Discovery, Innovation, and Risk

New Liberal Arts Series

Light, Wind, and Structure: The Mystery of the Master Builders, by Robert Mark, 1990

The Age of Electronic Messages, by John G. Truxal, 1990

Medical Technology, Ethics, and Economics, by Joseph D. Bronzino, Vincent H. Smith, and Maurice L. Wade, 1990

Understanding Quantitative History, by Loren Haskins and Kirk Jeffrey, 1990

Personal Mathematics and Computing: Tools for the Liberal Arts, by Frank Wattenberg, 1990

Nuclear Choices: A Citizen's Guide to Nuclear Technology, by Richard Wolfson, 1991

Discovery, Innovation, and Risk: Case Studies in Science and Technology, by Newton Copp and Andrew Zanella, 1992

This book is published as part of an Alfred P. Sloan Foundation program.

Discovery, Innovation, and Risk

Case Studies in Science and Technology

Newton Copp and Andrew Zanella

The MIT Press
Cambridge, Massachusetts
London, England

© 1993 Massachusetts Institute of Technology

All rights reserved. No part of this book may be reproduced in any form by any electronic or mechanical means (including photocopying, recording, or information storage and retrieval) without permission in writing from the publisher.

This book was set in Bembo by Asco Trade Typesetting Ltd., Hong Kong, and was printed and bound in the United States of America.

Library of Congress Cataloging-in-Publication Data

Copp, Newton.
    Discovery, innovation, and risk : case studies in science and technology / Newton Copp and Andrew Zanella.
      p. cm.—(New liberal arts series)
    Includes bibliographical references and index.
    ISBN 0-262-03199-X.—ISBN 0-262-53111-9 (pbk.)
      1. Science—Methodology. 2. Engineering—Methodology.
3. Research—Methodology—Case studies. 4. Technological innovations—Case studies. I. Zanella, Andrew. II. Title.
III. Series.
Q175.C734   1993
500—dc20
                                                                                   92-22206
                                                                                         CIP

Contents

Preface ... vii
Acknowledgments ... xi

Introduction ... 1

1
Science Joins Engineering ... 3

I Discovery

2
Telegraphy: The Beginnings ... 13

3
Hydroelectric Power: The Irony of Los Angeles ... 36

4
The Flying Machine Problem: The Wright Stuff ... 78

II Innovation

5
Fossil Fuels, Steam Power, and Electricity: Los Angeles Revisited ... 129

6
Gasoline: From Waste Product to Fuel ... 146

7
Bridge Design: Concrete Aesthetics ... 192

III Risk

8
Vaccines: Good Intentions Are Not Enough　　　245

9
The Greenhouse Effect: Revolution Involves Risk　　　289

10
Atomic Power: Difficulty in Estimating Cancer Risks　　　321

Conclusion

11
An Intricate Web　　　377

Abbreviations, Symbols, and Units　　　385
Glossary　　　389
Bibliography　　　405
Index　　　417

Preface

The New Liberal Arts Program

In 1982, the Alfred P. Sloan Foundation began the New Liberal Arts program to foster an understanding of technology and quantitative reasoning in students attending liberal arts colleges. The Sloan Foundation appealed to thirty-six colleges and fourteen universities around the United States to develop materials for the undergraduate curriculum that would further the goals of the program. The New Liberal Arts series, which includes this book, represents part of that effort.

Focus of this Book

Some things are more easily defined by stating what they are not than by trying to determine exactly what they are. This book is not an introduction to the natural sciences; neither is it an introduction to engineering—many very good books have already been written in each of these areas. This book represents our attempt to look simultaneously, cross-eyed if you will, at both science and engineering in the historical context of particular technological developments. Our major purpose in creating this unusual amalgam is to provide a meaningful technological framework within which major scientific principles in biology, chemistry, and physics may be discussed. Detailed treatment of the relevant scientific principles, however, is left to the instructor.

The impulse to write this book came from a realization that most people do not relate directly with the practice and findings of science but instead interact indirectly through technology. We live not so much in an age of science as in an age of scientific technology. The blossoming of the natural sciences that has been going on for roughly three hundred years appears to most people through the technological achievements of the last century, a period overlapping much of what historians have named the Second Industrial Revolution. In this brief period, an increasingly complex interplay between science and engineering within our culture has produced the airplane, modern pharmaceuticals, gasoline and the automobile, telecommunications, commercial electricity, modern highway systems, and other technological marvels that have shaped our culture. Modern technology thus provides the most meaningful context for understanding the principles and impact of modern science, and yet many general science courses lack material on modern technology and the interplay between science and engineering that has produced it. We hope this book will fill that gap.

Each chapter focuses on a specific technology, and the most conspicuously relevant scientific principles are introduced as they apply to each case. A full appreciation of the technological advances discussed in this book includes a quantitative view of the subject matter. In keeping with this notion, simple calculations are included where appropriate to complete the story and provide a level of detail that we hope increases the reader's appreciation for the elegance of the achievements. Various chapters will read differently depending upon whether the mathematics are more central to the arguments, as in chapters 4 and 7, or less central, as in chapter 10. Likewise, the study of new material implies that a new vocabulary is encountered; to facilitate understanding, a glossary of technical terms is provided at the end of the book for handy reference. When a technological advance has a particular bearing on urban life, we have chosen Los Angeles to illustrate its impact. There are several reasons for this choice apart from the fact that we live near Los Angeles. A massive westward redistribution of the population in the United States occurred during the Second Industrial Revolution and redefined the southwestern states as a desirable region in which to live rather than as a forbidding desert. The desert environment has placed a premium on innovative technological solutions to the problems associated with supporting a large population. Los Angeles represents a particularly good illustration for these solutions, as well as for the attending problems, because its growth has been unusually rapid. Finally, the lessons that can be learned from studying Los Angeles apply to many other large cities around the country.

Our intent is not to provide a thorough treatment of the scientific principles but to develop a technological context in which various scientific principles are applicable. We hope that this approach offers material relevant to students and faculty interested in science without excluding those whose interests lie in the history of science and technology and without duplicating unnecessarily much of the material commonly offered in general science courses. The examples were chosen to illustrate events in each of the major divisions of engineering including civil, electrical, mechanical, chemical, and medical as well as to touch on principles in physics, chemistry, and biology. Not all of the examples represent spectacular achievements, but they do represent undeniably significant contributions to our modern culture that arose during the Second Industrial Revolution.

Each of the chapters, except the first and the last, falls into one of three sections: discovery, innovation, and risk. The section on discovery emphasizes technological developments that sprang directly from the discovery of new scientific principles (chapters 2 and 3) or from the fresh application of scientific reasoning to an engineering problem (chapter 4). Scientific discovery and reasoning does not always foster technological development as directly as described in these early chapters. The cases in the section on innovation show that a great deal of technical innovation can proceed without a full understanding of the underlying science (chapter 5), or that it may grow in a climate of scientific inquiry without direct support provided by recent scientific discoveries (chapters 6 and 7). Regardless of the relationship between science and technology that spawned a major development, new developments often bring risks. Three case studies are presented in chapters 8, 9, and 10 to indicate the scientific and technical aspects of the danger as well as to suggest the power and limits of science and engineering in reducing risks in our technological society. By placing each topic in its historical context, we hope to communicate a sense of how human understanding and control of nature have increased in fits and starts as engineers and scientists struggled to solve specific problems. As small as the glimpse may be, we also hope that these studies help the reader share our growing appreciation for nature and what human beings manage to do with it.

Acknowledgments

We would not have undertaken this project without the stimulation and support provided by the Alfred P. Sloan Foundation's New Liberal Arts Program and the many energetic, talented people involved with it. We are especially grateful to David P. Billington, Samuel Goldberg, John Truxal, and Marion Visich for encouraging and advising us at every step.

Much of the material in this book was developed in a series of workshops sponsored by the Sloan Foundation and conducted at Princeton University. Four of our colleagues in these workshops generously agreed to share the results of their work and cooperate with us in converting their materials into chapters. Alfonso Albano (Bryn Mawr College) wrote the original draft of the chapter on telegraphy. The chapter on gasoline was developed from a manuscript written by J. Nicholas Burnett (Davidson College). William Case (Grinnell College) contributed the analyses and discussion of the Walnut Lane and the Felsenau bridges to the chapter on prestressed concrete. The chapter on flight grew out of a manuscript written by Robert Prigo (Middlebury College). We are indebted to these four people for unselfishly contributing their time and expertise to this project.

We also want to thank Peter Myers, Ken Stunkel, David Billington, John Truxal, John Moeur, our colleagues in the Joint Science Department of Claremont McKenna, Pitzer, and Scripps Colleges of The Claremont Colleges, and several anonymous reviewers for carefully evaluating portions of this book and helping us correct and clarify the manuscript. Finally, we thank Helen Wirtz for her help in designing and drawing many of the illustrations. Portions of this book have appeared in several monographs.

Introduction

Science Joins Engineering

1

The Pace of Change

Anyone born in the United States at the start of the Civil War and fortunate enough to live eighty-four years until the end of World War II would have experienced a pace of change remarkable, perhaps unparalleled, in human history. In 1861, we were a nation of thirty-four states with only a few wagon trails connecting the eastern seaboard with the vast, sparsely populated and largely unincorporated reaches of the west. Within a single lifetime, America was transformed socially, culturally, politically, militarily, and technologically into a world-power nation of forty-eight states linked increasingly tightly by telecommunications networks, nearly two million miles of improved roads, airline routes, and electrical power networks. The fastest way to get mail from New York to Sacramento, California early in 1861 involved putting it on a train for a two-day trip to St. Joseph, Missouri, then transferring it to the saddle bags of Pony Express riders who took at least eight more days to carry it the remaining 2,000 miles. The speed of mail delivery by this route astonished people at the time, but it quickly paled in comparison to the virtually instantaneous communication made possible by the first transcontinental telegraph line erected in the same year. This event established the first truly high-speed, long-distance communication channel between the two coasts of the North American continent and was a harbinger of the important technological changes soon to come.

During the first thirty years of someone's life spanning this eight-four-year period, the steam engine served as the major source of machine power with energy drawn from burning wood and coal. But by the person's fortieth birthday, the technological seeds had been sown that would develop rapidly into extensive networks of electrical power lines. Dramatic improvements in the ability to transmit electricity long distances allowed its consumers to locate their homes and businesses far from the oil deposits, coal fields, or rivers that supplied the energy. This valued aspect of electricity played a central role in opening the American Southwest for settlement and industry after the turn of the century.

In the person's forty-second year, the ancient dream of human flight became a reality as the first powered, piloted airplane took to the air near Kitty Hawk, North Carolina. Supported by advances in the refining of oil to produce gasoline, the aircraft industry developed so rapidly that the person's seventy-fifth birthday could have been celebrated by flying coast to coast in a little more than eighteen hours, a trip that took at least five days before the advent of airplanes. At about the same time as the first powered airplane flight, Americans stood a 33 percent chance of dying from an infectious disease. Pneumonia, tuberculosis, diphtheria, and gastrointestinal infections such as cholera ranked among the ten major causes of death in the United States. A battery of vaccines and antibiotics developed mostly after 1920, along with improvements in sanitation and the average standard of living, produced a 75 percent cut in the mortality from these causes by 1935 and eliminated all infectious diseases except pneumonia from the top ten list within the next thirty years.

A life that began in the same year as transcontinental telegraphy and lasted eighty-four years ended in the year that atomic fission first promised abundant power even as it destroyed two Japanese cities. This event, perhaps more than any other, symbolizes the uncertainty rooted in the technological changes that unfolded between 1861 and 1945. The technological developments of this period have become so integrated into the fabric of our lives that we are hard pressed to find a single facet of the nation's well-being that does not depend on them. Telecommunications, transportation, medicines, and energy do not exhaust the list of technological fields crucial to us, but they comprise a significant part of it. Advances along these fronts have come so rapidly that we appear to be stunned by the pace of change even as we take it for granted. Coping with the consequences of change, the benefits as well as the problems, now commands a considerable amount of attention from writers, social critics, politicians, and others.

Science, Engineering, and Technology

The obvious question is how did all of this happen in such a short amount of time? A complete answer to this question, if one could be obtained at all, would necessarily refer to the almost unimaginably complex interactions among the myriad components of a society in a growing nation. One portion of this web of interactions, however, deserves special attention: the relationship between science and technology. By this we mean the relationship between engineering and the natural sciences that grew notably closer over the past 100 to 150 years.

Modern science gradually separated from its ancestral field of natural philosophy but retained the understanding of nature as its overriding goal. Scientists seek general patterns in nature that can be summarized in models and theories. Theories are then used to formulate predictions that are tested experimentally, leading to further support or refutation of the theory. Many people, including some scientists, would take exception to this highly simplified view of the scientific enterprise, but it serves to point out that science, or basic research, is not explicitly concerned with practical applications. Nuclear physicists in the 1920s wanted to understand the atomic nucleus. They had no intention of building a bomb.

The profession of engineering, on the other hand traces its major roots to craftsmanship and works toward increasing our control over nature for human benefit. The engineer may be helped by an understanding of general patterns in nature, but that is not the ultimate goal of the profession. Engineers have come to use much of the methodology of science, but the nature of the questions differ from science as engineers seek better machines, structures, systems, chemicals, or processes. Problems of scale become important to the engineer as do problems of cost and other aspects that influence utility. While general scientific theories set the broad constraints, final designs must be informed by more detailed considerations than general descriptions of nature can provide.

Modern technology results from engineering, but this has not always been the case. Fabrics, dyes, medicines, pottery, metal tools and weapons, the plow, the stirrup, other useful items, and even dozens of crops were developed over the course of human history without benefit of professional engineers or scientists. The developers of the early water wheels, for example, were not physicists or mechanical engineers, but they knew about tapping the power in falling water and had the skill, the tools, and the materials to do it. Likewise, the first steam power enthusiasts were simply trying to remove water from mines. They probably had no idea that they were launching a technological revolution.

The trial-and-error approach to technology should not be demeaned—it fostered the First Industrial Revolution in the eighteenth century. This transition in western society sprang from the intelligent efforts of British inventors, craftsmen, machinists, toolmakers, and businesspeople. The antecedents of the revolution go back to the Middle Ages when wind and water power were harnessed to run hammers, saws, and grain mills, thus beginning the emancipation of people and animals from being the main sources of such power. The eighteenth century witnessed a great leap forward when the steam engine was developed to run water pumps, initially, then looms and other machinery. Mills and factories, no longer constrained to locations near running water, were built on sites closer to raw materials and became centers for the growth of large cities.

The technology of the First Industrial Revolution developed more from the practical experiences of the craftsmen than from sound theoretical foundations. New structures, machines, and materials reflected common sense and creative insights, rather than scientific data and principles, because scientific information was largely unavailable at the time. This is not to say that scientific principles were ignored when they were understood. When James Watt designed steam engines in the late 1700s, he certainly knew that the pressure of a gas depends on its temperature, but he had no theoretical knowledge of the laws of thermodynamics. These laws were not formulated until the nineteenth century and eventually proved crucial in designing safer, more efficient engines. Iron, the signature material of the First Industrial Revolution, also was developed without much reference to science. Iron had been produced many centuries before, but innovators in the eighteenth century found ways to improve its quality and produce it in quantities sufficient for use in bridges, boilers, and other large structures. They could determine its bulk chemical composition, but they knew nothing about its atomic structure nor about the beneficial effects of changing the proportion of various impurities in the iron. Such information later proved crucial in the development of modern steel.

The Second Industrial Revolution

The trial-and-error method of developing technology gradually gave way to more systematic approaches such that a Second Industrial Revolution was under way by the last quarter of the nineteenth century. Practitioners in the emerging profession of engineering began paying careful attention to the methods and knowledge being developed by scientists at an increasingly rapid rate. Innovations that began to flow from the new science-based approach to engineering contributed the technological underpinnings of our modern society. Whereas the previous industrial revolution had emphasized coal, steam, and iron, this period witnessed the rapid introduction of electricity, steel, telecommunications, synthetic chemicals, and new modes of transportation including the automobile and the airplane.

The first telegraph and electric generator, ancestors to our telecommunications and electrical power industries, grew directly from the fundamental understanding of electricity provided by scientific researchers such as Hans Oersted, Joseph Henry, and Michael Faraday. Scientific methodology proved critical in solving the flying machine problem. Advances in organic chemistry supported creation of the petroleum industry that continues to feed industry, automobiles, and airplanes. A biological understanding of microorganisms led to the taming or eradication of diseases that had plagued humans for centuries. These and many other examples testify to the bountiful cross-fertilization between science and engineering that began in earnest during the Second Industrial Revolution.

The connection between science and technology established during this period is not a one-way street, however. Scientists do not simply inform the engineer who then applies scientific principles in useful ways and then waits for new concepts. Engineering is itself a creative enterprise that influences science in return. In addition to contributing useful tools for scientific research, engineering helps shape basic questions in scientific research. Careful investigation of steam engines revealed the laws of thermodynamics, which in turn led to improvements in engine design. Smallpox vaccine eventually helped spark scientific interest in the immune system and eventually promoted further development of vaccines. This positive-feedback effect of the interaction between science and technology no doubt accounts in large part for the remarkably rapid pace of change described at the beginning of this chapter.

The Social Context and Risk

Science and technology exert an influence at more than the material level. Our industrialized culture has become permeated by the ideas of "progress" and "technological fixes." In this climate of optimism, all problems seem conquerable if only we spend enough money and effort on them. NASA's Apollo Project of the 1960s provides the most common reference for this optimism: everyone has heard the phrase that begins, "If we can land on the moon, why can't we ...?" and ends with any one of a variety of current problems. This question fails to recognize the factors that can limit progress toward solving certain problems. The task of rocketing people to the moon and getting them safely back to earth involved tremendously difficult technical problems, but the scientific basis for their solution was already well in hand when the project was undertaken. The same can not be said about the "war on cancer" undertaken in the 1970s. Gaping holes in our basic understanding of cells and genes hinder progress toward cures for this group of all-too-often lethal conditions. Even when the scientific understanding is at hand, a solution to a problem may not be forthcoming. The chemistry of air pollution from internal combustion engines is fairly well understood, for example, but the pollution-free engine has not yet been developed. Factors outside the realms of science and engineering frequently limit our options as well. We could use solar cells to reduce our dependence on coal-fired power plants, but the cost would be prohibitive and people might not be willing to settle for less reliable service from their electrical utility companies or to cover the deserts with solar panels.

Limits to our understanding, as well as limits on technology imposed by society, often mean that we assume risks as we reap the benefits of new technology. Most of us lead less risky lives than our ancestors did, but that realization does not remove the fear of nuclear power plant accidents, pollution, or airplane wrecks. We somehow learn to live with these risks as we work to reduce them. An understanding of science, engineering, and technology, taken in their proper social and historical context, may help us in both regards.

Discovery

I

Modern technology reflects a complex interplay between the processes of scientific discovery and engineering innovation, the uncovering of new knowledge about nature, and the clever application of knowledge. The following three chapters describe technological developments in which scientific discovery played a key role in stimulating subsequent engineering innovations. Chapters 2 and 3 present engineering developments that grew from discoveries of the relationship between electricity and magnetism. It is fitting to begin this book with these cases because electricity became one of the hallmarks of the Second Industrial Revolution and remains crucial to modern life in a staggering variety of applications. But scientific discovery may also mean the employment of scientific methods and reasoning to uncover solutions to engineering problems. The development of the airplane illustrates this role of science in contributing to the technological world. Chapter 4 on the Wright brothers thus stands at the interface between scientific discovery and engineering innovation.

Telegraphy: The Beginnings

2

Dawn of the Information Age

On January 8, 1815, the Battle of New Orleans ended with an American victory and thousands of casualties. The next day a message arrived containing the good news that the war was over—two weeks before! The Treaty of Ghent, signed in late December of 1814, had formally ended the War of 1812. One might well call this a classic example of the failure to communicate, although the mail ships crossed the Atlantic as fast as they could. By 1860, things were not much better, because the fabled Pony Express took eight days to carry mail two thousand miles from St. Joseph, Missouri, to Sacramento, California. But in 1861, the time required for news to traverse the entire breadth of the United States plummeted from days to minutes. This unprecedented increase in the speed of communication came from the installation of the first transcontinental telegraph system in October of that year.

This watershed in communications was comparable to the linking of the Atlantic and Pacific Oceans by railroad just eight years later, and both provided the bases for immense and vital industries. Even more significantly, together these two achievements brought to a practical reality the dream of a nation from the Atlantic to the Pacific.

Over a century later, in 1977, the U.S. Department of Commerce reported that over 45 percent of the country's gross national product and nearly half of its work force were involved in the production, analysis, storage, and distribution of information. The "information industry" has become so "pervasive and influential that it is now becoming clear that the U.S. is moving into a new era—the information age" (Dizard 1985).

Telegraphy is the primary technological innovation that made the information age possible. Once telecommunications networks were established, it became possible to transfer vast quantities of information with tremendous speed over great distances. In the following sections, we will examine the development of telegraphy in order to understand its scientific and technical underpinnings as well as to put our modern communications systems into perspective.

Familiar as we are with long-distance telephone calls, two-way radios, live television pictures from across the world, fax machines, telexes, and such, we might view the telegraph as a bit archaic and unsophisticated. But it was the telegraph, after all, that first made the "instantaneous communication of intelligence" a worldwide reality. Modern communications technologies can be legitimately viewed as descendants of telegraphy. In contrast, the telegraph represented a dramatic break from the preceding methods of communication. A piece of electronic mail sent by one computer to another along an optical fiber is more like a telegram sent on a copper wire than a telegram is to a message sent by semaphores, smoke signals, stagecoach, or railroad.

The telegraph gave rise to the large "wire services" which, to this day, dominate the worldwide gathering and dissemination of news. Ninety percent of the international news going to the world's newsrooms in the early 1980s was supplied by four companies: Reuter's, Agence-France Presse (AFP), United Press International (UPI), and Associated Press (AP). Electronic transfer of funds is the same as the old "wire transfer," a service that Western Union, the first large U.S. telegraph company (1856), still advertises. Indeed, many of the details that characterize telecommunications today were defined by telegraphy less than 150 years ago. The birth of the telegraph marked the birth of the information age.

Challenge of Signaling over Distance

Attempts to build systems of rapid communications over long distances date back to ancient times. News of the fall of Troy, Homer claims, was transmitted to Agamemnon's fortress in Mycenae, some

250 miles away, by a series of signal fires (ca. 1200 BC). The same technique played a key role in history some two and-a-half millennia later when beacons of fire warned Elizabeth I's England of the approach of the Spanish Armada (1588). These highly limited signals transmitted only one piece of information to confirm the occurrence of an anticipated event. They probably used a prearranged code like "one if by land, two if by sea" as with the lights that signaled to Paul Revere from Boston's Old North Church in 1775.

Smoke signals used by native Americans and made popular by innumerable western movies represented a more sophisticated communications technology, capable of transmitting a larger variety of messages. Semaphores further increased the information content by using wooden arms or flags to wave signals from high towers to receiving stations as far away as vision permitted, later aided by telescopes. The relative orientations of the arms or flags denoted letters of the alphabet, specific syllables, phrases, or sentences. Scouts and lifeguards still practice a variation of this form of communication. The French revolutionary government developed a network of semaphore stations that included over 200 stations and stretched more than 1000 miles by the end of the Napoleonic Era.

Anyone who tries to communicate over long distances using a system entirely dependent on vision faces serious difficulties—"rain, or hail, or snow, or gloom of night," as well as the curvature of the earth, interfere with the process. Even under optimal conditions, the receiver must be facing the sender for the message to be communicated. These limitations placed a premium on alternative communications systems and, when it became clear during the eighteenth century that electrical signals could be transmitted along wires, numerous attempts to develop electrical communications systems followed.

Quest for the Electrical Telegraph

George Louis Le Sage, a Swiss inventor, constructed the first well-documented electrical communications device in 1774. This instrument utilized the fact that two objects carrying the same electrical charge repel each other. The receiving end consisted of two pith balls suspended by strings from the end of a wire. When the other end of the wire, the transmitting end, was charged, the pith balls at the receiving end jumped. Le Sage's telegraph consisted of twenty-four wires, one for each letter of the alphabet, except for the combinations i, j and u, v which shared one wire per pair. Most of the other early devices were equally cumbersome, some with as many as thirty-six wires to accommodate the alphabet as well as numbers.

Other communications instruments built as early as 1823 measured electric currents with galvanometers, devices similar to modern-day meters, in which the deflection of a needle indicates the magnitude of the current in the circuit. A galvanometer telegraph consisting of five independent circuits was constructed in Russia by Paul Ludovitch Schilling, Baron of Cromstadt, in 1823. In 1833 a mathematician, Karl Friedrich Gauss, and a physicist, Wilhelm Eduard Weber, constructed a device that used a single galvanometer needle to allow rapid communication between the physics laboratory at the University of Goettingen and the university observatory, a mile and a half away. They used this system to synchronize experiments performed at the two locations. It is not clear if Gauss and Weber developed a code for the alphabet, but they transmitted complete sentences. Improvements of their design by C. A. Steinhall of Munich eventually included an automatic recording device. When the first trans-Atlantic cable was laid in 1858, the telegraph was of the Gauss-Weber type.

In 1836, a London professor, Charles Wheatstone, constructed another type of telegraph, a five-needle device. He collaborated with William Cooke, and together they instituted modifications that later resulted in an automatic recording telegraph. In the 1840s Wheatstone performed some of the earliest successful experiments on the transmission of telegraphic signals through underwater cables.

The best known name in the history of telegraphy is, however, Samuel Finley Breese Morse. Although there were various other people who constructed telegraphs based on electric or electromagnetic principles before him (box 2.1), it was Morse who is best remembered as its inventor, especially in the United States. This seems all the more amazing when his background and, particularly, his lack of experience with electricity are considered. Nevertheless, Morse's perseverance and imagination served him well when the U.S. Congress set up a committee in 1837 to study the issue

Electromagnetism

The scientific and technological developments that made the invention of telegraphy possible were at least as significant and profound as the effects of telegraphy on society. From the time of the ancient Greeks, electricity and magnetism had been studied as separate phenomena, but scientists had long sought a connection between them. This was especially true towards the end of the eighteenth century and the beginning of the nineteenth when electrical phenomena were finally beginning to be understood. Oersted's discovery of the connection between electricity and magnetism not only led to the scientific basis of telegraphy, it also led to the possibility of electric power generation that made our twentieth-century electrical technology possible.

Joseph Henry, the American, and Michael Faraday, the remarkable British chemist and physicist, separately discovered the basic components of the electric generator and motor in the 1830s. In addition to practical applications, these discoveries led to the unification of the theories of electricity and magnetism by the great Scottish physicist, James Clerk Maxwell (1831–1879). Maxwell's work provided the basis for understanding the interactions between electric and magnetic fields.

One consequence of Maxwell's unified theory of electromagnetism was the demonstration that light behaves like a wave, an electromagnetic wave. Light is just a small part of a large spectrum of electromagnetic waves ranging from the gamma rays emitted by atomic nuclei to those used in radio and television broadcasting. The unification of electricity and magnetism inspired Albert Einstein (1879–1955) many years later to seek a similar unification of electricity, magnetism, and gravitation. The goal eluded Einstein and still eludes physicists to this day.

Box 2.1

of telegraphy, especially how to overcome deficiencies of visual signaling methods, which were ineffective during bad weather.

Morse: An American Leonardo?

Samuel Morse was called "the American Leonardo" (da Vinci) by an adoring biographer. The comparison may seem a bit outlandish, but the biographer's enthusiasm may be forgiven if one considers Morse's multifaceted career, perhaps almost as varied as da Vinci's. Morse (figure 2.1) was a well-known portrait painter and a cofounder and president for many years of the National Academy of Design. He took up the post of professor at New York University, helped found the *Journal of Commerce*, a New York newspaper, and wrote for other New York journals as well. Active in politics in New York City, Morse continued his political involvement in

Figure 2.1 Samuel Finley Breeze Morse. (Reproduced from an engraving from a photograph by M. Brady in *The Life of Samuel B. Morse*, by Samuel I. Prime, D. Appleton & Co., New York, 1875.)

Poughkeepsie, N.Y. In addition, he promoted the daguerreotype, an early version of photography, in North America. And then, of course, there was the telegraph.

Morse was born on April 27, 1791, in Charlestown, Massachusetts. His maternal grandfather was one of the first presidents of Princeton University, and his father was a clergyman who later authored a widely used textbook in geography. He was educated at Phillips Academy in Andover, Massachusetts and at Yale College. Although he attended some lectures in physics and chemistry, including electromagnetism, while at Yale, his primary interest apparently was in painting.

In 1811, a year after graduating from Yale, he sailed to England to study painting under Benjamin West, one of the best known painters in England. Morse returned to the United States in 1815 and embarked on a successful career as an artist. In 1829, he went on a second trip to France and Italy.

The possibility of using electricity for instantaneous communication reportedly occurred to him in 1832 while sailing home from England on board the packet ship, *Sully*. At that time, and for many years thereafter, he seems to have been unaware that just such a possibility had been contemplated by a number of people since the mid-eighteenth century. In spite of his original enthusiasm, Morse apparently did not do much work on the telegraph until the fall of 1835. At that time, he had just been appointed Professor of the Literature of the Art of Design at the University of the City of New York (now known as New York University).

The Invention of the Telegraph: Science Applied

"The electric telegraph was the first great application of the growing body of electrical research" (Dibner 1967) by using electricity and magnetism for truly rapid long-distance communication. All of the early electrical communication devices depended on batteries and electromagnets. Batteries provided sustainable electric currents that made electromagnets possible. Electromagnets, in turn, allowed the motion of a piece of iron at one location to be controlled by opening or closing a switch at another position, possibly at a considerable distance from the first.

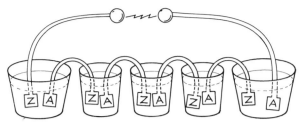

Figure 2.2 Volta's Battery, which he called "a crown of cups." Here Z represents zinc, and A stands for copper. Note that the figure as drawn is an "open circuit," and it would be necessary to connect the first and last cup to produce a current. (Adapted from a figure accompanying a letter from Volta to Sir Joseph Banks of the Royal Society and published in the "Philosophical Transactions" of the Royal Society for 1800, p. 405. Reproduced in Magie, 1935.)

The batteries available in Morse's time resembled the original battery (figure 2.2) made in 1800 by Alessandro Volta, then professor of physics at the University of Pavia. Volta's battery, to which he gave the almost poetic name, "crown of cups," consisted of a series of cups (he highly recommended crystal goblets!) containing water, or lye, or brine. He immersed a copper plate and a zinc plate into each cup, and the cups were joined together by metallic wires so that the zinc plate of one cup was connected to the copper plate of the next cup. The series of cups was arranged so that the copper plate of the cup at one end and the zinc plate of the cup at the other end were not connected. These unconnected plates became the terminals of the battery. In modern terminology, we would say that each of the cups was a cell (a voltaic cell) consisting of copper and zinc electrodes. The water, or brine, or lye solution served as an electrolyte to conduct electricity between the immersed electrodes. The cells were connected in series to form a battery of electrochemical cells.

The main difference between Volta's crown of cups and later versions of the battery is that the later versions typically used an acid solution for the electrolyte. The batteries used by Morse contained a sulfuric acid solution. Modern automobile batteries also use sulfuric acid electrolytes but have a combination of lead and lead oxide electrodes.

Chemical reactions between the electrolyte and each of the electrodes result in a situation where electrical charges have greater energy (more strictly, electrostatic potential energy) at one terminal than at the other. We may describe this situation by saying that there exists a potential difference between the two terminals of the battery. This difference is also referred to as a voltage and is measured in volts (V) to honor Volta. If an external conducting pathway, a circuit, connects the two terminals, charges will move from the terminal where they have higher energy to the other. This motion of electrical charges is an electrical current, whose magnitude depends on the potential difference across the terminals of the battery and on a property of the pathway called its resistance. The conducting pathway can be a metal wire, a piece of graphite, or an ionic solution, among others.

The relation among voltage, current, and resistance was defined in 1826 by Georg Simon Ohm, a German physicist who showed that when a voltage is applied to the ends of an electrical conductor, the resulting current is directly proportional to the voltage that caused it. In symbols, the voltage difference, V, and the resulting current, I, measured in amperes (see below), are related by:

$$V = IR$$

where R, measured in units of ohms, is the resistance of the object. This equation is known as *Ohm's Law*.

In the construction of a telegraph line or circuit, it was clearly important to be able to apply the above relationship. Inventors obviously needed to test their apparatus, but they also tried to estimate as accurately as possible the strength of the batteries required and the amount of copper needed for the wires. In order to minimize costs, they did not want to overdesign the apparatus, and copper was the most expensive of the materials. A major concern was the resistance of the copper wire, which depended on both its length and thickness.

The electrical resistance, R, of an object of length L in the direction of current flow, and cross-sectional area, A, is given by,

$$R = \rho L / A$$

where ρ (Greek rho), called the resistivity, is characteristic of the material of which the object is made. Resistivities vary greatly depending on the material's ability to conduct electricity or to "insulate." Copper, a very good conductor, has a resistivity of 1.7×10^{-6}

ohm-cm at 20°C, while glass, a very good insulator, has a resistivity in excess of 10^{15} ohm-cm at the same temperature. Thus, low resistivity means that a substance is a good electrical conductor.

Using the resistivity of copper given above, a one-meter (10^2 cm) long copper wire with a cross-sectional area of one square millimeter (10^{-2} cm²) has a resistance of:

$$R = \frac{(1.7 \times 10^{-6} \text{ ohm-cm}) \times (10^2 \text{ cm})}{10^{-2} \text{ cm}^2} = 1.72 \times 10^{-2} \text{ ohm}$$

If this wire was connected to the terminals of a 1.5-volt flashlight battery, the resulting current in the wire would, by Ohm's law, be:

$$I = \frac{1.5 \text{ V}}{1.7 \times 10^{-2} \text{ ohm}} = 88 \text{ amp}$$

This is an extremely large current that would produce enough heat in the wire to melt it (a short circuit). For comparison, ordinary house fuses or circuit breakers are set to blow at 15 or 30 amps.

A further important discovery related to electricity laid the final groundwork for the telegraph. In addition to producing heat as in the heating element of an electric stove, or light as in an incandescent lamp, electrical currents also produce magnetic fields. The fact that magnetism is a consequence of the motion of electrical charges is one of the more profound scientific insights of the first half of the nineteenth century.

In 1820 the Danish physicist Hans Christian Oersted reported having observed that an electric current in a wire caused the deflection of a nearby magnetized needle. Within months of this report, André Marie Ampere of the École Polytechnique in Paris and his fellow members of the Académie Française confirmed and amplified Oersted's work and obtained an almost complete theoretical description of the phenomenon. These advances led to the development of the electromagnet.

Henry: the American Faraday

From the discovery that currents have magnetic effects to the construction of practical electromagnets, we go from Oersted in Copenhagen to Joseph Henry in Albany, N.Y. In the early 1820s,

Figure 2.3 Joseph Henry. (From a photograph in *Joseph Henry: His Life and Work*, by Thomas Coulson, Princeton, 1950. Reprinted by permission of Princeton University Press.)

Henry (figure 2.3) was enrolled in the Albany Academy, a secondary school that still exists. In Henry's time, it must have been more than just an ordinary secondary school for while he was a student there, Henry learned mathematics through integral calculus, chemistry, and natural philosophy (physics). He had no formal education beyond that.

After graduation, Henry was appointed as an assistant to help the chemistry professor at the academy with the professor's demonstrations. In 1826 he was appointed professor of mathematics and natural philosophy. A year later, he started his experiments on electricity and magnetism. Working only in the summer when he had no teaching duties and could use part of a classroom for a laboratory, and often using his own funds for equipment and supplies, he nevertheless produced a body of work that eventually gained him international recognition.

In a paper published in 1831, Henry discussed the construction of powerful electromagnets, which he called "intensity magnets,"

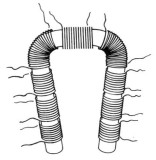

Figure 2.4 One of Henry's "intensity" magnets. Adapted from a figure in Henry's 1831 paper, reproduced in Coulson, 1950. The whole magnet weighed 21 pounds, and was used to lift a maximum weight of 750 pounds.

by using high-voltage batteries ("intensity batteries") and many coils of shellac-coated wire wrapped around soft iron cores (figure 2.4). The shellac served to insulate the wire from itself and prevented shorting out. We now know that the magnetic fields produced in electromagnets depend crucially on the current used and the number of turns of the coil. We may summarize these by the equation:

Magnetic field $= KIN$

where I is the current in the coil, N is the number of turns of the coil, and K is a parameter that depends on the material of the core, its size, and shape. Henry obtained large values for the current by using batteries with large voltages. Recall Ohm's law ($V = IR$, or $I = V/R$), so for a given resistance, R, the larger V is, the larger I will be. The value of N could be increased by merely winding more turns on the coil but only up to a point—K decreases as the radius of the coil increases. Also, K can vary very dramatically from one material to another. If, for an air-core magnet, the value of K is of the order of one, the same magnet with a soft iron core could have a K of as much as 5000. A core made of one of the modern magnetic alloys such as Permalloy, an alloy consisting of about 80 percent nickel and 20 percent iron, could give a K between ten thousand and one million (or of the order 10^4 to 10^6, in scientific notation). Henry was able to build an electromagnet that was powerful enough to lift several thousand pounds. This was used to demonstrate the electromagnet's capability, but smaller ones were used in the telegraphic equipment.

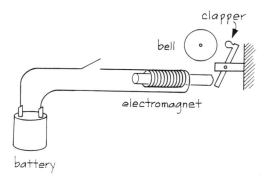

Figure 2.5 A schematic diagram of Henry's telegraph showing the core attracting the clapper (i.e., the circuit is open).

Henry recognized that the electromagnet could be used as the basis for a telegraph, and he built the first "acoustic electromagnetic telegraph" (figure 2.5) in Albany. Around his classroom-laboratory, he suspended one mile of copper wire that was part of a circuit that included a switch, a battery, and an electromagnet with a soft iron core. The core normally attracted one end of a magnetic iron clapper pivoted at its center. Whenever the switch on the electric circuit was closed, current flowed in the coil and induced an opposite magnetism in the core, which then repelled the clapper causing its other end to strike a bell. The typical electrical doorbell operates on a similar principle. Thus, in 1831, five years before Morse built his first apparatus, a working model of what we now know as the Morse telegraph was produced. That is, it contained the basic elements Morse used in his device to solve the problem of communicating signals over a relatively large distance.

Henry became professor of natural philosophy at the College of New Jersey (which later became Princeton University) in 1832. It was in Princeton where he produced his most fundamental scientific contributions. He shares with Michael Faraday, the English scientist (1791–1867), the credit for discovering that changing magnetic fields may be used to induce or generate electric currents. It is this scientific principle that underlies the operation of machines that generate electrical power by mechanical means—electrical generators. He left Princeton in 1846 to become secretary of the newly established Smithsonian Institution in Washington where he remained until his death.

The Morse Telegraph

Henry's scientific precedence in constructing a working telegraph should not, however, diminish the credit due to Morse. For it was Morse's singlemindedness and dedication that eventually led to the development of a practical system of telegraphy. In Henry's words, "... I may say that I was the first to bring the electro-magnet into the condition necessary to its use in telegraphy ... and to illustrate this by constructing a working telegraph ... To Mr. Morse, however, great credit is due for his alphabet, and in bringing telegraphy to practical use" (From a letter to S. B. Dod, as quoted by Harlow 1971).

Unlike later versions, or Henry's earlier version, that used sound to convey the transmitted message, Morse's original receiver (figure 2.6) was meant to make a permanent record (hard copy) of the message. A pencil point attached to the bottom of a pendulum touched a strip of paper moved by rollers. The pendulum was hung from a canvas stretcher, which must have been readily available since Morse built this first model in his studio. On the pendulum was an iron bar near an electromagnet mounted on the canvas stretcher. Whenever there was sufficient current in the electromagnet, it would attract the pendulum causing the pencil point to move in a direction transverse to the motion of the paper. By closing the electromagnet circuit only momentarily, the pencil would perform

Figure 2.6 Morse's first telegraph. (Adapted from an 1837 drawing reproduced in Andrews 1989.)

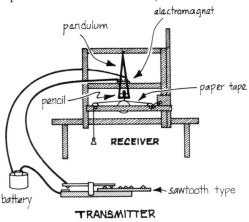

a quick zigzag, marking a V with a sharp bottom. Keeping the electromagnet circuit closed a bit longer would cause the pencil to zig, pause, and then zag—marking a V with a flat bottom—the predecessors of the dot and the dash. For as long as the pendulum was undisturbed, the pencil marked a straight line on the moving piece of paper.

Morse's original intention was to transmit only numbers. Different combinations of numbers would stand for phrases or words according to a vocabulary that he was going to develop. Each numeral was represented by a series of dots and dashes, or sharp-bottomed and flat-bottomed Vs. Borrowing from the printing technology of the time, he cast some "sawtooth type" (Harlow 1971; Hindle 1983)—flat pieces of lead that were grooved on one side with the appropriate sequence of sharp-bottomed or flat-bottomed Vs. This was to be assembled on a printer's composing stick and the coded message would pass under one arm of a lever. As the lever rose and fell according to the grooves on the type, it opened or closed a circuit, transmitting the message. Compared to this cumbersome system, the simple telegraph key that was later developed was indeed a marvel of simplicity and elegance.

Eventually operators (telegraphers) realized that by listening to the noise pattern of the "dots and dashes" they could interpret letters and words. This made telegraph use much faster, since they could translate as the message came in and no longer had to wait to decode the paper tape. By 1860, skilled telegraphers could send and receive at the remarkable rate of sixty words per minute.

A Practical System

With his first instrument, Morse tried to transmit a message through forty feet of wire using a single voltaic cell. It did not work so he sought the advice of a colleague, Leonard D. Gale, then professor of chemistry at New York University. Gale advised him to use many cells in series, increase the number of turns of wire in his electromagnet, and to read Henry's paper on the electromagnet, published in the *American Journal of Science* in 1831.

In this paper Henry discussed, as Volta did three decades earlier, how higher voltages could be attained by connecting several cells in series. Using a battery of twenty cells in series and a coil of a few hundred turns, Gale eventually transmitted messages over ten miles of wire strung around his lecture room.

One of Gale's lecture room demonstrations was witnessed by Alfred Vail, at that time a recent graduate of New York University. Vail was so impressed that he volunteered his labor, materials from his father's ironworks in New Jersey, and money. Vail introduced many improvements to Morse's device—he replaced the sawtooth type with the now familiar key and increased the sensitivity of the instrument so that batteries of smaller voltages could be used. By the late 1830s, the "Morse telegraph" retained few of Morse's original design elements and was, by then, essentially Vail's design. There are even claims that it was Vail who invented the Morse code. Morse, Gale, Alfred Vail, and Alfred's brother George later became partners, jointly owning the patent rights to the telegraph.

In February, 1838, the partners went to Washington to demonstrate the new device to federal officials and members of Congress who were seeking new technologies for signaling over long distances. Morse proposed that the government have exclusive control over the use of the telegraph, granting franchises for the construction of private lines under appropriate restrictions. He made this proposal to Congress and included a request for funds to build an experimental line between Baltimore and Washington to show the capabilities of the instrument. The proposal was referred to the House Commerce Committee, then chaired by Francis Ormand Jonathan Smith of Maine. Smith obviously appreciated what the telegraph had to offer, to the world in general, and to Smith in particular. He persuaded Morse, Gale, and Vail to take him on as a partner. After becoming a full partner, he favorably reported to the House a bill appropriating $30,000 for the construction of an experimental line. He came to be known as "Fog" Smith and, as a historian of the era put it rather indelicately, "came to be still more noted for his lack of scruple" (Harlow 1971).

The bill appropriating the $30,000 eventually passed, and a line between Washington and Baltimore was constructed under the supervision of Ezra Cornell. Cornell was initially hired by Morse to lay electric cables underground between tracks of the Baltimore & Ohio Railroad. The cables were for two complete telegraph circuits and consisted of a total of four wires (color coded yellow, red, green, and black, just as they are now in telephone cables), individually wrapped in cloth, covered with shellac, and encased in lead. Cornell suspected that this system was susceptible to short circuits, and testing—after nine miles of cable had been laid– proved him right. On Cornell's suggestion, the wires were separated and each wire was attached to glass insulators mounted on crossbars attached to poles, a practice that has survived to this day. Cornell's scheme assured the success of the line, and his financial success resulted in the founding of Cornell College in Iowa and Cornell University in New York.

Even before the completion of the line, messages were being sent from intermediate points to Washington. Among the first messages to be transmitted were reports about the national convention of the Whig Party that was being held in Baltimore. However, some people did not believe in the new technology's accuracy and they asked for confirmation of the nominee by mail! On May 24, 1844, the Washington-Baltimore line was formally opened. Transmitting from the Supreme Court room in the Capitol, Morse sent to Vail, who was forty-four miles away at the Mount Clare Station of the Baltimore & Ohio Railroad in Baltimore, the biblical quotation, "What hath God wrought!" (Numbers, 23:23).

Controversy: Who Invented the Telegraph?

As with many inventions and discoveries, there often exists a climate that stimulates the development of a successful device or process. The time becomes ripe and many inventors compete to produce the best solution to a particular problem or challenge. There were some early attempts at constructing electrical telegraphs in Europe, but strikingly parallel events occurred in both England and America, presumably independent of each other.

William Cooke was a young English artist who saw an experimental electric telegraph in Heidelberg in 1836. He was impressed with this instrument and thought about designing a useful apparatus based on its principles. On his return to England, Cooke consulted one of the foremost physicists there, Professor Charles Wheatstone, who by coincidence was also experimenting with telegraphy. Together they developed a system for monitoring railway trains (box 2.2). In 1837, one of their telegraphs using underground wires was

Telegraphs, Railroads, and Time Zones

As citizens of the late twentieth century, we are familiar with different time zones dividing large countries: an early afternoon ball game on the West Coast becomes late afternoon viewing for a fan in Philadelphia, but a soccer fan in Moscow has to get up before dawn to watch a match from Vladivostok because of the C.I.S.'s eleven time zones. This makes television broadcasting complicated, and new stories break before a news program can be screened at the "same" hour in other locations.

This was not always so. Official time was often a local matter in the United States until the late ninteenth century. With the advent of the railroads, however, the problem of constructing rational and standardized schedules arose. You could not, for example, have a train travel 100 miles west and arrive at a time earlier than it departed, no matter how swift the old "iron horse" seemed to move. Railroad people quickly appreciated the utility of the telegraph because they now had a handy way of informing stations down the line as to the timing of trains, the number of passengers, the amount of freight, and information on weather conditions (Oliver 1956).

Another enduring legacy of the partnership between telegraphy and railroads in the United States was the establishment of "public time" within the country. With the introduction of the telegraph, messages could travel thousands of miles almost instantaneously. Therefore, it seemed reasonable that an office in Cleveland might be at an earlier time than the sender's office in New York. Before 1883, however, there was a profusion of local times determined by the position of the sun. These times sometimes differed from each other by minutes and seconds. For instance, when it was noon in Washington, D.C., it was 12:13:13 in Albany, New York, and 10:53:31 in Des Moines, Iowa (Stephens 1989). On November 18, 1883, most U.S. and Canadian railroads voluntarily adopted Railway Standard Time and the time zones that exist to this day. The telegraph made it possible for the railroads to synchronize the entire North American continent.

Box 2.2

installed by the London and Birmingham Railway, and the Great Western Railway did likewise in 1839. However, like Morse and Vail, they experienced problems with the insulation failing and had to string wire on poles. In contrast to the Morse telegraph, the Cooke-Wheatstone device depended on the principle, discovered by Oersted, that a magnetized needle will be deflected by a wire carrying current. Instead of moving a pen to write, however, the magnetic field of the current caused the needle to move one way or another depending on the direction of current flow. Using a complicated system of five needles, they were able to actually point to individual letters of the alphabet arranged on a grid. This system required more wires than the Morse telegraph and eventually proved too cumbersome and expensive to operate. They simplified it to a version more like Morse's, and later England adopted the Morse telegraph and code.

On this side of the Atlantic, the allocation of credit for the invention of the telegraph was surrounded by controversy almost from the start. Soon after Morse's early demonstrations at New York University, Dr. Charles T. Jackson, a fellow passenger of Morse on the *Sully* in 1832, claimed to be part inventor of the telegraph. It was Jackson's shipboard accounts of recent experiments in electromagnetism that had apparently inspired Morse to think of using electromagnetic phenomena for telegraphy. And Dr. Jackson wanted part of the credit and the anticipated profits, as did many others.

There were lawsuits and charges and counter-charges in pamphlets and newspapers. Before it was all over, there were sixty-two claimants to the invention of telegraphy! Morse was kept busy defending his precedence and his patent rights. One of the controversies that developed involved Joseph Henry. "The one highly deplorable controversy of his lifetime, in which he was quite unforgivably wrong," historian Allan Nevins wrote "was that with Joseph Henry, a scientist as gentle as he was great, whose friendship and aid had been invaluable to Morse" (Mabee 1943).

In one of the lawsuits, Henry was asked to testify as an expert witness. Morse took exception to some parts of the depositions made by Henry at the trials. In 1867, Morse published a pamphlet in which he wrote, "I am not indebted to [Henry] for any discovery in science bearing on the telegraph; and that all the discoveries of principles having this bearing, were made, not by Professor Henry, but by others, and prior to any experiments of Professor Henry in the science of electro-magnetism" (from Morse, "A Defence Against the Injurious Deductions Drawn from the Deposition of Prof. Joseph Henry," as quoted by Coulson 1971). This was clearly indefensible. A contemporary biographer wrote, "[Morse] believed himself an instrument employed by Heaven to achieve a great result, and having accomplished it, he claimed simply to be the original and only instrument by which that result could be reached" (Prime, as quoted by Coulson 1971). A later biographer, the one who had called him "The American Leonardo," was kinder: "the controversies ... had warped his perspective" (Mabee 1943).

But Morse was no stranger to controversy. He had, in fact, sought it many times in his life. His participation in the founding of the National Academy of Design was a rebellion against the American Academy of Fine Arts. He had espoused a sufficient number of unpopular causes that he had been called "... a snob, a bigot, a charlatan avid for fame ..." (Mabee 1943). In the 1830s he wrote polemics against the Irish and other "foreigners," against Catholics, and immigration, and naturalization. He was a candidate for mayor of New York City on an antiforeigner platform and lost. During the Civil War, he was president of a proslavery society, and cofounder of an organization that the poet and then editor of the *New York Evening Post*, William Cullen Bryant, claimed to have been involved in an "unscrupulous campaign against the government ... and in behalf of a body of rebels now in arms." (Mabee 1943). While he considered himself a "Jeffersonian Democrat," he did not quite agree with Jefferson's notions of equality and the "inalienable character of liberty" (Mabee 1943).

Even after more than a century, the controversies in which Morse was involved and the causes he espoused still aroused much emotion. These issues also confuse the allocation of credit for the development of telegraphy. It is true that others built telegraphs of various forms before he did, that Henry's scientific work and personal assistance were invaluable to him, and that the mature form of the telegraph was due mainly to Vail. It may even be true that Vail had a lot to do with the development of Morse code. But it was

> ## Edison and the Telegraph
>
> Say "Thomas Edison" and most people think of the electric light bulb, the phonograph, or, in some regions, the electric utility company. The legendary American inventor was also initially linked to the telegraph: his first profession, beginning at age 16, was that of a telegrapher on the railroad. He worked for Western Union at one stage, but he also kept up an ardent interest in inventing. He was especially fond of experimenting with chemicals, which often got him into trouble when things went awry.
>
> In the late 1860s, Edison moved from the midwest to Boston and began work on an ingenious application for telegraphy—a stock market ticker. In order to market this invention more effectively, he moved to New York and set up shop to manufacture stock tickers for Western Union's network. He also made improvements on other telegraphic equipment. With the money he earned from these devices he was able to fund the construction of his famous research and development laboratory ("invention factory") in Menlo Park, N.J. There, he and his skilled staff worked tirelessly to develop a system of lights powered by electricity and numerous other technologies.
>
> Like Samuel Morse with the telegraph, Edison clearly envisioned what devices he wanted to construct, knew that they would be important new technologies, and had every intention of realizing profits from them. In particular, the light bulb, as with the telegraph, was envisioned as part of a large system that would change people's daily lives. (Abbot 1932)

Box 2.3

Morse who had the vision to bring all these elements together. He successfully used electromagnetism to transmit information in the form of signals using a code consisting of only three symbols (dot, dash, pause). He took advantage of what scientific knowledge was available, as does every inventor. It was his singlemindedness and entrepreneurship that led to the development of a practical system of telegraphy.

Growth of a Communications Network

Slowly at first, and then rapidly, often chaotically, the American telegraph network grew and telegraphic technology improved, one providing impetus for the other. By 1861 transcontinental U.S. service was introduced. By 1866, the U.S. network included 100,000 route miles.

Although the first experimental line was laid on railroad right-of-way, it took a while before railroads and telegraph companies joined in a mutually beneficial partnership. Eventually, however, the advantages of cooperation became obvious to both sides. Telegraph lines paralleled railroad tracks, the railways provided easy access to the lines for installation and maintenance, and the railways depended on the telegraph for scheduling trains. This latter function was crucial, especially for single-track railroads. This partnership between the railroads and the telegraph networks helped to facilitate the settling of the American West and contributed to the forging of a sense of nationhood among people scattered from one coast to the other.

The electric telegraph paved the way for the development of the telephone (by Bell in 1876) and inspired the invention of wireless communication (by Marconi in 1897) and eventually radio and television. Today satellite and fiber optic cables cover the globe. For example, the trans-Atlantic cable laid in 1988 can carry forty thousand simultaneous conversations. Morse would certainly be pleased that he pioneered the beginnings of the information age.

Was Morse an "inspired gadgeteer" who doggedly pursued one insight and then retired on the profits it garnered? It seems too simple a description for this multifaceted and probably not very likable genius. He was not, by any means, as original, or as creative, or as brilliant as Leonardo da Vinci. While the biographer's enthusiasm is understandable, the comparison that he suggests by calling Morse "the American Leonardo" is rather inappropriate. But the system of telegraphy that Morse and his associates developed, and the communications networks that grew from it, has certainly affected more lives than DaVinci's work. By the time Morse died in 1872, his vision of the "instantaneous communication of intelligence" across telegraph lines girding the globe had become a reality. Morse developed a practical system of telegraphy. This system included not only transmitting and receiving devices but also a code utilizing signals that could reliably and easily be transmitted with the existing technology. Just as important, he founded a network that eventually grew to make possible, for the first time, rapid worldwide communication.

Discovery

Exercises

1. Some time ago, you probably learned that Morse invented the telegraph. How would you consider that assumption now?

2. (a) A wire commonly used in homes is No. 10 copper wire, which has a radius of 0.129 cm (1.29×10^{-2} m). What is the resistance of 12 m (approximately 40 ft.) of this wire? (Resistivity of copper $= 1.7 \times 10^{-6}$ ohm-cm $= 1.7 \times 10^{-8}$ ohm-m).
   (b) If Morse had used the above wire for his initial 40-ft transmission test, what current would he have generated in his circuit using a single 1.5 V cell?

3. In the case of the development of the telegraph, what was the relationship between the "basic science" and the "technology"? What scientific principles helped the inventors in their pursuit of a practical telegraph?

4. (a) In some of his demonstrations, Gale used twenty 1.5-volt cells in series to get a 30-volt battery. He claims to have been able to transmit through 10 miles (16 kilometers). Assuming that he also used something similar to No. 10 copper wire, with what current was he operating?
   (b) Gale's demonstrations used the same receiver that Morse used for his (failed) 40-ft attempt except that Gale increased the number of coils in the electromagnet. If Morse used only ten coils, how many coils were needed in Gale's improved version to generate the same magnetic field in the electromagnet?

5. Why were the railroad and telegraph systems linked? What did this have to do with the institution of standard time zones?

6. (a) Estimate the time needed for a minimum type of signal to be transmitted across France using the semaphore system built in the late 1700s.
   (b) Estimate the time required for a telegraph signal to go from coast to coast in the United States (Hint: electromagnetic signals travel at about 3×10^8 meters per sec.)

Hydroelectric Power: The Irony of Los Angeles

3

An Unlikely Spot for a City

If you conducted a poll just over a century ago asking people to name the ten likeliest spots for a major American city, few if any would have answered "the Los Angeles basin." Los Angeles housed only ten thousand people in 1880. Residents of the dusty town lived under spare conditions by today's standards. Most people still suffered the odor and risk of kerosene in order to illuminate their homes. Virtually no industries had seen fit to locate in the basin. Most income came from cattle ranching and agriculture. Horses and feet were the common modes of transportation.

The climate of Southern California is attractively mild but also dry. The Los Angeles basin nearly qualifies as a desert, receiving only fifteen inches of rain per year on average. Lack of rainfall robbed people of a source of energy commonly used elsewhere, flowing water. The Los Angeles River was small and supplied barely enough water for domestic and agricultural needs. To make matters worse, the area lacked abundant supplies of other common energy sources of the day such as wood and coal. Oil had been discovered in the basin about the time of the Civil War, but it was low-grade material that refused to submit to the refining techniques in use at that time. Unlike cities in the eastern United States, historical patterns of settlement in the Los Angeles basin led to a widely dispersed population, which increased the difficulties of distributing people, water, and power.

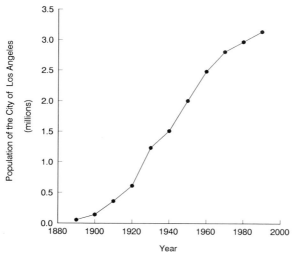

Figure 3.1 Population growth in the city of Los Angeles, California.

A less likely place for a major metropolis might have seemed hard to find in the United States during the latter part of the nineteenth century. Yet extension of transcontinental railroad service to Southern California in 1876 encouraged immigration from the midwest as land developers offered train tickets to Los Angeles for as little as one dollar. The city's population passed the half million mark shortly after the turn of the century and exceeded 1.2 million ten years later. Now more than 3.4 million people live in the city of Los Angeles (figure 3.1), and four times that number live in the Los Angeles basin. Los Angeles is currently the second largest city in the United States with an industry-based economy that rivals that of many countries.

Something dramatic must have happened to transform an apparently forbidding region into the land of freeways, suburbs, Hollywood movies, and Disneyland. The small town of Los Angeles was swept up in a tide of engineering innovation that marked the Second Industrial Revolution. The renewed sense of opportunity, growth, and progress that accompanied the new technologies helped people see the rigorous environment of Southern California not as an insurmountable barrier to development but as a challenge to American ingenuity and determination. Perhaps no challenge was greater than supplying energy to a rapidly growing population.

Electricity, that mysterious phenomenon, proved key to meeting this challenge and became a symbol of the new era. Los Angeles provides an especially useful illustration of how major electrical technologies gained prominence because the rigors of the local environment demanded innovation in power generation. These innovations depended, in turn, on nineteenth-century scientific discoveries regarding the relationship between electricity and magnetism. The major lessons gleaned from the electrification of Los Angeles apply generally to large cities across the nation.

The First Electricity in Los Angeles

People who witnessed the modest beginning of electrification in Los Angeles could hardly have realized what was to come. On December 31, 1882, electric current flowed through twenty-one outdoor arc lamps arranged on seven street corners in the center of Los Angeles. The power for the arc lamps came a short distance from a nondescript building that contained two small electrical generators powered by coal-fired steam engines. This first electric power plant in Los Angeles, located at the corner of Banning and Alameda Streets in what is now Little Tokyo, produced up to 300 kilowatts (kW)—300,000 watts—of direct current electricity, less than 0.1 percent the output of a typical modern electric power plant.

The new arc lamps provided much brighter light over a wider area than was supplied by the more common gas lamps, but the arc lamp system experienced difficulties. At the outset, acquiring fuel for generating electricity posed a significant problem in Los Angeles. Most of the coal used in the Banning-Alameda steam plant had to be imported from Australia and New Zealand because local coal deposits were small and poor quality, and transporting coal from Utah or Colorado by train was surprisingly expensive.

Also, not everyone greeted the new technology with open arms. Arc lamps emitted an intense glow as electricity flowed between two slightly separated carbon rods heating the carbon until it produced an incandescent brilliance equal to four thousand candles per lamp. Some people feared that the unprecedented brightness of the lamps endangered ladies' complexions. Chicken owners, not a small fraction of the city's populace at the time, worried that the arc lamps would keep their chickens awake. Other complaints that electricity was an unnecessary and costly experiment probably originated with investors in local gas companies who stood to lose money if the new technology succeeded. Despite high fuel costs, fears of a new technology, and resistance from supporters of existing lighting systems, the great utility of electricity was beginning to be realized. The ability to transmit electrical energy over thin cables allowed industries to dispense with noisy systems of belts, pulleys, and axles or dangerous gas pipes. Once an electrical distribution system was completed, electricity proved much cheaper to transport over long distances than other forms of energy. The great speed at which electricity could be moved from one place to another offered another distinct advantage over other energy forms, an advantage also appreciated in communication systems. The ability to quickly alter the "behavioral" properties of electricity by changing its voltage and current ultimately supported its utility in a staggering array of applications. Despite these advantages of electricity, the technology initially used to produce electrical power in Los Angeles proved insufficient to accommodate the city's rapid growth.

By the mid-1890s, Los Angeles had grown from a town into a small city of nearly a hundred thousand residents, and the uses of electricity had spread from arc lamps to the operation of trolley cars and telephones. Widespread availability of incandescent light bulbs accelerated the use of electricity for indoor lighting in replacement of gas, but the trolley companies accounted for the lion's share of electrical power development in the early days. One of the many ironies of Los Angeles, famous as it now is for traffic jams on freeways, is that it once sported the largest interurban rail transit system in the world. Electric trolleys carried thousands of people every day around the downtown area and out to the burgeoning suburbs.

Small, coal-fired steam plants such as the Banning-Alameda plant simply could not meet the city's growing demand for electricity. Put another way, the city could not grow rapidly unless new methods for generating electricity were found. Electric power companies, then as now, often operated under confused motives as they have tried not only to meet but also to create demand for their product. The only way to supply large amounts of electricity to Los Angeles in the late 1880s appeared to involve burning more coal and building ever larger steam engines. Coal was too expensive to make this a very attractive option to power company executives, however, and simply duplicating small steam plants, like the Banning-Alameda plant, presented none of the economies of scale offered by large centralized power plants. In addition to these economic constraints, serious technical problems awaited anyone who attempted to use existing steam engine technology to build a much larger power plant. A new source of energy had to be found.

Water Power in a Desert

Ironically for Los Angeles, water provided the new energy source. Although the basin itself remained as dry as ever, the surrounding mountains abounded with potential sources of energy in streams and rivers. People in the late 1880s understood well enough how to extract electrical power from falling water, but they did not know how to transmit that electricity long distances. As long as that problem remained unsolved, the mountain streams surrounding Los Angeles lay beyond the reach of electrical technology.

The solution to Southern California's early energy problem first appeared in Germany in 1891 at the International Electrical Exhibition in Frankfurt on the Main. On August 24 of that year electricity began to flow from a hydropower plant in Lauffen, near Stuttgart, 109 miles north to Frankfurt. This demonstration that electricity could be transmitted a long distance from a river to a city was made possible by a new type of electrical generator called a *three-phase AC generator*.

It might seem that electrical engineers in Los Angeles would have gobbled up the new technology in order to begin tapping new energy sources for their growing city. But major technological advances often get introduced as solutions to small, specific problems rather than as means for achieving some grand design. This proved to be the case for three-phase AC generators in Los Angeles. The new form of generator eventually became the standard for commercial production of electricity, but it arrived in Los Angeles by way of a small town that needed inexpensive ice for its oranges and lemons.

Citrus, Ice, and Three-phase AC Electricity

Citrus growers in Redlands, a small town seventy miles east of Los Angeles, faced escalating costs in the 1880s that threatened to drive the prices of their fruit to exorbitant levels. It had become very expensive to make ice for the refrigerated rail cars that carried the fruit to eastern markets because the Union Ice House in Redlands, supplier of ice to local citrus growers, depended on coal-fired steam engines to generate its electrical power. Just as the high cost of coal affected the electric power plants in Los Angeles, it affected ice production in Redlands as well. The citrus growers needed cheaper ice, and that required cheaper electricity.

Henry Harbison, a local businessman, led a group of citrus growers and entrepreneurs in Redlands to form the Redlands Electric Light and Power Company, which raised funds for a hydroelectric plant on Mill Creek about twelve miles east of Redlands. The chief engineer for this project, Almarian W. Decker, was aware of the demonstration at the Frankfurt Exhibition and elected to install three-phase AC generators at the new facility. A transmission distance of twelve miles was substantial in southern California at that time, and Decker reasoned that the new type of generator would be best for the job.

The generators went into service at the new Mill Creek Station #1, as it was called, on September 7, 1893, making it the first three-phase AC electrical power plant in the United States. Mill Creek Station #1 supplied electricity to Redlands so effectively that two other stations were soon added upstream from #1 to generate power for a hotel elevator, city lights, and pumps that supplied irrigation water to crops. Mill Creek Station #1 has produced electricity continually since 1893 when the first generators were installed. One of the original generators remains on display at the station although it is no longer in service.

Virtually all commercial electricity now takes the form of three-phase AC, but you would never guess the historical importance of Mill Creek Station #1 from its appearance. The small, tan, stone building sits by itself in a nearly dry creek bed below San Gorgonio Mountain. Mill Creek flows year round carrying snow melt from the southern slopes of San Gorgonio Mountain to the

Santa Ana River. This flow of water, small as it is, represents a source of energy for the generation of electricity. An understanding of how electrical energy can be extracted from falling water proved crucial for the development of Los Angeles. The Mill Creek power plant illustrates the basic principles especially clearly. Large-scale energy production by a hydroelectric power plant required more than an understanding of the basic scientific principles, however, as described when we discuss the Hoover Dam.

Energy and Power in Falling Water

Water wheels have been used for hundreds, perhaps thousands, of years to convert the energy in falling water into mechanical (rotational) energy (figure 3.2). Lifting water to a height above the water wheel gives it the potential to do this work. That is, lifting the water invests it with potential energy. The amount of potential energy is equivalent to the amount of work done in lifting the water. Since work is the product of the force applied to an object and the distance that the object is moved, then the amount of potential energy in a body of water is the product of the water's weight and the height to which it has been raised. Typically, the sun supplies the energy needed to do this work by heating the water until it evaporates and rises into the atmosphere. The water then rains down to fill the lakes behind dams. But we can store energy in water as well. According to an increasingly attractive option, for example, the excess electrical energy generated by nuclear power stations or wind turbines during hours of low consumption can be used to pump water behind a dam and store the electrical energy as potential energy for future use. *Pumped storage*, as this process is called, overcomes one of

Figure 3.2 Transformation of energy in a hydroelectric power plant.

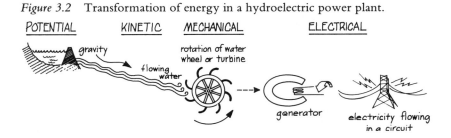

the nagging problems with electricity—the inability to store it for extended periods.

The weight of water is the product of its mass, m, and the acceleration due to gravity, g: one cubic foot of water weighs 62.4 pounds. The amount of potential energy (PE) in one cubic foot of water above Mill Creek Station #1 can be determined by using the following equation in which h stands for head, or the height to which the water has been raised behind the dam:

potential energy, $PE = (mg) \times h$

$= 62.4 \text{ lbs} \times 510 \text{ ft}$

$= 31{,}800 \text{ ft-lbs}$

The potential energy of water above a hydroelectric plant sets the upper limit on the energy available to the facility. That energy is extracted for the purpose of doing work only when the water is allowed to fall. Once that happens, the potential energy is expressed as *kinetic energy*, or the energy of motion. The faster a given amount of water moves, the more kinetic energy it has and the more work it can do. As water is lifted higher above a water wheel and its potential energy increases, it will strike the wheel at a higher velocity and thus will be capable of more work.

The engineer's main concern, in this case, is how rapidly that work can be done, that is, the amount of power that can be developed. More electrical power will be produced if the water wheel and generator rotate rapidly. Water may be piled high behind a dam, but if it trickles out at a low rate, it is not very useful for doing work because it does not generate much power. The two major contributors to power in a stream are the rate at which a given amount of water falls and the vertical height, the head, through which it drops to the water wheel. One cubic foot of water falling 8.81 feet in one second represents one horsepower or 550 foot-pounds per second (ft-lbs/sec).

At Mill Creek Station #1, water falls 510 feet through a pipe, the *penstock*, at a carefully regulated rate of 28 cubic feet per second (ft^3/sec). Multiplying the flow rate (q) by the density of water (d) gives the rate at which water falls in pounds per second. Multiplying this value by the head (h) gives the maximum amount of available power (P_a), the amount that would be extracted from the stream if the water wheel functioned with 100 percent efficiency.

$$P_a = qdh$$

$$= \frac{28 \text{ ft}^3}{\text{sec}} \times \frac{62.4 \text{ lbs}}{\text{ft}^3} \times 510 \text{ ft}$$

$$= 891{,}000 \frac{\text{ft-lbs}}{\text{sec}}$$

You get a feeling for this relationship every time you stand under a shower. If a small amount of water falls slowly through a short distance before it hits your head, the experience is pleasantly refreshing. A large amount of water falling a great distance before pounding you on the head produces a less pleasing experience.

The calculation just described reports the power available to Mill Creek Station in foot-pounds per second. Although this unit supplies a perfectly good measure of power, electrical engineers prefer the kilowatt (kW). One kW equals 737 ft-lbs/sec. In deference to electrical engineers, the power available at Mill Creek Station #1 is converted to kW:

$$P_a = 891{,}000 \frac{\text{ft-lbs}}{\text{sec}} \times \frac{1 \text{ kW}}{737 \text{ ft-lbs/sec}} = 1200 \text{ kW}$$

This is sufficient power to meet the average needs of 1,200 households in the United States, although it would not satisfy peak demand. It also amounts to about 2 percent of the power delivered by the smallest nuclear power plant currently operating in the United States.

Capturing the Power: The Water Wheel

Almarian Decker learned from similar calculations that Mill Creek could supply sufficient power for the Union Ice House if enough of the water's power could be captured. Decker, undoubtedly like every other engineer of a hydroelectric plant, would have loved to convert all of the power available in the falling water into electrical power. Unfortunately, nature does not allow us to be so efficient. Some energy that might otherwise be used to do work is lost in friction as water falls along rocks and pipes and as the water wheel turns in its bearings. This places a burden on the engineer to design or choose a water wheel that minimizes unnecessary energy losses and ultimately maximizes electrical power production.

Figure 3.3 One of the original Pelton wheels and generators in the Mill Creek Station #1 hydroelectric facility near Redlands, California. Water emerged at high velocity from the jets to strike the buckets of the Pelton wheel (foreground). The wheel's rotation turned the rotor of the generator on which were mounted the wire coils (background). The coils rotated within a circle of electromagnets, six of which are visible in the photograph. (Photograph by N. Copp.)

Decker selected water wheels made by Lester Pelton. Pelton, an engineer working in California, produced some of the best water wheels made in the late nineteenth century. Pelton wheels featured twenty to thirty buckets distributed around the perimeter of a disc several feet in diameter (figure 3.3). Each bucket consisted of two spoon-shaped surfaces separated by a ridge or "splitter." A jet directed water from the penstock to the splitter of a bucket. The stream of water, divided by the splitter, swept across the two sides of the bucket before being discarded into the tailrace below the wheel. As the wheel turned in response to this collision, a new bucket swung into the stream and the process continued.

The action of the Pelton wheel illustrates the law of conservation of momentum that can be derived from Isaac Newton's laws of motion. Newton, in the late seventeenth century, departed radically from the ancient teachings of Aristotle in arguing that bodies in motion possess an intrinsic "quantity of motion." We now call this quantity *momentum*, but we preserve Newton's definition of it as the

> ## Why a High Flow Rate Is So Useful
>
> The amount of kinetic energy in a given mass (m) of flowing water can be increased by a factor of four simply by doubling the velocity (v) of its flow:
>
> kinetic energy $= 1/2 mv^2$
>
> The key factor in the operation of a Pelton wheel is the change in kinetic energy as a given mass of water flows over the scooped buckets, that is, the difference in kinetic energy between the water that enters the bucket rapidly (v_1) and the water that leaves the bucket less rapidly (v_2):
>
> change in kinetic energy $= 1/2 m(v_1^2 - v_2^2)$
>
> The jet at the end of the penstock assures a high value for v_1. Because water is delivered to the pipe behind the jet under pressure and at a regulated flow rate, its velocity increases as it passes through the jet's constricted opening. This is similar to the way you increase the velocity of water flowing from a garden hose by using your thumb to reduce the opening at the end of the hose. At Mill Creek Station, v_1 was about 160 ft/s, and v_2 was about half that.
>
> The high flow rate creates a large amount of kinetic energy which, in turn, allows a great deal of work to be done in turning the wheel rapidly.

Box 3.1

product of a body's mass and its velocity. The *conservation of momentum* means that, in systems with no net external force acting on them, the amount of momentum remains constant. The interaction of water with a Pelton wheel can be considered such a system for the purposes of this chapter.

The jet placed at the end of the penstock in Mill Creek Station #1 served to increase the velocity of the water and thereby enhanced the conversion of potential energy into kinetic energy of the stream that struck the buckets on the wheel (see box 3.1). The collision of the water with the wheel's buckets reduces the water's velocity, hence its momentum. The conservation of momentum requires that the reduction of the water's momentum be accompanied by an increase in the momentum of the water wheel. Since the water wheel's mass does not change in this interaction, its velocity must increase, and so it does.

The transfer of momentum from the water to the wheel continues as long as the water retains some velocity and remains in contact with a bucket. This transfer occurs so quickly in the case of Pelton wheels that it is called an *impulse*. Not surprisingly, Pelton wheels are classified as impulse wheels. The magnitude of an impulse is the change in momentum or the total amount of force applied to the wheel over the time that the water contacts the bucket. The scooped design of the buckets reflected Pelton's attempt to transfer as much momentum as possible from the water to the wheel. His innovation nearly doubled the momentum transfer because the bucket nearly reversed the direction of the water's flow and because the water remained in contact with the bucket's surface for a long time. (Henry Aaron repeatedly applied this scientific principle in becoming the greatest home run hitter in professional baseball. High-speed photographs reveal that, by snapping his wrists as the bat passed over home plate, Aaron managed to keep his bat in contact with the ball for a slightly longer period of time, creating a larger impulse than other hitters.)

Pelton sought to make a water wheel that captured as much energy in falling water as possible. Pelton wheels gained rotational energy as the stream of water lost kinetic energy. If Pelton could have designed a bucket that reduced the water's kinetic energy to zero, he would have come quite close to the unattainable efficiency of 100 percent. Water with no kinetic energy is not moving, however, and would not exit the bucket, thus making the water wheel useless. Despite nature's constraint, Pelton succeeded admirably and designed water wheels that operated with an efficiency near 90 percent over a wide range of flow rates. Chief Engineer Decker wisely selected Pelton wheels for Mill Creek Station #1 because of both their remarkable efficiency and their suitability for the conditions of high head and low flow rate characteristic of Mill Creek.

Electricity from Magnetism: The Scientific Discovery

A hydropower plant brings together in coordinated service two technologies, the water wheel and the generator, that could hardly be more different in terms of their origins. Pelton wheels evolved gradually as an ancient technology was fine-tuned and improved by application of basic physical principles that had been understood for

Figure 3.4 Michael Faraday. (Reproduced from *Faraday's Diary*, vol. 1, G. Bell and Sons, Ltd. publishers.)

over two hundred years. The electric generator, on the other hand, grew fresh from a scientific breakthrough in our understanding of nature. The new discovery did not lead to improvements in a preexisting technology but to an entirely new idea for controlling a part of nature.

Michael Faraday, a slightly built but tireless experimental physicist (figure 3.4), worked in England during the 1820s and 1830s to understand the relationship between electricity and magnetism. He knew such a relationship existed because, in 1820, Hans Christian Oersted had observed in one of his lecture demonstrations that a magnetized needle could be deflected simply by running an electric current through a nearby wire. The needle's behavior contradicted what Oersted had told his class to expect, as happens all too frequently in lecture demonstrations, but Oersted had the presence of mind to admit that something must have been wrong with the prediction instead of blaming the anomaly on faulty equipment. Oersted's discovery of electromagnetism quickly found application in the telegraph and established the basis of the telephone, as de-

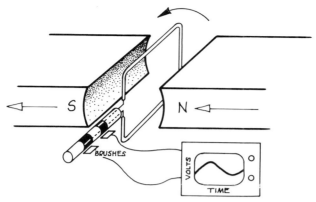

Figure 3.5 Electromagnetic induction. Turning a wire coil within a magnetic field produces an electric current with a sinusoidally varying voltage.

scribed in chapter 2. The American Association of Physics Teachers later honored Oersted by naming their award for "notable contributions to the teaching of physics" after him.

Guided by an unshakable faith in the symmetry of nature, Faraday reasoned that, if electricity can create a magnetic field, then magnetism must be able to create electricity. He worked on and off for the better part of ten years arranging bits of wire, metal, delicate voltmeters, and the best permanent magnets in England to produce every conceivable configuration in his search for this expression of the relation between electricity and magnetism. Finally, on October 17, 1831, Faraday made the crucial observation that the magnet must *move* with respect to the coil for an electric current to be induced in the wire. In all his previous fiddling, the various coils and magnets had remained stationary relative to each other, and no sustained current emerged. Only when he thrust a bar magnet into the cavity of a wire coil or withdrew it did he observe a powerful response by his homemade voltmeter. Faraday's faith was vindicated: magnetic fields could indeed be used to induce electric currents in a phenomenon called *electromagnetic induction* (figure 3.5). Joseph Henry, the American scientist and coinventor of the telegraph (see chapter 2), discovered induction at almost exactly the same time as Faraday, but Faraday supported his ideas with more extensive experimental evidence than Henry and published his work first, so credit that has gone almost exclusively to Faraday has been justified.

After testing various materials and configurations in his characteristically thorough way, Faraday replaced the wire coil with a copper disc that rotated between the poles of a permanent magnet and, in doing so, built the first electric generator. Faraday showed no interest, however, in producing generators for practical purposes. His desire, typical of so-called pure scientists, remained entirely to understand the nature of electricity and magnetism. He spent the rest of his career performing variations on his electromagnetic induction experiments, studying the magnetic properties of various materials, conducting experiments in electrochemistry, and attempting to relate gravity to electricity, a task that Einstein subsequently undertook and also failed to complete.

Electricity from Magnetism: Application

A discovery as fundamental as Faraday's rarely lies around very long without being put to use. Many people quickly realized the practical significance of Faraday's work and began considering how to build electrical generators for commercial use. Faraday's prototype produced pitifully little electrical power and served only for experimental work on induction (see box 3.2). A useful generator would have to be much larger to produce enough electricity for commercial benefit. The first practical application of the generator came in 1853, twenty-two years after Faraday's crucial observation, when a device much like Faraday's was used to power an arc lamp. This device proved too feeble for widespread use. The problem of creating a useful generator proved more difficult than simply scaling up Faraday's prototype.

Nearly fifty years separated Faraday's description of electromagnetic induction from its application in the most direct ancestor to modern generators. The slow pace with which the generator was developed may seem surprising given the clarity of Faraday's discovery. But simply knowing the scientific principle proved insufficient to spark immediate production of broadly useful generators. Problems in engineering design appeared independently of problems in scientific understanding and required the creative efforts of a great many people interested much more in making a commercially successful device than in understanding electricity and magnetism. Also, developers of the generator may not have been driven by a

> ## A Web of Consequences
>
> Just as Oersted's discovery of electromagnetism led in two distinct directions, one toward telecommunications and the other toward a greater understanding of electricity and magnetism, Faraday's description of electromagnetic induction led to significant scientific advances as well as engineering developments. Faraday was averse to modeling his ideas mathematically and so never produced a mathematical description of electromagnetic induction. James Clerk Maxwell, a Scotsman forty years younger than Faraday, generalized from Faraday's experimental observations to produce four mathematical statements that became the core of electromagnetic theory. Maxwell's famous equations, first published together in 1864, worked out the interactions between electrical and magnetic fields and enabled people for the first time to predict electromagnetic behavior. This small group of equations remains one of the cornerstones of physics. The engineering and scientific outcomes of a major discovery do not remain distinct, however. Maxwell's equations, in addition to describing nature, provide the basis for designing electrical transmission systems, radio and television broadcasting systems, computers, and a myriad of other electrical devices. In this way, threads of engineering and science become woven in a complex web.

Box 3.2

sense of urgency. Batteries had been reasonably satisfactory suppliers of steady current for the meager variety of electrical devices in existence at the time. The apparent delay between scientific discovery and application may be somewhat illusory, however, because the generator did not simply spring on the technological scene fully formed. It evolved as understanding improved and as the demands of newly developed electrical devices and systems changed.

Progress toward a useful generator did not depend entirely on trial and error but was guided by Faraday's detailed understanding of the induction phenomenon. Faraday had learned from his experiments that the size of the electrical difference (i.e., the voltage) produced by a generator depends primarily upon the number of turns of wire in the coil (N), the strength of the magnet field (B), and the rate at which the coil rotates through the magnet field (indicated by the quantity $\omega \cos \omega t$ where $t = $ time). These quantities are now related in a single equation with a fourth factor, the area swept out by the wire coil (A), to give the voltage that a generator will produce. Knowing the generator's voltage helps predict its current because voltage and current are related in these systems by Ohm's Law (see chapter 2): the larger the voltage created by the generator, the larger the current. Equations such as the following have accelerated the development of electrical generators by providing a sound rationale for design decisions.

$$V = NBA(\omega \cos \omega t)$$

A German company directed by Werner von Siemens considerably increased the generator's power production in 1856 by replacing the rotating disc with a series of densely wound wire coils (an armature) and substituting powerful electromagnets for the less effective permanent magnets. These changes increased factors N, B, and A in the preceding equation and so increased the voltage and current produced by the generator.

The Siemens generator was well suited to the arc lamp systems of the day, but it proved inadequate for the more modern incandescent lamps developed by Thomas Edison. Edison resolved the problem in 1879 by designing and building his own generators that produced more power than the earlier ones and became the forerunners of modern electrical generators.

The basic components of electrical generators, coils of wire and a magnet, have remained unchanged since 1831 when Faraday conducted his key experiments. The type of electrical current produced, however, depends on how contact is made with the wire coils in which the current appears. The strength of the current in the coil actually fluctuates as the coil rotates through different positions relative to the lines of the magnetic field. By cleverly designing a series of contacts between the transmission wires and the generator's coil, however, Edison was able to draw off a current of nearly constant strength. This type of electricity is called *direct current*, or DC, because current flows in only one direction through the circuit. Edison built this type of generator because it best suited the system of incandescent lamps that he designed and later installed around his generating station on Pearl Street in New York.

The first electrical generating station in Los Angeles followed Edison's lead and employed DC generators (figure 3.6). In these generators, multiple coils of wire wound on an iron loop rotated through the fields of two sets of magnets. Each coil was electrically

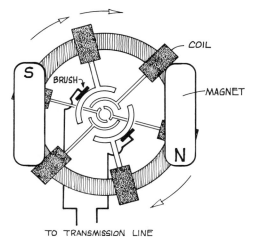

Figure 3.6 Schematic diagram of a Brush generator looking along the axis of rotation. Brushes at the ends of the transmission lines make electrical contact with coils on opposite sides of the rotor. The contacts from the coils are shown in concentric rings for the sake of clarity. They actually extend the same distance inward to the shaft of the rotor. The rotor rotates behind the electromagnets.

connected to a small metal strip running along the shaft of the rotor. Each brush contacted only one metal strip at a time as the shaft rotated under it. A brush would thus pick up current only when the coil and its corresponding metal strip rotated into a certain position relative to the brush and the magnet. The magnitude of that current changed only slightly as the metal strip rotated under the brush before being replaced by the next strip on the shaft. The magnitude of current "seen" by each brush thus remained nearly constant. Because the two brushes picked up current from different, fixed points within the coil's cycle, the electrical difference or voltage that appeared between them also remained nearly constant.

As the Current Flows: AC or DC?

Thomas Edison relentlessly championed the virtues of DC electricity. George Westinghouse, another prolific American inventor, took a decidedly different position. Not long after Edison built his Pearl Street Station, Westinghouse brought into commercial production a different type of generator, one that produced electricity with a periodically varying voltage. This form of electricity is termed *alter-*

nating current or AC because the voltage and current regularly alternate directions as the generator runs through its cycle. Edison and Westinghouse were soon embroiled in a prolonged and spirited debate, often called the battle of the currents, over which type of electricity should come into standard use. The arguments for each were complex and varied, but proponents of DC cited the availability of DC motors in boosting their case whereas proponents of AC pointed out its advantages in terms of long-distance transmission. Unscrupulous opponents of AC, with Edison's support, encouraged the public to associate that form of electricity with death by lobbying the New York State legislature to use Westinghouse generators to electrocute capital criminals. The first electrocution as capital punishment was indeed carried out in New York in 1890. This gruesome publicity ploy had little effect, however, and AC electricity gradually came into general use.

In selecting an AC generator for the Mill Creek Station, the young engineer Decker audaciously contradicted the wisdom of Edison who had done more than any other single person to establish commercial electricity and electrical power systems. Even Edison recognized, however, that it was prohibitively expensive to transmit DC electricity over distances greater than a few miles. Three-phase AC generators offered Decker the long-distance transmission capability he needed.

The problem of transmitting electricity over long distances stems from the power losses that occur during transmission. Electrical power losses increase with the amount of power transmitted, and the two can be described with the same equation. Ohm's Law relates voltage (V) to current (I) and resistance (R).

$$V = IR$$

Electrical power is given as:

$$P = IV$$

where V refers to the voltage of the electricity leaving the generator. Using Ohm's Law to substitute for V in the power equation gives:

$$P = I(IR) = I^2R$$

These statements also hold true for the electrical power lost as heat along the transmission cable except that in calculating power losses V refers to the voltage drop along the transmission line, not the voltage at the generator. It behooves electrical utility companies to reduce power losses to a minimum. The equations for power show that power losses in a transmission cable increase as the square of the current. Power delivered to the customer, however, is proportional to IV, the product of current and voltage. Transmitting electricity at low current helps utility companies deliver to their customers large proportions of the electricity they generate. In Edison's time, there was no device for reducing the current (and simultaneously raising the voltage) of DC electricity for efficient transmission and then reversing that process for the customer's use. Consequently, Edison and other operators of DC power plants were required to transmit their electricity at high current to ensure that their customers obtained sufficient electrical power. This constraint limited the distance over which DC could be transmitted and remain affordable. Longer transmission distances required longer cables, of course, but longer wires meant higher electrical resistances. Higher resistances, in turn, brought greater losses of electrical power. As transmission distance increased, this cascade of effects quickly raised the amount of current needed from the generators beyond acceptable limits. Some utility companies in Edison's time coped with this problem somewhat successfully by installing storage batteries at intervals along the transmission lines. Less expensive answers to the DC transmission problem were eventually developed but not before most engineers turned to AC for the solution.

AC: Three Phases Are Better Than One

After the Frankfurt Exhibition of 1891, AC became the preferred form of electricity for long-distance transmission. The two most important developments that led to this significant advance were the three-phase AC generator and a transformer for AC electricity. Considerable disagreement exists over who deserves credit for developing the polyphase AC generator, the type now commonly used to generate commercial electrical power. Nikola Tesla, a brilliant and eccentric pioneer in electrical engineering, developed the idea of producing and using several AC currents simultaneously. He won several court battles to gain patents for the AC motor and polyphase generator and is widely recognized as the most important figure in their development. Other people made significant contributions as well, however.

Alternating current is created by electromagnetic induction, just as DC is, but instead of producing electricity with a constant voltage and current, AC generators produce a smooth wave of oscillating voltage and current. When a coil of wire rotates in a magnetic field, the induced current gains strength as the coil moves through a position perpendicular to the magnetic field lines and weakens when the coil moves through a position parallel to the magnetic field lines. If electricity is drawn from the coil continuously during its rotation, the voltage and current increase and decrease cyclically at a rate dependent on the coil's rotation rate, and the current will alternate directions. The same effect occurs if the magnet rotates instead of the coil. Most modern generators employ the latter design.

A three-phase AC generator simultaneously produces three electrical currents in this way with each current's cycle lagging behind the preceding one by a precisely fixed length of time. The original generators at Mill Creek Station #1 performed this feat in a beautifully simple way (figure 3.7). Numerous wire coils were wound on a frame attached to the rotating part of the generator, called the *rotor*. Each coil was insulated from its neighbors and connected to one of three metal cuffs encircling the shaft of the rotor. Three adjacent coils led in sequence to three adjacent cuffs on the rotor's shaft. This pattern was repeated many times around the perimeter of the rotor.

The rotor with its coils was mounted on the same axle as the Pelton wheel and moved within a stationary cylinder of electromagnets, called the *stator*. Water striking the buckets of the wheel thus turned the rotor also. The current induced in any coil (for example, coil A in figure 3.7) increased as that coil rotated into the strongest region of one of the magnetic fields and subsequently decreased as the coil rotated out of the magnetic field. A similar alternating current ebbed and flowed in coil B as it followed A in the rotation, but B arrived at the strongest region of the magnetic field and thus developed its peak voltage later than A. The same argument applied to coil C. Each of the three phases of current escaped from the rotor's shaft to a copper conductor by way of a small carbon rod pressed against the corresponding cuff on the shaft.

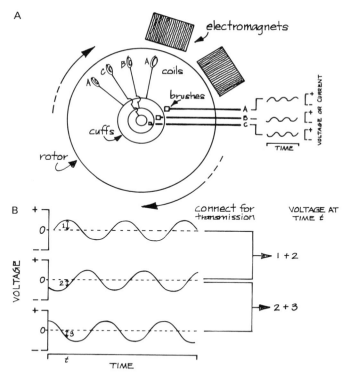

Figure 3.7 Three-phase AC electricity. (*A*) Schematic diagram of the generator at Mill Creek Station #1. Each coil (A, B, C) produces a sinusoidally varying current slightly out of phase with the other two. (*B*) By connecting transmission lines across two of the generator's three outputs, larger voltages are obtained than if each output was simply connected through the transmission wires to ground.

By carefully choosing the space between the coils or electromagnets and adjusting the rotation rate of the water wheel, the engineers at Mill Creek Station #1 produced three-phase AC current that went through sixty cycles each second with an offset of 120 degrees between each pair of phases. Most commercial electricity takes this form today.

Three-phase AC generators offer greater efficiency in the production of electrical power than single-phase AC generators. If a generator produces only one phase of AC—that is, only one electrical wave—the only available voltage is the fluctuating electrical difference between the conductor and the ground. The amount of current, and hence power, produced will be related to this voltage. In a three-phase system, however, large voltages exist not only between each electrical wave and the ground but also between one conductor and another because their electrical waves are not in phase with each other (figure 3.7b). The outputs of a three-phase AC generator can be connected in three pairs (e.g., in figure 3.7b: 1 + 2, 1 + 3, 2 + 3). Each conductor produced by one such pair offers a source of electrical power. Thus, a three-phase AC generator can produce three times as much power as a single-phase AC generator and do so without requiring three times as much material or more water power.

An Unsung Hero

The new generators may not have attracted much attention if transformers had not been developed to increase the voltage and reduce the current for efficient transmission and then reverse the process for use. Faraday established the basis for the transformer when he observed in 1831, just four months before his discovery of induction, that a fluctuating electric current flowing in one coil of wire induces a secondary current in an adjacent, independent coil of wire. The first coil became, in effect, an electromagnet, and it induced an electrical current in any wire within its magnetic field as long as the strength of that field continued to change. This phenomenon is essentially identical to the induction that underlies the operation of a generator.

William Stanley, while working during the early 1880s for George Westinghouse, used Faraday's discovery and expanded upon the work of European contemporaries in developing a transformer to convert electrical currents from high to low values and vice versa. He realized that if the coil on the input side has fewer windings than the output coil, then the output voltage will be larger than the input voltage. A device that increases voltage in this way is called a *step-up transformer*. It might appear that step-up transformers amplify electrical power as they increase voltage because electrical power is the product of voltage and current. Unfortunately, however, nature does not allow us to gain something for nothing, and as the voltage increases in a step-up transformer, the current decreases in the same proportion so that the electrical power leaving the transformer is the same as the power entering minus a small percentage for inevitable losses to heating. A *step-down transformer* performs the inverse task of decreasing the voltage by using an output coil that is smaller than the input coil. Stanley's success in designing transformers contributed significantly to making AC electricity a commercially attractive commodity and helped convince Westinghouse to begin large-scale production of AC generators.

The entrepreneurs backing construction of the Mill Creek Station invited the Westinghouse corporation to bid on the contract for the facility's generators, but the company refused arguing that, although AC was beginning to look useful, the future of three-phase AC generators remained too uncertain to justify an investment. The General Electric Company, major competitors of Westinghouse, jumped at the challenge and supplied four three-phase generators that produced electricity at 750 volts. Step-up transformers at the plant increased this to 12,000 volts for transmission. Step-down transformers on the power poles near a customer typically reduced the value to 120 volts for local use.

Efficiency: An Advantage of Hydro- Over Steam Power

The elegant process of converting the energy in falling water into electricity became widespread in Southern California during the early part of the twentieth century because it was more efficient and less expensive than other sources of electricity in use at the time. The total efficiency of a hydroelectric station is the ratio of the electric power produced to the power available in the water that falls through the station. Mill Creek Station #1, as originally configured, housed four three-phase AC generators, each of which normally produced 250 kW giving a total capacity of 1000 kW or 1 megawatt. The power available to Mill Creek Station #1 was calculated to be 1200 kW. The efficiency of the station was thus:

$$\text{efficiency} = \frac{\text{power generated}}{\text{power available}}$$

$$= \frac{1000 \text{ kW}}{1200 \text{ kW}} \times 100 = 83\%$$

This is a remarkably high efficiency considering all the pathways by which energy can be lost in a hydroelectric station, such as through turbulence in the penstocks and friction in the bearings of the machinery. The efficiency as viewed by the customer would be lower by about 10 percent if the losses during transmission and transformation were counted. In comparison, steam plants in use at the time managed to achieve efficiencies of only 15 percent.

The economy of hydroelectricity proved irresistible in Southern California. Construction of new hydroelectric plants was sure to provide an ample supply of inexpensive electricity to a growing populace, but what form should these facilities take? Replication of small stations like the one at Mill Creek could hardly meet a large demand for electrical power in an economical fashion. Given the factors that govern hydroelectric power, the obvious alternative was to build facilities either with much higher heads or much faster flow rates than at Mill Creek Station #1. Some hydroelectric plants built in the Sierra Nevada mountains of California utilized heads of over 1,000 feet, but the most spectacular achievement in hydropower before World War II explored the other option.

Controlling the Treacherous Colorado River

In the mid-nineteenth century, the Colorado River had a reputation as an unpredictable, treacherous river in a "profitless locality." Major John Wesley Powell led several historic explorations of the Colorado River between 1869 and 1877 and came away with a somewhat more positive appraisal of the river's utility, but remained skeptical that the rigorous environment would ever support large settlements. These pessimistic attitudes later gave way in the face of technological developments that opened the desert to large numbers of people and to agriculture.

Controlling the Colorado for the purpose of irrigation posed the first major challenge for those who wanted to see the desert settled. The river flooded frequently and violently. Seven large floods were recorded between 1825 and 1870. Six more struck between 1904 and 1919. President Theodore Roosevelt, in a speech to Congress in 1907, described the earlier floods as a matter of national concern. Also during this period, the value of Colorado River water for irrigation was becoming increasingly apparent. Construction of small flood control projects and the Imperial Canal just after the turn of the century encouraged farmers to develop the 600,000 acres of fertile lands in the Imperial Valley of Southern California. By 1927, 75,000 people lived in the Imperial Valley, produced crops valued at $40 million, and helped create one of the most valuable agricultural regions in the country.

In 1922 the Secretary of the Interior, Albert Fall, and the Chief Engineer of the Bureau of Reclamation, A. P. Davis, advised that a large dam be built on the Colorado River to control flooding and supply irrigation water to farmlands downstream. This project fell under the jurisdiction of the Bureau of Reclamation that had been created in 1902 for the purpose of making western deserts suitable for agriculture and settlement. The proposed project involved unprecedented engineering and political difficulties. The Colorado River collects water from an area covering nearly one quarter of a million square miles and extends into seven states: Arizona, California, Colorado, Nevada, New Mexico, Utah, and Wyoming. The river winds in a southwesterly direction for 1,700 miles before emptying into the Gulf of California. The onerous task of distributing the river's water among the seven claimant states fell to a distinguished engineer, Herbert Hoover. His commission produced the Colorado River Compact in 1922 that allocated water rights to an upper basin covering four states and a lower basin composed of three states (see box 3.3). Congress followed this action in 1928 by passing the Boulder Canyon Project Act, which was named for the location specified in a preliminary proposal.

Congress placed an interesting stipulation on the project. The federal government agreed to set aside $165 million to pay the project's expenses provided that this amount be repaid with interest within fifty years of the dam's completion. Fall and Davis had recognized earlier that a large dam on the Colorado River could generate enough electricity to pay for its own construction, but a suitable market for the electrical power had to be found.

The city of Los Angeles offered one such market. Ironically, the city's guarantee came from its need for water rather than electricity. In 1925, the city authorized a bond of $2 million to study the feasibility of constructing a canal to carry water from the Colorado River to Los Angeles. Regardless of the proposed route, considerable amounts of electricity would be needed to pump water over the mountains that intersected the canal's path. Approval of the bond indicated to Congress that enough electricity could be sold

The Colorado River Compact

The dispute over the proposed dam on the Colorado River grew fierce as the seven states in the river's watershed vied for rights to its water. Hoover's commission neatly sidestepped the thorniest features of this issue by allocating water to an upper basin covering Colorado, New Mexico, Utah, and Wyoming and a lower basin composed of Arizona, California, and Nevada. Despite this tactical maneuver, representatives from California and Arizona strenuously objected to the plan, each claiming that it deprived them of the water they needed for their states to grow. The California delegation demanded that the water storage capacity behind the proposed dam be increased from 5 million acre-feet to 20 million acre-feet in order to meet Southern California's projected needs. Arizonans saw this demand as further evidence that "California was a water-thirsty vampire intent on sinking its fangs into the Copper State's jugular and sucking it dry" (Stevens 1988). California won: the dam's final water storage capacity amounted to almost 29 million acre-feet. The Arizona legislature initially refused to ratify the 1922 compact, an impediment that was overcome by deciding that only six of the seven states needed to ratify the compact for it to become binding. Arizona ratified the compact many years later but continued to battle in the courts over water rights. A 1962 court case finally established specific state-by-state water allocations in the lower basin, assigning approximately 27 percent of the lower basin's water to Arizona, 59 percent to California, and 4 percent to Nevada. The remaining lower basin water, plus any water left over from the upper basin, was to go to Mexico.

Neglect of Mexico in the compact's original water allocations was not unintentional. Conflict between the southwestern United States, especially Southern California, and Mexico over Colorado River water extended back at least to the turn of the century when farmers in the rapidly growing Imperial Valley grew concerned that Mexico would divert all of the river's water for its own use. The Californians' desire to secure at least a large portion of the Colorado River helped fuel the political drive that produced the compact. The fact that most of the Colorado River basin lay north of the border gave the United States tremendous leverage in negotiating water allocations. Negotiation, however, is perhaps the wrong word in this context. At one of the commission's meetings in 1922, Hoover declared, "We do not believe that [Mexico] ever had any rights" to Colorado River water. Despite the obvious dependence of the Mexicali Valley, one of Mexico's richest agricultural regions, on water from the Colorado River, Mexico received only "surplus" water from the river until the

> Mexican Water Treaty, ratified in 1945 over California's objections, guaranteed 1.5 million acre-feet of water to Mexico, one third of the compact's allocation to California. The controversy did not end there, unfortunately, because most of the water that Mexico received was salty runoff from American farms. Not until 1973 was Mexico finally assured of "good quality" water from the Colorado River.

Box 3.3

to pay for Hoover Dam's construction. Accustomed as we have become to delays and cost overruns on federal projects, it is noteworthy that the dam was finished ahead of schedule and the loan was repaid on time.

When the federal government invited cities to apply for the electricity to be produced at Hoover Dam, Los Angeles applied for all of it. In the final allocation, however, Los Angeles and neighboring communities won rights to approximately 65 percent of the electricity produced by the new facility. Given this allocation and the capacity for generating electricity at Hoover Dam, it is not surprising that the dam supplied a large proportion of the electricity used in the Los Angeles area between 1936 and 1950. Hoover Dam supplied as much as 75 percent of the energy sold by the Department of Water and Power and 24 percent of the energy sold by Southern California Edison, companies that supply electricity to the city of Los Angeles and surrounding communities respectively.

The electricity produced at Hoover Dam remained less expensive than electricity produced by steam plants, although the cost differential was reduced considerably relative to the difference prevailing shortly after the turn of the century. The City of Los Angeles and the Metropolitan Water District paid about 0.4 cents per kilowatt-hour for electricity from oil-fired power plants and less than half that amount for electricity from Hoover Dam. A cost advantage that had once favored hydropower by a factor of forty was now reduced to a factor slightly greater than two, at least for the hydropower produced at Hoover Dam. Nevertheless, consumers of electricity in Los Angeles saved $1,300,000 in the first year of the dam's operation.

The dam was initially called Boulder Canyon Dam although it was actually built in Black Canyon. It was later renamed in honor of Herbert Hoover. At the dedication of the dam in 1935, however, Secretary of the Interior Harold Ickes insisted that the dam should not be named for any one person, and the name reverted to Boulder Dam. The political backstabbing ended in 1947 when an act of Congress established the current name and restored recognition to Hoover.

Scaling Up: Hoover Dam

Nothing compared to the scale of Hoover Dam in 1936 when it was completed (figure 3.8). Hoover Dam was then the largest structure ever built (3.25 million cubic yards of concrete), the highest dam in the world (726.4 feet), and the most powerful hydroelectric facility in the world (1,323,500 kW). In addition, the lake it created, Lake Mead, was the largest artificial lake, and the transmission lines that carried electricity from the dam's power plant to Los Angeles set world records for voltage and length.

Building a facility as enormous as Hoover Dam requires more than an understanding of the basic scientific principles underlying its operation. Hoover Dam uses nature in exactly the same way as its modest counterpart on Mill Creek, but it could not have been built simply by scaling up the smaller plant. Tiny Mill Creek was dammed by mixing a few bags of cement with water, sand, and rock and pouring the concrete to a height of about five feet across the stream bed. The Colorado River did not submit so meekly. With the larger river came larger forces that required unusually strong materials and careful designs if the challenge was to be met. Whereas Mill Creek Station #1 was notable for its early application of scientific discoveries in electromagnetism, Hoover Dam represented the frontier of civil engineering.

Hoover Dam required new methods for pouring concrete to ensure that the finished structure would not crack. The penstocks had to be constructed from unusually strong steel to withstand the tremendous pressures they would experience. Construction of the transmission lines across a forbidding desert was in itself recognized as a major engineering achievement. A small city, Boulder City, was built to house thousands of workers and their families.

Figure 3.8 Hoover Dam with Lake Mead behind. (Photograph courtesy of the Bureau of Reclamation.)

Work on such a large engineering project is always treacherous, but normal risks of construction were exacerbated by the fierce desert environment. President Hoover, in order to provide employment in relief of the Great Depression, ordered work on the dam to begin in March, 1931 although adequate housing for the workers would not be completed until October. The average daily high temperature reached 119 degrees Fahrenheit by late July of that year, and the average low was 95. Workers unlucky enough to be assigned the task of digging the diversion tunnels sweltered as the temperature in the tunnels rose to 140 degrees. Deaths from heat prostration became so common that summer that a physiologist was called in to examine the situation. His advice to the workers was simple and effective: drink more water. Contrary to the popular but gruesome myth, no one was entombed in the concrete of the dam.

In one unfortunate accident, a worker died in an avalanche of concrete, but his body was recovered.

Most of the public acclaim for the Hoover Dam project has gone to Elwood Mead, Commissioner of the Bureau of Reclamation, and Herbert Hoover, for whom the lake and the dam are named respectively. The person most deserving of praise for turning the proposal into a finished structure, however, is Frank Crowe, the superintendent of construction. He admitted spending ten years of his life, from 1925 to 1935, obsessed with the Hoover Dam project. "I was wild to build this dam ... the biggest dam ever built by anyone anywhere anytime" he recalled. In working on previous dam projects, Crowe demonstrated his overriding concern for efficiency and speed by developing methods for rapidly pouring large amounts of concrete. His coworkers on the Hoover Dam project nicknamed him "Hurry Up." Crowe's diligence helped bring Hoover Dam to completion two years ahead of schedule. His work on the Hoover Dam project earned him the reputation as the "finest field engineer in the world."

Harnessing the Colorado's Power

When the dam was finished, water began accumulating in Lake Mead and eventually created an average head of 520 feet and a maximum head of 590 feet. Water flows from the lake into the four intake towers, through the penstocks and into the spiral casings that deliver water to the turbines at an astounding rate of 32,000 cubic feet per second. The water then flows over the blades of the turbines and into the tailrace below the dam.

The Hoover Dam power plant produces electricity according to exactly the same scientific principles that govern operation of the much smaller Mill Creek Station #1. The maximum amount of power available to the turbines at Hoover Dam is given by:

$$P_a = qdh$$

$$= 32,000 \frac{\text{ft}^3}{\text{sec}} \times 62.4 \frac{\text{lbs}}{\text{ft}^3} \times 590 \text{ ft}$$

$$= 1{,}178{,}110{,}000 \frac{\text{ft-lbs}}{\text{sec}} \text{ or } 1{,}600{,}000 \text{ kW}$$

Discovery

The head is not much higher behind Hoover Dam than at Mill Creek, but there is about a thousand times more power available to the larger facility's generators because the water flow is about a thousand times greater.

The tremendous flow of water through the power plants at the Hoover Dam called for a different kind of water wheel than employed at Mill Creek Station #1. James Francis, an American engineer and contemporary of Pelton, had developed a water wheel or turbine in the late nineteenth century that was well suited to the new conditions. These turbines serve the same purpose as Pelton wheels in converting energy in flowing water into rotational energy, but they do so in a different way.

Unlike Pelton wheels, Francis turbines function while completely immersed in water. Water, under high pressure, flows in a "scroll casing" around the perimeter of the rotor and passes guide vanes before shooting across the blades on the rotor (figure 3.9a). The adjustable guide vanes direct the water at a shallow angle across the surface of the blades (figure 3.9b). Because water is not shot directly at the blades, a Francis turbine does not work purely by impulse as do Pelton wheels.

The blades of a Francis turbine function more like wings on an airplane than buckets on a Pelton wheel. A single blade on a Francis turbine can be imagined as a narrow section cut from a wing, twisted slightly along its length, and then stuck, trailing edge first, into the hub of the turbine. The leading edge of the wing segment now juts out from the hub toward the scroll casing and into the fierce current of water. The current passing over the blade acts in the same way as wind passing over a wing and creates lift according to a concept generated by Daniel Bernoulli (see box 3.4). The pressure is lower over the more curved of the blade's two surfaces, and so the blade is pushed from the other side. Lift can also be explained as an application of Newton's third law of motion, commonly abbreviated "action/reaction," in which lift is considered a reaction to a downward deflection of the air. Hence, Francis turbines are called *reaction* turbines. After flowing over the blades toward the center of the rotor, the water drops into the draft tube below the turbine and passes into the tailrace.

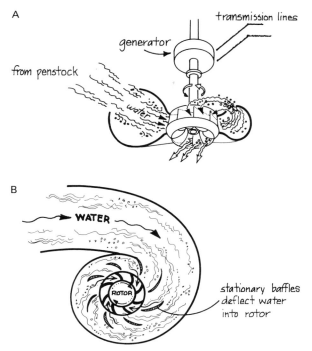

Figure 3.9 The Francis turbine. (*A*) Cut-away view. Water descends the penstock into the scroll casing to flow around the perimeter of the turbine and enter it at numerous points. (*B*) View of the penstock, scroll casing, and Francis turbine from above.

Efficiency with Megawatt Generators

The generators installed in the power houses at Hoover Dam operate in the same way as their modest kin at Mill Creek Station #1 but produce megawatts (MW) of power (one MW equals 1000 kW). The original installation at Hoover Dam included fifteen three-phase AC generators that varied in capacity from 40 MW to 130 MW. The total installed capacity was 1,323.5 MW (1,323,500 kW) or 1,300 times greater than Mill Creek Station #1. Since the dam's completion, two generators have been added, and most of the original generators have been improved either by rewinding the coils or replacing old blades on the turbines. The total generating capacity of the Hoover Dam power plants now stands at 1,900,000 kW and will grow to over 2,000,000 kW when the improvements are completed in 1992.

> ## The Bernoulli Effect
>
> Daniel Bernoulli, an eighteenth-century Swiss mathematician and one in a long line of distinguished mathematicians, made the somewhat counterintuitive determination that as the flow rate (v) of a fluid increases, its pressure (P) decreases. Under certain idealized conditions this relationship can be expressed as follows:
>
> $P + 1/2\rho v^2 = $ constant   where $\rho = $ density
>
> Although the conditions necessary to make this equation strictly true rarely appear in actual practice, Bernoulli's equation provides one reasonable, general description of how an airplane wing or a Francis turbine blade works. Pressure is lower over the more curved surface of these two objects because the fluid (air in the case of the wing) flows faster over that surface. The application of this theory in flight is described in chapter 4. You can demonstrate Bernoulli's principle with a piece of notebook paper. Hold the long edge of the paper against your lower lip so that it droops out away from you and toward the floor. Blow a stream of air directly out, parallel to the floor. The paper should "react" and lift.

Box 3.4

The immense generating capacity of Hoover Dam relative to Mill Creek was not obtained by increasing the efficiency of the power plant's operation. The original configuration of the generating system at Hoover Dam permitted the following efficiency under conditions of maximum output.

$$\text{efficiency} = \frac{1{,}323{,}500 \text{ kW}}{1{,}600{,}000 \text{ kW}} \times 100 = 83\%$$

Recent increases in the maximum flow rate of the water and the aforementioned improvements in the generators now permit an efficiency of 90 percent. This value represents the efficiency of power production when all generators are on-line and operating at their designed levels under the maximum flow rate, conditions that do not occur often. Engineers regulate the flow rate according to the need for water downstream from the dam, and this rate is usually less than the maximum. A computer now governs which generators are brought on-line when the flow rate is reduced in order to maintain a high operating efficiency. The average operating efficiency is currently 84 percent.

The Stretch for Power

The history of the electrification of Los Angeles has been one of reaching ever further for sources of electrical power. Before Hoover Dam began sending its power 266 miles to Los Angeles, electricity had never been transmitted over such great distances in the United States. Transmitting electricity over long distances meant transmitting at low currents and correspondingly high voltages. When electrical power first traveled from Mill Creek Station #1 to Redlands 12 miles away, the then-record level of 12,000 volts was required to keep the transmission efficient. The much larger distance separating Hoover Dam from Los Angeles presented a serious challenge to the engineers by calling for unprecedented voltages in the transmission lines. Not surprisingly, the engineers on the Hoover Dam project met the challenge in 1936 by transmitting at the record level of 287,000 volts.

Many technological systems develop as a cascade of solutions to problems that present themselves one after another. This proved to be the case for the transmission of electricity from Hoover Dam to Los Angeles. The high voltage in the transmission cables created serious risks of arcing between power lines, between the power lines and the ground, and between the power lines and the towers. These risks prompted scientists and engineers to study high-tension lines carefully and to develop methods for reducing both the danger and the power losses created by arcing. These studies produced new designs for cables and ceramic insulators and new ideas about how to space transmission cables on a tower. Largely as a result of these changes, Southern California utility companies have been able to extend their reach for electrical power even further than Hoover Dam. Los Angeles now receives some of its electricity by way of an 800,000-volt line that originates in a hydropower plant on the Columbia River nearly 900 miles away in Oregon.

An Unexpected Problem

A dazzling show of electric lights in downtown Los Angeles on October 9, 1936 heralded the arrival of electricity from Hoover Dam. The significance of the dam as a source of electricity grew during World War II as its electrical power went to steel mills and aircraft manufacturing plants that produced 20 percent of the nation's aircraft between 1941 and 1945. Hoover Dam accounted for approximately 75 percent of the electricity sold in the city of Los Angeles during 1945, but that year marked the zenith of hydroelectricity for Southern California (figure 3.10). Although other large hydroelectric projects were to follow Hoover Dam, the major period of hydroelectric development for Los Angeles had ended. One reason was that many of the largest sources of hydropower had been tapped. Another reason was that rivers had proven to be an unreliable source of electricity for transportation systems.

Los Angelenos once boasted of the world's largest interurban rapid transit system. More than 1,100 miles of track carried electric trolleys with their passengers between the downtown area and outlying communities in the 1920s. Henry Huntington, the "trolley king," owned two of the largest transit companies, the Los Angeles Railway Company with its yellow cars, and the Pacific Electric Company fondly remembered for its big red cars. These trolleys operated entirely on electrical power, and in expanding the territory served by his companies, Huntington played a major role in developing inexpensive sources of electricity for Los Angeles. His financing enabled the construction of a number of large hydroelectric

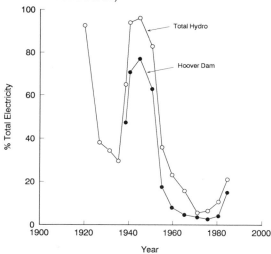

Figure 3.10 Production of hydroelectricity for the city of Los Angeles from all sources and Hoover Dam. (Data provided by the Los Angeles Department of Water and Power.)

plants in the Sierra Nevada Mountains northeast of the city. These facilities supplied electricity for transportation and other enterprises.

The partnership between hydroelectricity and transportation was fragile, however, as revealed by a severe drought during the early 1920s. Several years of below-average rainfall in Southern California forced sharp reductions in trolley service in 1924. Lower than normal amounts of snow in the Sierra Nevada Mountains meant less runoff to the dams that provided hydroelectric power for Huntington's trolleys. Trolleys ran less frequently and stopped at fewer places. The electric railways in Los Angeles never recovered their predrought level of activity as indicated by the gradual reduction in the amount of electrical power used by trolley companies after 1923. People turned increasingly to automobiles and buses to meet their transportation needs.

The proportion of electricity in Los Angeles that was contributed by the Hoover Dam declined rapidly after 1945 (figure 3.10). Its share of the electric power market dropped to 17 percent within ten years after World War II. Twenty years later, that value was reduced to 2 percent. Other hydroelectric facilities followed a similar trend. Today hydroelectricity from all sources accounts for less than 10 percent of the electricity sold in the Los Angeles basin. The rapid reduction in the importance of hydroelectricity was not produced by sharp cuts in the amount of power generated by hydroelectric facilities. Instead, fossil-fuel fired plants outstripped the lagging hydropower industry in meeting the rapidly growing demand for electricity in Los Angeles during its postwar population boom.

The story of hydroelectricity in Southern California is not yet over. As the costs of fossil fuels rise, the Department of Water and Power and the Metropolitan Water District look with increasing interest at the benefits of installing microhydro stations on their water distribution lines. These very small facilities will help reduce the department's expenses for electricity by reducing the need to burn fossil fuels to supply electrical power for the department's water pumps.

Symbolic Power

Despite its diminished importance as a source of electrical power for Los Angeles, Hoover Dam remains a compelling symbol of an

Figure 3.11 Suspended Power, painting by Charles Scheeler, 1939, Dallas Museum of Art, gift of Edmund J. Kahn. (Photograph reproduced with permission from the Dallas Museum of Art.)

earlier optimistic belief in modern technology as the foundation of a new social order. This monumental structure not only embodies the heroic accomplishments of the people who built it but also the conviction of many influential people in the 1920s and 1930s that large-scale public works projects can produce dramatic social improvements. Hydroelectricity enjoyed a special status in this belief. Henry Ford claimed that hydroelectric plants offered a way to improve the country by dispersing manufacturing centers and eliminating the ugly congestion and pollution of urban industry. A number of American artists began to incorporate an almost reverent view of technology into their work. Charles Sheeler's 1939 painting *Suspended Power* (figure 3.11) depicts a turbine being lowered into place in a hydroelectric plant and reveals his perception that technological structures represent a virtual substitute for religion in providing people with a sense of order. Margaret Bourke-White's famous 1936 photograph of the Fort Peck Dam in Montana that appeared on the cover of the first *Life* magazine also testifies to the powerful symbolic element of these huge structures.

The widespread view of modern technology as a benevolent tool for social engineering may have accounted for the apparent lack of concern that the Hoover Dam would seriously affect the environment of the lower Colorado River. Attitudes about large civil engineering projects and the environment have changed considerably since 1936 as indicated by a 1989 newspaper article in the *Los Angeles Times* proclaiming that "Environmentalists Triumph as Dam Project Is Scuttled." A proposal for a 615-foot dam on the South Platte River in Colorado was vetoed by administrators of the Environmental Protection Agency because the predicted adverse effects of this project on the habitats of trout and whooping cranes as well as on a scenic waterway were judged to outweigh the intended benefit of securing a water supply for Denver and other cities along the Rocky Mountains. An attorney for the Audubon Society commented on the broad significance of the EPA's decision by proclaiming that, "It says once again that the era of building these enormous projects is over." In a social climate diametrically opposed to that which produced Hoover Dam, opponents of the Two Forks Dam project, as it was called, suggested that Denver's need for water should be met by conservation measures and small water diversion projects. This represents a shift in our relation with nature. Evaluations of beauty, biological habitats, and recreational opportunities now compete against, and often supersede, evaluations of rivers as sources of water and power.

Science and Technology

Hydropower plants stand on a broader scientific foundation than supported invention of the telegraph. Newton's laws, Bernoulli's principle, and hydrostatics all came into play along with electromagnetism as engineers grappled with the tasks of improving waterwheels and increasing the extraction of energy from falling water. The path that links these principles to their applications in this context is a tortuous one, however, obscured by years of intervening work. The scientific principles certainly provided the groundwork on which improvements in waterwheel and generator design were based, but they could not suffice in the absence of extensive testing, even trial-and-error work, that produced improved, working devices.

The key scientific development for hydropower, and all other modern methods for generating electricity, remains the discovery of electromagnetic induction. Here, the link between scientific understanding and its application is especially clear and direct as in the case of the telegraph. Faraday revealed in his diaries that he understood at least the immediate practical implications of his discovery, but he left it to others to apply the principles, preferring instead to continue his basic scientific research. The astonishing rate at which the discovery of electromagnetic induction transformed electricity from a mysterious phenomenon known most widely through dangerous parlor tricks to a major source of power suggests that the gap between scientific discovery and commercial application was quite small.

In contrast to scientists such as Faraday or Maxwell, engineers like Tesla strove not to understand nature but to find ways to put nature to work, to control it for our benefit. Of course in the development of hydroelectric power, a broad understanding of nature proved to be of immense help to the engineers, but their first task was to create a working device or system and then make it work better. The science produced by Newton and Faraday no doubt shortened the search or even indicated what to search for, but it took the work of people like Edison, Tesla, Westinghouse, and Stanley to produce the workable solutions.

Exercises

1. Palm Springs and Cathedral City are the two Southern California communities closest to the San Gorgonio windfarm, the largest array of wind-powered generators in the United States. These communities at one time had an opportunity to buy as much electricity from the windfarm as they wanted or could get. (They turned down the offer.) A total of 70,000 people live in these two cities. If we assume that four people make up a household, on average, then there are 17,500 households in the two cities. An average household in Southern California now uses approximately 5,000 kWh of electricity in one year. The most cost-effective wind tur-

bines at the San Gorgonio windfarm are those that generate 100 kW of electricity.

(a) How many 100-kW turbines would it take to meet the total domestic need for electricity in Palm Springs and Cathedral City during the months of June, July, and August (when winds are at their peak)? Assume that (i) all electricity is used during the day when the wind is highest, (ii) one third of a household's annual demand for electricity occurs in these three months (it may well be higher), and (iii) wind turbines presently generate on average only 10 percent of the electricity that they could have if they worked at full capacity all the time.

(b) Each turbine is normally placed a distance of three times its rotor diameter from the next turbine in the same row, and rows are separated by a distance of seven to eight times the rotor diameter. The 100-kW turbines have rotors that are sixty feet in diameter. Approximately how many square miles would the facility in (a) require, assuming that the turbines are not placed in a single row?

(c) How many days would Mill Creek Station #1 have to run to meet the demand for electricity as described in (a)? Years? Make the same calculations for Hoover Dam. (Assume that each hydroelectric plant operates at peak capacity twenty-four hours per day. This is a more reasonable assumption for hydroelectric plants than for wind turbines because wind turbines experience mechanical failure frequently.)

(d) How many barrels of oil would have to be burned to meet the same demand for electricity as described in (a)? (One gallon of #2 fuel oil has an energy equivalent of 1,740 kWh of electricity, and there are forty-two gallons of oil in each barrel.)

(e) What problems do you foresee in trying to replace fossil fuels with wind energy on a large scale in the United States?

2. The Department of Water and Power (DWP) is the sole supplier of electrical power for the city of Los Angeles. As the city grows, the DWP must either add to its generating capacity to meet increasing demand for electricity or require customers to demand less. What should the strategy be?

(a) More generators? The generating capacity of the DWP must be large enough not only to meet average hourly needs for electrical power but also to meet sharp increases in demand ("peak demand") while retaining sufficient buffer capacity to compensate for the loss of one or more generators because of unexpected malfunctions or planned maintenance. In 1991, the DWP had a total generating capacity of 7,263 MW. The average (per hour) demand was for 3,170 MW in that year. Hot weather has led to a peak demand as high as 5,100 MW. Officials at the DWP assume that peak demand will increase 2 percent each year for the next twenty years. They do not project further than twenty years into the future because the results are too uncertain to be useful.

According to this projection, how many years will pass before the DWP will need to increase its generating capacity? (Officials at DWP strive to keep the peak demand at least 16 to 20 percent less than the total capacity.)

(b) Do you think your answer to (a) underestimates or overestimates the number of years before new generating capacity must be added? Why?

Discovery

(c) Conservation? The DWP served 3,400,000 people in 1991 when the residential use of electricity averaged 5,000 kWh per household. (A household is assumed to mean four people.) The population of Los Angeles will be 3,800,000 in the year 2007 according to recent projections.

How much will each household in 2007 have to reduce its consumption of electricity below the 1991 level to keep the total residential use equal to the 1991 value?

Look at the table below and propose at least two ways to achieve the needed savings.

Typical electricity consumption by household appliances

Appliance	Consumption*
central air conditioner	5.3 kWh per hour
electric blanket	0.17 kWh per hour
coffee maker	0.2 kWh per hour
dishwasher	1.0 kWh per load
garbage disposal	0.08 kWh per day
microwave oven	1.5 kWh per hour
range with oven	3.2 kWh per day
hair dryer	0.4 kWh per hour
shaver	0.02 kWh per hour
radio/phonograph	0.1 kWh per hour
solid state color TV	0.2 kWh per hour
refrigerator/freezer (14 cubic feet)	2.1 kWh per day
100-watt light bulb	0.1 kWh per hour
clothes dryer	3.6 kWh per load
iron	1.0 kWh per hour
automatic washing machine	0.25 kWh per full load
electric water heater	13.2 kWh per day

*Data taken from Los Angeles Department of Water and Power publication; indicates energy expended during average use of appliances.

The Flying Machine Problem: The Wright Stuff

4

I sometimes think that the desire to fly after the fashion of the birds is an ideal handed down to us by our ancestors who, in their grueling travels across trackless lands in prehistoric times, looked enviously on the birds soaring freely through space, at full speed, above all obstacles, on the infinite highway of the air.
Wilbur Wright

Disaster and Success

On December 8, 1903, Charles Manley crawled into the sling hung below an aircraft that looked like a "giant dragonfly" sitting atop a houseboat floating on the Potomac River south of the nation's capitol. The aircraft, dubbed the *Great Aerodrome* (figure 4.1), culminated Samuel Pierpont Langley's efforts to realize the dream of powered flight. The craft's four wings spanned 52 feet from tip to tip and reflected Langley's best estimate of how the craft's considerable weight of 850 pounds could be kept in the air. A powerful and remarkably light-weight internal combustion engine promised to supply the thrust necessary to sustain flight.

With Manley in place, the *Great Aerodrome* rested at the end of a short wooden runway along which it would be catapulted into flight, much as jets are now launched from aircraft carriers. On cue, the catapult was released and the *Aerodrome* sped to the end of its plank. Shortly after clearing the houseboat, the *Great Aerodrome's* tail

Figure 4.1 Samuel Langley's *Great Aerodrome*. (Photograph reproduced from *The Invention of the Aeroplane 1799–1909*, © Charles H. Gibbs-Smith, with permission.)

section crumpled disastrously, spilling the craft and its pilot into the frigid river water. A courageous assistant dove in and pulled Manley from the wreckage.

No one was seriously hurt in the accident, but the dramatic failure of Langley's aircraft appeared to be a major setback to aeronautical engineering and progress toward flight. The recent debacle was the *Aerodrome*'s second failure, following the first unfortunate trial by two months. Major N. W. Macomb, one of the many observers hoping to witness a great moment in history, despaired that, "We are still far from the ultimate goal, and it would seem as if years of constant work and study by experts, together with the expenditure of thousands of dollars, would still be necessary before we can hope to produce an apparatus of practical utility on these lines." Although Macomb could have been more concise, his sentiment carried some justification.

Samuel Pierpont Langley (1834–1906), a self-taught astronomer, had risen to become the preeminent American scientist of his day. His position as secretary of the Smithsonian Institution placed him at the top of the most prestigious scientific research center in the United States. After years of measuring the lifting properties of various surfaces and testing different configurations of model airplanes, Langley constructed a series of small, unpiloted, steam-

Figure 4.2 The *Flyer* just after take-off at 10:35 A.M. on December 17, 1903 at Kill Devil Hills, N. C. Orville Wright is the pilot. Wilbur Wright runs near the right wingtip. (Photograph courtesy of the Library of Congress.)

powered "aerodromes" with financial and technical assistance provided by the Smithsonian. On May 6, 1896, he achieved the first sustained, powered flight of a heavier-than-air machine when one of his unpiloted aerodromes was catapulted from the top of a houseboat on the Potomac and flew a distance of one half mile in one and a half minutes. This accomplishment represented a major step ahead of hot-air balloons and gliders.

Langley's early success, combined with support from Secretary of the Navy Theodore Roosevelt, encouraged the United States War Department to grant Langley $50,000 for the purpose of turning his small-scale models into full-scale, piloted flying machines. His second and more highly publicized failure appeared to dash the best hopes for a solution to the flying machine problem. But other minds and hands remained hard at work.

Just nine days after the crash of the *Great Aerodrome*, on December 17, 1903, Orville Wright guided a frail looking aircraft into the air and flew 120 feet before landing on the sand of a little-known beach four miles south of Kitty Hawk, North Carolina (figure 4.2).

This modest flight marked the first time that a "heavier-than-air craft left the ground under its own power, moved forward through the air, and landed safely all under the control of a pilot" (Anderson 1984). The Wright Flyer, as it came to be called, made three more trips that day with Orville and his older brother, Wilbur, taking turns as pilots. On the fourth flight, Wilbur remained aloft for 59 seconds as he flew a distance of 852 feet. Reaction to this historic event was strangely muted. Frank Tunison, a reporter for the Dayton *Journal* who responded to a slightly incorrect report of the longest flight, skeptically remarked, "Fifty-seven seconds, hey? If it had been 57 minutes, then it might have been a news item" (Crouch 1989).

Despite a lingering disbelief among many that the flying machine problem had been solved, a new mode of transportation had been born and progress followed rapidly. The first commercial airliners appeared within twenty years of the Wrights' momentous achievement. War planes were developed even earlier. Only twenty-four years after the Wrights' first powered flight, Charles Lindbergh flew across the Atlantic in about thirty-six hours, a trip that took six days by steamship. The rapidest way to cross the country from New York to San Francisco in the 1890s required five days on a train. In 1936, The DC-3 airplane reduced cross-country travel time to eighteen hours, not counting the several refueling stops it made. A mere forty-four years after the first flights at Kitty Hawk, Chuck Yeager broke the mythical sound barrier flying the jet-powered Bell XS-1. By the late 1950s, jet airplanes began carrying passengers coast-to-coast in five hours. These astounding increases in the speed of commercial travel redefined our world, doing for people what the telegraph, telephone, and radio had done earlier for information: they transformed the world into a smaller place with repercussions that have affected virtually every aspect of our lives since.

Why the Wrights?

Why did Orville and Wilbur Wright (figure 4.3) succeed where many other better-known scientists and engineers had failed? One key lies in the Wrights' ability to see all aspects of the flying machine

Figure 4.3 The Wright brothers: Wilbur on the left and Orville on the right, at their home in Dayton, Ohio, 1909. (Photograph courtesy of the Library of Congress.)

problem. They recognized that a successful flying machine must include a source of lift to counteract the weight of the aircraft and pilot, a propulsion system sufficient to launch the aircraft and counteract the drag to sustain its forward movement, and a suitable control mechanism to enable the pilot to control the aircraft in flight.

A second, and equally important, key to the Wrights' success lay in their realization that the old trial-and-error method of design would have to give way to a more systematic, scientific approach to the problem. In taking this approach, the Wright brothers became the best aeronautical engineers of their day, always practicing their scientific methods in the service of their goal, a practical flying machine.

A Systematic Study of Flight: The Wrights Begin

The Wright brothers began considering flight in 1894 apparently after they read a magazine article about the glides being made in Germany by a daring man named Otto Lilienthal. Oddly enough, it was the news of Lilienthal's death two years later that caused them to think more deeply about the flying machine problem. Lilienthal had designed and built gliders according to his observations of birds in flight and his experiments with surfaces as they moved through the air. Firmly convinced that the only way to learn about flying was to get into the air, Lilienthal courageously extended his experiments to include nearly two thousand glides before crashing to his death in August 1896. Orville and Wilbur spent a good part of the next three years furiously debating the cause of Lilienthal's accident and other topics related to flight. Finally, in May 1899, Wilbur wrote the Smithsonian Institution requesting all the information they had on aeronautics, explaining that, "I am about to begin a systematic study of [flight] . . . ." At this stage, Wilbur was much more interested in pursuing the flying machine problem than his younger brother, but Orville "caught the fever" soon enough.

Their review of the documents sent by the Smithsonian, including papers by Langley and Lilienthal and discouraging comments by Thomas Edison on the feasibility of powered flight, convinced them that much of the best aeronautical science of the time was little more than speculation. Lilienthal's work stood out as the best and helped guide their early efforts, especially their first attempts to solve the problem of lift.

Old Roots to the Problem of Lift

The problem of generating lift extends its roots into the broad area of fluid dynamics, which includes a wide variety of phenomena associated with the flow of fluids and the movement of objects through fluids. What the Wright brothers needed to solve the lift problem, indeed hoped to obtain from papers sent to them by the Smithsonian, was an accurate theoretical framework within which the problem of lift could be resolved.

Newton's Three Laws of Motion

Why do objects move? Aristotle believed that objects continue to move only as long as they are pushed. It took more than two thousand years to replace this idea with less intuitive but more correct answers. In the early seventeenth century, Galileo created the first major crack in Aristotle's monumental body of reasoning by claiming that moving objects possess an intrinsic quantity of motion. Isaac Newton soon expanded on Galileo's notion and generated the three laws of motion that remain the foundation of our modern understanding of motion, at least on a macroscopic scale. We now use the term *momentum* in describing a moving object's intrinsic quantity of motion: momentum is the product of an object's mass and its velocity.

The first law A body moving in a straight line at constant speed tends to remain moving in a straight line at constant speed or to remain at rest if at rest, unless acted on by a net outside force.

The second law The rate at which an object's momentum changes, which is directly related to its acceleration, is determined by the net force acting on the object. The most common mathematical expression of this law, $F = ma$ (force equals mass times acceleration), is among the most famous and useful equations in physics.

The third law When two objects interact, they always exert equally strong and oppositely directed forces on each other.

Despite Newton's remarkable work, many people continue to follow Aristotle in their intuitive interpretations of nature. Test yourself: if a Walkman falls from your right hip as you are running, where will it land relative to your right hip? In front? At the same place? Behind?

Adapted from Wolfson, R. and J. M. Pasachoff. 1987. *Physics*. Little, Brown and Company, Boston, pp. 70–72.

Box 4.1

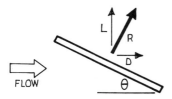

Figure 4.4 Isaac Newton's concept of resistance experienced by an object in a flowing fluid. The flat plate, seen end-on, is tilted at an angle, θ, to the flow. The resistance vector, **R**, represents the sum of two other vectors, lift (**L**) and drag (**D**).

A theoretical framework of a sort was actually available to them: Isaac Newton had provided it in his *Principia* two hundred years earlier. In the *Principia*, Newton analyzed problems of fluid flow in terms of the laws of motion that he had developed (see box 4.1). He had rejected Aristotle's idea of air or any fluid as exerting a propelling influence on moving objects and agreed with Galileo that air acts only as a resisting medium. Newton sought to understand the nature of this resistance and eventually created an expression for resistance felt by a surface placed in a stream of a hypothetical, frictionless fluid. He defined the resistance, R, as a force acting in a direction perpendicular to the surface (figure 4.4) and dependent on the density of the fluid (ρ), the square of the velocity of the incoming fluid stream (V^2), the surface area (S), and the angle that the surface makes relative to the initial flow direction (called the angle of attack and represented by θ).

$$R = \rho V^2 S \sin^2 \theta$$

Although Newton's use of the term *resistance* to describe this force survived until the early twentieth century, it will be less confusing if we substitute the modern term *reaction force*. Several aspects of this equation follow directly from Newton's second law of motion, usually expressed as $F = ma$ (force equals mass times acceleration). A larger force is required to deflect (accelerate) fluid of more mass per unit volume, so the reaction force increases with fluid density. Newton's second law also explains the effect of surface area on reaction force: a larger area deflects more air mass in a specified period of time, giving a larger force. The dependence of the reaction force on the angle of attack is explained similarly: increasing the angle of attack causes the fluid to be deflected through a larger

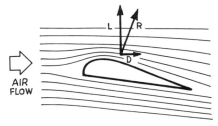

Figure 4.5 Wing section in an airstream. Resistance, lift, and drag are indicated as in figure 4.4. As air flows over the curved upper surface of the wing, the streamlines are seen to get closer together, indicating a region of higher velocity and lower pressure compared to the air flowing under the wing.

angle. Consequently, the acceleration and corresponding force are larger.

Newton based his theory of fluid dynamics on certain naive assumptions about how fluid actually flows around a surface. He assumed, for example, that the fluid particles collide directly with the surface. This assumption and others led him to think that the influence of the angle of attack on the reaction force should be adjusted by the factor "$\sin^2$". This "sine squared law" later proved inadequate to describe the behavior of wings flying at low speeds, although later still it proved correct in the context of supersonic flight.

Despite its shortcoming, Newton's equation for the reaction force provided subsequent investigators a way to estimate the lift and drag on a surface. A wing in a normal horizontal flying position, for example, deflects the airstream downward (figure 4.5). Then, according to Newton's third law of motion, the air pushes back on the wing with a reaction force of the same strength. The reaction force can be conveniently broken down into two components. The one that extends perpendicular to the direction of fluid flow is the lift, L. Lift in level flight opposes gravity and keeps an aircraft aloft. Unfortunately, lift is always accompanied by the other vector, drag, D, that extends parallel to the direction of fluid flow and impedes forward movement. Because the third law represents an expression of the law of conservation of momentum, this analysis of lift reflects an emphasis on momentum, the intrinsic quantity of motion that Newton described as the product of mass and velocity.

Another Way to Explain Lift

The next giant step in the history of fluid dynamics was taken by Daniel Bernoulli (1700–1782). He and his friend Leonhard Euler provided an alternative to the momentum-based explanation of lift. Euler, in 1791, realized that fluid particles do not actually strike the object's surface, as assumed by Newton, but follow pathways that bend just before reaching the object and course along its surface. Euler showed that the reaction force accompanying the movement of an object through a fluid varies as the sine of the angle of attack ($\sin \theta$), rather than as the sine squared as Newton had claimed, at least at low speeds.

The analysis of idealized fluid flow completed by Bernoulli and Euler revealed a relationship between the flow velocity and pressure that remains as surprising as it is important: the faster the flow velocity, the lower the fluid pressure. For horizontal flow of a fluid, this relationship can be expressed in the following equation where P is the fluid pressure, ρ is the fluid density, and V is the fluid velocity:

$$P + \rho V^2/2 = \text{constant}.$$

This relationship follows from the law of conservation of energy: the $\rho V^2/2$ term represents kinetic energy per unit volume and the pressure term indicates potential energy per unit volume, so if one increases, the other must decrease. When expressed this way, Bernoulli's theorem assumes that the density of the fluid does not change during its flow. This may not hold true for air, especially at supersonic speeds, but the changes in density that occur in air flowing at slower rates are small enough to be ignored, at least in this context.

Bernoulli's principle forms the cornerstone of modern fluid dynamics. It provides us with a framework for understanding lift: if air can be made to flow faster over the upper surface of a wing than over the lower surface, the pressure will be less above the wing than below, and the wing will be lifted (figure 4.5). Even at zero angle of attack, some lift will be generated by an upwardly curved wing in keeping with Bernoulli's principle. It is interesting to note that a complete scientific understanding of this phenomenon did not come until the early twentieth century, *after* the Wrights' historic flight.

This general understanding of a wing's behavior grows from the theories created by Newton, Bernoulli, and Euler, yet none of these great scientists ever considered the problem of flight. Not surprisingly then, this body of theory failed to supply the specific information needed to design a wing. How large should a wing be to create a specified amount of lift? How can a wing be made to generate much more lift than drag? How much curvature should the wing have? Where should the peak of the wing's curvature be located relative to the leading edge of the wing? How do wings behave in the air where flow does not conform to the ideal conditions assumed by Bernoulli? All of these questions bear directly on the design of a flying machine, yet Newton's and Bernoulli's equations offered frustratingly little specific guidance. Two paths led toward answers to these questions. One was followed by theorists and experimental scientists; the other path was taken by the Wright brothers. The two paths of theory and practice ran largely parallel courses during the first decade of this century before converging in the second decade, but it was the Wrights' approach that produced the first flying machine.

Not from Theory but from Practice

Undisturbed by the elegant theories developed by Newton, Euler, and Bernoulli, many flight enthusiasts continued to pursue their dream by mimicking birds. The sight of wing-clad men foolishly launching their bodies into the air became so commonplace that they were given a name: tower jumpers. One calamity after another followed this haphazard approach to the flying machine problem. The Wrights attacked the problem in a decidedly different way by using systematic, scientific analyses to inform their designs and improve their practical tests. This is not to say that they followed basic physical principles closely as they worked—they probably relied very little on the theories of Newton, Euler, and Bernoulli, although they included Euler's $\sin\theta$ factor in their calculations of lift. The systematic nature of their work was unusual, but an emphasis on the practical problems of flight, as opposed to the theoretical aspects, had a substantial history.

One of the early contributors to a practical understanding of lift was John Smeaton (1742–1792), an Englishman interested not in flight but in the performance of waterwheels and windmills. He used a windmill-like device in examining the behavior of flat surfaces as they moved through the air. Smeaton concluded from these experiments that the pressure over a moving surface varies as the square of the relative velocity multiplied by a coefficient, k. Smeaton estimated k to be 0.005 for a flat surface tilted 90 degrees to the airstream. Smeaton's coefficient returned a hundred years later to assist the Wright brothers in designing their first wings, but it also haunted their first failures.

A young, wealthy Englishman named George Cayley picked up Smeaton's methods in the first years of the nineteenth century and extended the earlier investigator's observations in important new directions. Cayley dissected Newton's reaction force into its two components of lift and drag and found that the amount of lift generated by a wing in an airflow increased with the wing's angle of attack and its surface area as implied by Newton's equation for the reaction force. Like Smeaton, he also observed that lift increased with the upward curvature, called the *camber*, of the wing. This follows from Bernoulli's principle, but a detailed understanding of this effect would not come for many years.

Cayley followed these experiments by designing an aircraft with features so like modern aircraft—including a single fixed-wing, a fuselage, a horizontal stabilizer in the rear, and a vertical tail—that some historians have called Cayley the father of modern aeronautics and inventor of the modern airplane. These appellations underestimate the role played by Orville and Wilbur Wright, but Cayley's contributions helped set the stage for the Wrights' success.

Cayley tried to interest others in flight research, but public ridicule of the wing-flapping tower jumpers discouraged many people from taking the new enterprise seriously. Finally, in 1866, Francis H. Wenham presented a theoretical paper to the Aeronautical Society of Great Britain that supported many of Cayley's views and helped enlist the efforts of a small but growing number of serious flight enthusiasts. Wenham emphasized a point that Cayley had tried to make earlier; namely, that generating experimental data on wing design was of real value to researchers attempting to build an airplane. He followed his own advice in 1871 by introducing the wind tunnel as a tool for the systematic study of wing design. Wenham found that most of the lift of a wing is generated nearer the leading edge than the trailing edge, indicating that a long, narrow wing should be preferred over a short, broad one. He also concluded that a wing's upward curvature should be deepest near the leading edge for maximum lift. Finally, he argued from his data that two wings stacked one on top of the other in a biplane configuration may be necessary to create sufficient lift to support a flying machine and its pilot.

Wenham's work represented a turning point in flight research that would profoundly influence the Wright brothers. Wenham and his coworkers splendidly represented a growing class of new professionals, the engineers, that appeared on the scene after 1860 and applied the methods and knowledge of science to technological problems. Aeronautical engineering societies began forming all over the world, and aeronautics became established as a respectable field of research for the engineer.

By 1890, a group of technicians interested in flight had been established in the United States. While they did not organize themselves into official professional aeronautical societies like their European counterparts, they did form a loose network of professionals interested in the problems of flight. The leading spokesperson for this community was Octave Chanute, a French-born, highly respected civil engineer. Chanute became interested in aeronautics in 1875 and quickly made contact with major aeronautical figures all over the world. His book, *Progress in Flying Machines*, established his reputation in aeronautics and served as the major text on aeronautical research, one to which the Wright brothers turned repeatedly during their years of study. Chanute's efforts helped establish a framework that allowed researchers to share and evaluate information systematically and, eventually, to give direction to promising avenues of attack. In other words, the kind of empirical and practical knowledge that would be of use to the Wright brothers could only be obtained from an organized, coordinated, and sustained approach to flight.

The public credibility of human flight received a significant boost from Otto Lilienthal, the German pioneer in aviation. Although he valued the experimental work of his predecessors and conducted important experiments of his own on the lifting prop-

Figure 4.6 Otto Lilienthal and one of his many gliders, 1894. (Photograph reproduced from *The Invention of the Aeroplane 1799–1909*, © Charles H. Gibbs-Smith, with permission.)

erties of various wing shapes, he firmly believed that one must be "on intimate terms with the air" to learn to fly. This became known as the "airman's" approach to flight. The key to this approach was that the aircraft ought to be sensitive to the pilot's efforts to control it. That is, it should be inherently unstable. Accordingly, no other activity than gliding could give a hopeful pilot the feeling needed to control an inherently unstable aircraft. A few people, notable by their failures, had tried another tactic called the "chauffeur's" approach in which the focus was on using massive power plants to propel an inherently stable craft through the air. These two approaches to flight differed primarily in the particular aspect of the flying machine problem they chose to emphasize: the airman's approach emphasized control, whereas the chauffeur's approach concentrated on stability and propulsion.

Until his death in 1896, Lilienthal flew a wide variety of gliders made from cloth and willow rods and shaped like bat wings (figure 4.6). After leaping from a hill built expressly for the purpose, he controlled his gliders by shifting his weight much as pilots of hang gliders do today. He made nearly two thousand glides up to an altitude of 1,150 feet in the period from 1891 to 1896. Beautiful photographs of these glides appeared all over the world and inspired a skeptical public that flight need not be restricted to the birds.

Flight Is a Delicate Balance

Within three months of receiving the materials from the Smithsonian, the Wright brothers had produced a design for a glider and decided the time had come to test it in flight. They moved their base of operations in September 1900 from their home in Dayton, Ohio, to a beach four miles south of Kitty Hawk, North Carolina. Their choice of this site reflected the characteristically thoughtful approach they took to all aspects of the flight problem. Wilbur calculated that the glider they had designed in Dayton would require a large area of open, level ground as well as steady winds of no less than 15 miles per hour (mph) if it was to fly and land safely. The brothers sought advice from the United States Weather Bureau and Octave Chanute before they selected an area around the Kill Devil Hills on a long, thin spit of land just north of Cape Hatteras. The average wind velocity at this location was not as high as at Chicago, but then they would not have a huge city and all its people to contend with in North Carolina.

After closing their business—a bicycle shop—for the remainder of the fall and winter, Wilbur Wright traveled to Kitty Hawk and camped below the Kill Devil Hills, a location so little known that residents of Elizabeth City only 35 miles to the northwest could not direct Wilbur to it on his first visit. Orville followed a short time later. Wilbur carried with him the design and parts for a large-scale glider. The problem of lift was only one of the obstacles confronting the Wrights on this trip. It is one thing to design a craft that generates enough lift to overcome its own weight and rise in a breeze, but a flying machine must also remain stable in the air. Wilbur believed that "the problem of equilibrium constituted the problem of flight itself" (Crouch 1978). According to Newton's first law of motion, when an airplane flies at a constant velocity in straight and level flight, the propelling force, called the thrust, balances the drag force, and the lift force balances the weight. Level flight means that the craft remains delicately balanced in aerodynamic equilibrium.

The Wright brothers knew from the work of their predecessors that the lift generated by a surface in a breeze increased with the relative velocity of the wind, the surface area of the wing, the wing's camber, and the angle of attack. They expressed these relationships in an equation previously used by Lilienthal, Langley, and Chanute where L is lift, k is Smeaton's coefficient, V is relative velocity, S is surface area of the wing and C_1 is the coefficient of lift, a factor that combines the effects on lift of the angle of attack and the camber.

$$L = kV^2SC_1$$

(Note that lift is a force and that this equation bears a strong resemblance to Newton's equation for the reaction force described previously.) Orville and Wilbur determined the value of the lift coefficient, C_1, for their first glider by consulting tables generated years earlier by Lilienthal. Although this equation proved useful to the Wright brothers, its combination of units is bewildering: velocity is given in miles per hour, surface area in square feet, and k is measured in $lbs(hr)^2/(miles)^2(ft)^2$ such that lift, L, comes out in pounds.

The Wrights' first attempt to solve the lift problem was with a wood and cloth glider weighing 52 pounds with a total wingspan of 17.5 feet and a wing surface area of 165 square feet. They had planned a larger wing, but Wilbur could not find in Kitty Hawk pieces of wood long enough to make wing spars 18 feet long, so he reduced the wingspan at the last minute. Following the example set by Cayley and Lilienthal, they added a camber to the wings, but they made it smaller than Lilienthal would have recommended and moved the peak of the curvature forward of the wing's center, also contrary to Lilienthal's practice. They thought these changes would reduce drag and increase the stability of the wing and prevent them from meeting the same fate as Lilienthal. The adventurous German had advised that the camber should rise above the wing's chord line a distance equal to one twelfth the width of the wing (figure 4.7). (The *chord line* is an imaginary straight line that runs across the width or chord of the wing and connects the wing's leading edge with its trailing edge.) Instead, the Wright brothers chose a camber only one twenty-third of the chord for their first glider. The Wright brothers constructed their wings as biplanes to create sufficient surface area in a light structure without sacrificing strength. The biplane with its struts and wires behaved as a truss, a characteristically strong form commonly used in bridges.

Figure 4.7 Wing sections showing a camber of one twelfth (1:12) (top) recommended by Lilienthal and a camber of one twenty-third (1:23) (bottom) used by the Wrights in designing the wing for their 1900 glider. Camber is the ratio of the wing's maximum depth to the length of its chord line, an imaginary straight line running from the leading edge to the trailing edge of the wing.

Out of respect for their father's concern and with some regard for their own safety, the brothers decided to fly their glider as a kite before piloting it themselves (figure 4.8). Their calculations suggested that the glider would lift a total of 192 pounds (a 52-pound glider plus Wilbur's weight of 140 pounds) if held at a 3° angle of attack in a 25-mph wind. It failed miserably. Their kite managed to lift only 75 pounds at the desired angle of attack. The feeble craft lifted the total weight of 192 pounds only when the wind blew at the unsafe velocity of 35 mph or the angle of attack was increased to 20°, an angle too steep for soaring flight. Their discouragement grew as they noticed that, under some conditions, the glider mysteriously lost all lift. The mystery deepened as they observed, happily, that their glider produced less drag than expected. They measured the amount of drag by using a fish scale to register the pull exerted on the tethers when the glider was flown as a kite.

Despite these problems, the two brothers realized, like Lilienthal, that if they were going to learn to fly, they would have to climb onto their machines and fly them. Wilbur elected to go first. The problem with the lift continued, but he found that he could glide short distances in safe winds as long as he maintained a steep angle of attack. He cautiously flew the glider within a few feet of the ground as he tested the stability of the aircraft. The decision to fly paid immediate dividends. Wilbur found right away that their design worked beautifully to maintain equilibrium for level flight. This achievement grew out of the brothers' understanding of a peculiar problem associated with lift.

They had correctly anticipated that as the wing's angle of attack increased, the craft would become unstable and its nose would

Figure 4.8 The Wright brothers' first glider flown as a kite, 1900. (Photograph courtesy of the Library of Congress.)

pitch upward uncontrollably because of the behavior of an abstract notion called the *center of pressure*. While the lift on a wing is a result of all the pressures acting on it, the overall or net lift on the wing can be treated as if it acts at a single point, called the center of pressure. The concept of center of pressure is analogous to the concept of center of gravity. Gravity acts on each mass element in an object, but the overall effect on the object is most conveniently determined by locating this overall effect (the weight of the object) at a specific point associated with the object (the center of gravity). In a similar way, the center of pressure is the specific point at which the overall effect of all the pressure forces acting on the wing can be considered to act.

    The Wright brothers knew that as long as the center of pressure coincided with the aircraft's center of gravity, the craft would remain level. But Langley had observed that this condition did not always occur: the center of pressure moved forward toward the

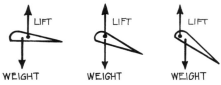

Figure 4.9 The position of a wing's center of gravity ("weight") remains constant, but the center of lift moves with the angle of attack. When the center of lift is behind the center of gravity, the wing will tend to tilt downward. When the center of lift moves in front of the center of gravity at high angles of attack, the wing will tilt upward—a potentially dangerous situation that could lead to a stall. Level flight occurs when the center of lift is directly over the center of gravity.

leading edge of the wing as the angle of attack increased. This movement presented the Wrights and other aeronautical engineers with a serious problem for longitudinal stability. Should a gust of wind increase the wing's angle of attack, the center of pressure or lift would move forward away from the center of gravity, causing the nose of the aircraft to tilt upward and making the problem worse (figure 4.9). When the wing reached a very steep angle of attack, the airflow over the upper surface of the wing would turn from smooth to turbulent, all lift would be lost, and the wing would stall. Lilienthal had died because of a stall.

We can now understand stall in terms of fluid dynamics. At a small angle of attack, the air flows faster over the top of the wing than past the bottom of the wing and the wing is lifted. Lift increases with the angle of attack, but so does the drag. Fortunately, for small angles of attack, the lift increases faster than the drag. Unfortunately, the drag begins to increase faster than lift past a certain angle. The angle of attack that provides the optimum lift-to-drag ratio usually lies in the range of 3.5° to 4° for propeller-driven aircraft. If the angle of attack becomes excessively steep (larger than about 15°), the air ceases to flow smoothly over the top of the wing and the wing loses its lift and stalls.

Wilbur and Orville coped with the center of pressure's movement and the threat of stalling by adding a small, flat, horizontal surface, called an *elevator*, to the front of their glider (figure 4.10). The pilot tilted the elevator upward or downward to counteract the effects of the shifting center of pressure. When the center of pressure moved forward, for example, Wilbur tilted the elevator downward to return the craft to level flight. In addition to offering stability for

Figure 4.10 The Wright brothers' 1901 glider being launched at Kill Devil Hills by Bill and Dan Tate with Wilbur as the pilot. The horizontal rudder is faintly visible in front of Wilbur. The 1901 glider lacked a vertical rudder. (Photograph courtesy of the Library of Congress.)

level flight, the forward placement of the elevator quite fortuitously saved the lives of the two brothers. The forward elevator kept their flying machines from nosing over and diving into the ground after a stall. Instead, the stalled craft floated leaflike to a bone-jarring but not lethal landing. Most modern airplanes carry the elevator attached to the horizontal stabilizer at the tail of the plane. A few small modern aircraft place the stabilizer in the front where it is called a *canard*.

A Warped Solution to a Control Problem

The elevator allowed the Wright brothers to control their flying machine in the pitch axis (figure 4.11), but they knew they would have to gain control over the aircraft in other directions as well to achieve true flight. Their first concern, after lift and equilibrium, was how to prevent the aircraft from rolling over. If they could control roll, they could initiate turns to either side because turns required only that one wing be made lower than the other. Lilienthal's method of initiating turns by shifting his body weight worked only for small wings, wings less than 155 square feet. The Wrights

The Flying Machine Problem

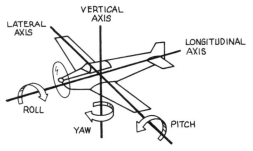

Figure 4.11 One problem with learning how to make a flying machine and then learning to fly it is having to cope with three axes of rotation shown here for a modern aircraft.

required much larger wings and so needed a new type of system for controlling roll.

Both Wilbur and Orville had explicitly rejected the tactic of solving the flying machine problem by copying birds. Too many of their predecessors had died from that strategy to make it an attractive one. Yet when it came to developing a roll control system, Wilbur depended heavily on his careful observations of buzzards and other soaring birds. One way these birds turned, he noticed, was by slightly retracting one wing to reduce the exposed surface area and lift of that wing. This caused the bird to tilt and rotate toward the retracted wing. Wilbur wisely recognized the difficulties involved in building such a mechanism into a flying machine. Fortunately, he observed another method of turning used by these birds: they rotate the tip of one wing slightly upward, changing its angle of attack and increasing its lift. The bird then tilts toward its unrotated wing and enters a banked turn.

Wilbur adapted this mechanism to the glider by adding a network of pulleys and cables that twisted the entire wing slightly when the pilot pulled on them. The brothers hoped that "warping" the wing in this way would increase the angle of attack on one side and decrease it on the other side, causing the craft to bank toward the side with less lift (figure 4.12a and b). That is exactly what happened when they tested the ingenious wing-warping mechanism during the kite-flying tests of their first glider.

Discovering ways to control pitch and roll counted as the greatest successes of that first trip to Kitty Hawk, but those achievements did not outweigh the Wrights' despair over their glider's

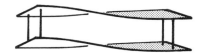

Figure 4.12 Wing warping. (*A*) Schematic drawing of a biplane wing configuration, such as that used by the Wright brothers, to show how twisting or warping the wing alters the angle of attack at both ends of the wing. (*B*) Wing warping is evident in this photograph that shows a right-hand turn of the Wrights' 1902 glider. (Photograph courtesy of the Library of Congress.)

failure to produce substantial lift. They returned to Dayton to think more carefully about the flying machine problem. Shortly afterward, a woman in a family who had befriended the Wright brothers cut the cloth from the wings of the abandoned glider and made dresses out of it for her daughters.

Progress and Blind Alleys: The 1901 Glider

Despite the mixed results yielded by their first glider, certain basic features of the Wrights' flying machines had emerged. The biplane-style wing, forward elevator, wing-warping method of roll control, and prone pilot position were all retained by the Wrights as they advanced from one experimental craft to another. Only after their most historic flight did they alter some of these features. The key advances, which carried them from their 1900 glider to powered flight in 1903, came only after a difficult struggle in which the Wrights boldly decided to throw out the old aeronautical data that had guided their first efforts and strike out on their own.

With their determination renewed, Wilbur and Orville hired Charles Taylor, a hard-drinking cigar-smoking mechanic, to manage the bicycle shop so that they could get an early start on their new flying machine. The brothers returned to Kitty Hawk in June of 1901 and expanded their encampment to include a hangar and a workshop. They shrugged off their previous failure and built an even larger aircraft, reasoning that the missing lift could be recovered by using a larger wing. The new glider's wings spanned 22 feet with a chord of 7 feet and surface area of 290 square feet, nearly twice the area of their previous glider. They also hoped to increase lift by drastically raising the wing's camber to one twelfth of the chord, instead of the one twenty-third value used the previous year.

They grew so confident in their new design that they broke with their earlier practice and flew it as a glider, with Wilbur as pilot, before testing it as a kite. The first eight attempts to fly the new glider produced disappointing results. The craft again showed depressingly little tendency to lift, and it kept turning its nose to the ground. On the ninth attempt, Wilbur managed to fly the craft a distance of 300 feet at an altitude of 1 to 3 feet, but he had to tilt the elevator upward as far as it would go to achieve this qualified success. The brothers knew that something was wrong with the design, but Wilbur insisted on taking it up again. This time the flying machine rose uncontrollably and dangerously to 30 feet where it stalled. Luckily, because of the forward elevator, the craft floated back to the sand without injuring its pilot.

Humbled by this near disaster, Orville and Wilbur disassembled the glider and flew each wing as a kite from the top of Big Kill Devil Hill. They watched carefully as the wing's behavior changed with wind velocity. At low wind velocities, the wings flew at a high angle of attack nearly overhead. At moderate wind speed, the angle of attack was reduced considerably and the "kite" flew level with the two brothers who were holding the tethers at the top of the hill. The kite actually flew below them when the wind rose to very high speeds.

Wilbur recognized the problem almost immediately. The center of pressure must travel forward at low wind speeds and remain over the center of gravity at moderate speeds, as others had observed. But at very high wind speeds, the center of pressure actually *reverses* its direction of travel, finally ending up behind the center of gravity and forcing the wing down (figure 4.9). No one before had thought that the center of pressure moved close to the trailing edge of a wing. Wilbur and his brother then took the counterintuitive step of reducing the wing's camber from one twelfth to one nineteenth in order to increase lift and improve stability. The design change worked so well that they confidently extended their flights to altitudes of tens of feet.

This modification did not solve the lift problem entirely, however. Just as in the previous glider, the new machine produced less lift, as well as less drag, than predicted by their equations, preventing them from flying their craft at the desired low angle of attack. Exasperated, they blamed the only thing left to blame: Lilienthal's data and lift coefficients. The prospect of scrapping the data that had supported a wealth of aeronautical research by so many respected people, including their close friend and advisor Octave Chanute, must have seemed rash, but the Wrights decided it had to be done. They would have to start over from scratch.

As if the problems with lift were not enough, a new aeronautical problem appeared unexpectedly. After the camber had been reduced and Wilbur began making "high-altitude" glides, he tested the craft's ability to turn. The wing warping worked to bank the aircraft, as he knew it would, but he sensed an instability during the turn that made him uneasy. The tilted aircraft slipped sideways through the air, losing altitude at a dangerous rate. One of these trials resulted in a crash after a wing tip struck the ground. Wilbur was unhurt.

The nature of this new problem can be seen in Newton's dissection of the reaction force into its two components of lift and drag. When Wilbur warped the wing to initiate a left turn, the right wing's angle of attack, and hence its lift, increased. But increasing the angle of attack also increased the drag on the right wing. As that wing rose it began to fall behind the left wing. This caused the glider to rotate in the horizontal (yaw) plane to the right as it entered a bank to the left! As a result, the leading edges of the wings no longer headed directly into the line of travel but at an angle to it, decreasing the lift considerably. The resulting lack of lift caused the glider to plummet sideways through the air. The mounting problems so discouraged Wilbur that he lamented to Orville on the train ride home, "Not within a thousand years would man ever fly!" (Crouch 1989).

The Engineers Return to Science

The Wright brothers were nothing if not tenaciously persistent. Rather than give up the flying machine problem, they conducted their own scientific investigations to generate new data on the lift properties of various wing shapes. They built a wind tunnel in their bicycle shop in Dayton (figure 4.13) and, from September 1901 to August 1902, tested about two hundred small models of wings using an ingenious balance they had developed (figure 4.14).

In a typical test by the Wright brothers, a piece of sheet metal was hammered into a cambered wing shape and mounted on a delicately arranged group of stiff wires. When air was blown across the wing, a force was transferred through the wires to a small gauge that read angles of deflection (figure 4.14). Flat pieces of hacksaw blade suspended below the airfoil counteracted the effects of drag. The Wrights used this apparatus to determine coefficients of lift for airfoils tilted at different angles in winds of various velocities. They

Figure 4.13 The wind tunnel built by the Wright brothers in 1901 and used to test the behavior of various airfoils. One of the Wrights would stand very still and peer into the test chamber, much as the woman is doing in this photograph, to watch the airfoil and the balance as wind blew across them from the fan on the right. (Photograph courtesy of the United States Air Force Museum.)

also determined the ratio of drag-to-lift, an important quantity that suggests how efficient the flight of an aircraft will be.

The thoroughly systematic nature of these wind tunnel studies was unprecedented in aeronautical engineering. Their experimental work made them the leading aeronautical engineers of their day. Contrary to their expectations, Lilienthal's data and lift coefficients proved to be essentially correct. The fault in their calculations must have been in one of the other factors. Wing surface areas were known precisely, and the wind velocity was measured accurately with an anemometer. The only remaining factor in the lift equation was Smeaton's coefficient of air pressure.

In a separate and surprisingly simple series of calculations, the Wright brothers reexamined Smeaton's coefficient. Using data from their 1900 glider for wing surface area, wind velocity, lift coefficient, and lift, they worked backward to estimate Smeaton's coeffi-

The Flying Machine Problem

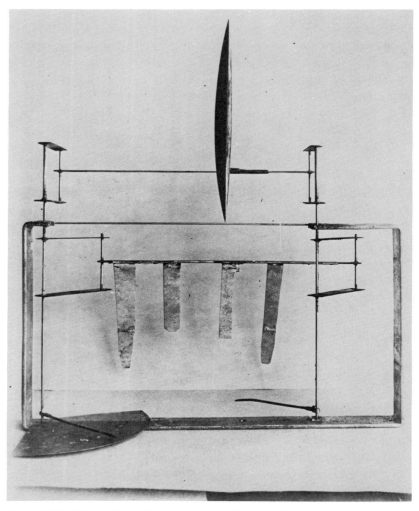

Figure 4.14 The balance designed and used by the Wright brothers, in concert with their wind tunnel, to measure the lift produced by model airfoils. The airfoil was attached in a vertical position to the uppermost wire as shown. An airstream blowing across the airfoil (i.e., along your line of sight as you look at the photograph) would cause the airfoil to move leftward as it experienced lift. The amount of lift could be measured by watching the rotation of the needle around the gauge at the lower left. The flat pieces of metal hanging from the lower wire counteracted the effects of drag. The balance is approximately one foot wide across its base. (Photograph courtesy of Wright State University Archives.)

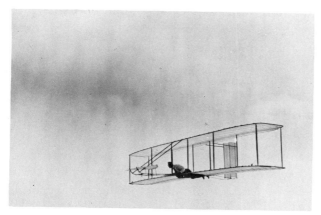

Figure 4.15 Wilbur flying the 1902 glider at Kill Devil Hills. The aircraft has a fixed, double rudder in the rear. (Photograph courtesy of the Library of Congress.)

cient. The correct value must lie between 0.003 and 0.0034, they reasoned, not 0.005 as Smeaton had concluded. Wilbur then decided that there was no particular reason to use a lift coefficient greater than 0.0033. (The actual value is 0.0026, or about 20 percent less than the value used by the Wrights, meaning that they continued to overestimate the lift that their aircraft would produce, but not by as much as before.) Both Langley and the United States Weather Bureau had suspected earlier that the actual value of Smeaton's coefficient must be in the neighborhood of 0.0032, but controversy had surrounded the issue and the Wrights had used Smeaton's value.

The error in Smeaton's coefficient explained not only why their gliders generated less lift than predicted but also why there was less drag: Smeaton's coefficient had entered into the calculations of drag as well as those for lift. Armed with new understanding about lift and a new value for Smeaton's coefficient, Orville and Wilbur returned to the task of designing a flying machine.

Light at the End of the Tunnel: The 1902 Glider

The glider the Wright brothers constructed below the Kill Devil Hills in August 1902 hardly resembled its predecessors in the important features (figure 4.15). The wing was made longer and narrower than in previous gliders: 32 feet by 5 feet, giving a length-to-width ratio of 6:4, a ratio more than twice as large as in the 1901 glider. The total wing surface area of 305 square feet slightly exceeded that of the 1901 glider. The camber was reduced again, this time to one twenty-fourth, and its peak was moved forward to a point a third of the way back from the wing's leading edge. The Wrights fixed two large, vertical rudders behind the wing to counteract sideslipping during turns. These design changes no longer reflected desperate shots in the dark.

The Wright brothers' persistence, intelligence, and reliance on solid information were rewarded by beautiful, soaring flights. The new aircraft generated at least as much lift as they had predicted and rode through gentle turns with ease. Orville finally tried his hand at piloting and quickly acquired the necessary skills.

As improved as it was over earlier models, all was not well with the new glider. It performed beautifully as long as the pilots kept the turns broad and gentle, but it again became unstable during more steeply banked turns. The fixed rudders were working but not well enough. Wilbur and Orville had thought that fixed rudders would keep the nose of the craft pointed in the direction of the turn by resisting sideways movement through the air and counteracting the undesirable yaw produced by wing warping. The 1902 glider with its fixed rudders did not yaw away from the intended turn as much as the preceding glider, but the rudder's corrective effect was not large enough to completely cancel the effects of wing drag. What was needed, Orville realized, was a *movable* rudder that the pilot could use to control the amount of sideslipping. By deflecting the rudder toward the left side of the aircraft during a left turn, the pilot would be able to increase the effectiveness of the rudder at counteracting rightward yaw. The brothers quickly replaced their two fixed rudders with a single movable one and the problems of sideslipping disappeared.

The addition of a third control system in the horizontal or yaw plane stands among the Wright brothers' landmark achievements. They were the first aeronautical engineers to recognize the necessity of controlling the aircraft in all three axes: pitch, roll, and yaw. They did not come easily to this conclusion, however. Until the tests with the 1902 glider, they had thought, like Langley and all other aeronautical engineers of the time, that pitch and roll control would suffice. Only after trying to fly their machines did they understand the need for a third type of control. The Wrights initially joined the rudder's control cables with the wing-warping cables so that the rudder would move automatically as the pilot warped the wings and turned the aircraft. This enforced coordination of yaw and roll control was abandoned by the brothers years later as they gave the pilot independent control over all three systems.

The modified glider was a total success. In October 1902, Orville and Wilbur made over a thousand glides to distances of 620 feet and with durations as long as 26 seconds. They set records for largest flying machine, longest distance covered in a glide, longest duration of a glide, and highest wind velocity successfully negotiated by a glider. The Wright brothers were rapidly becoming "intimate with the air". After three years of hard work, they were finally ready for the power plant and propulsion system.

The Last Steps to a Flying Machine

The Wrights followed their successes of 1902 with two important actions: they applied for patents on their inventions and set about to develop a propulsion system for their aircraft. The initial patent applications marked a turn in their view of aeronautics. Wilbur had written Chanute in 1900 that "I make no secret of my plans for the reason that I believe no financial profit will accrue to the inventor of the first flying machine ..." (McFarland 1953). The 1902 glider changed all that. In March 1903, Wilbur and Orville applied to the U.S. government for patent protection of the operation, construction, and, especially, the control systems of their gliders. Although the patent office rejected these applications on the grounds that the Wrights' inventions were not deemed original or workable, Orville and Wilbur had clearly begun to think of their machine as a marketable product.

The major technical problem they now faced concerned the propulsion system that would launch the aircraft and sustain its forward movement through the air. This turned out to be a greater challenge than either Wilbur or Orville anticipated. The key function of the propulsion system on an aircraft is to overcome drag. Lift counteracts gravity and propulsion counteracts drag; the two acting together make flight possible. Birds combine both functions in their flapping wings, but George Cayley years earlier had wisely decided that flying machines ought to separate them in different structures. The Wright brothers followed this advice.

The Wrights knew that adding an engine to their glider would complicate matters considerably. The added weight would have to be overcome by wings with greater lifting capabilities. But larger wings meant more drag, and more drag meant that a larger engine would be needed to produce more thrust. A larger engine would add more weight. The problem appeared to return on itself in a vicious cycle. The way out of this thicket lay in designing a lightweight aircraft that combined a large amount of lift with a small amount of drag.

One step the brothers took early on to reduce drag was to insist on a prone position for the pilot. As usual, this decision was far from arbitrary. There are two kinds of drag: drag that inevitably accompanies lift is called *induced drag* and is distinct from *parasitic drag*, which depends on the frontal area of the aircraft and indicates the difficulty of pushing (or pulling) the craft through the air. The Wrights calculated that a prone pilot would contribute 80 percent less parasitic drag than a seated pilot, an advantage they needed as they struggled to increase the ratio of lift to drag. Their close advisor, Octave Chanute, agreed with the brothers' decision, "provided that you do not plow the ground with your noses" (Combs 1979).

Generating power for the aircraft posed a problem in energy conversion somewhat similar to the one discussed in chapter 3. Chemical energy in the form of gasoline, another late-nineteenth-century development (chapter 6), would have to be converted to kinetic energy of a rotating axle. The kinetic energy must then be converted to thrust to propel the craft forward. Internal combustion engines were the logical choice for the power plant because they generated more horsepower per pound than the steam engines available at the time. Steam engines include an energy conversion step that internal combustion engines omit: the conversion of liquid water to steam. A railroad locomotive could afford to carry water and a boiler for a steam engine, but it is hard enough for a flying machine to get off the ground without adding all that extra weight. Thrust would be generated by propellers. Orville tackled the engine problem while Wilbur sought out a propeller. Each expected that they would simply be able to purchase useful devices and install them on their glider. Both tasks proved to be considerably more difficult than that.

The internal combustion engine had been invented in 1870, but when Orville canvassed ten manufacturers for a lightweight engine with sufficient power, he found none. The Wright brothers would have to build their own. Actually, Orville assigned the task to Charles Taylor, the mechanic who watched over the bicycle shop while the brothers were off at Kitty Hawk. Taylor's prior experience with internal combustion engines consisted of a single repair job. Orville calculated that their powered aircraft should fly at 23 mph, and that the engine required to achieve this velocity should generate eight horsepower, weigh no more than 200 pounds, and drive a propeller at 330 RPM to generate 90 pounds of thrust. These numbers are peculiarly specific. How did the Wright brothers arrive at them?

They estimated the necessary velocity by plugging their estimate for the aircraft's weight into the lift equation. Knowing that lift must equal weight for the plane to be in equilibrium, they then solved for velocity. Their previous experience with gliders suggested that a 285-pound aircraft would be needed to carry both a pilot and an engine. Wilbur and Orville each weighed about 140 pounds, and they estimated that the engine would weigh 200 pounds, bringing the total estimated weight to 625 pounds. Therefore, the aircraft must generate 625 pounds of lift. The other elements of the lift equation were either estimated or known. They judged from experience that the wings would need 500 square feet of surface area to produce sufficient lift. Values for the lift coefficient, C_1, were drawn from their wind tunnel data tables corresponding to airfoil 12 at angles of attack between 2.5° and 7.5°, the range they believed most desirable for soaring flight. At a 2.5° angle of attack, the wing shape they planned to use had a lift coefficient of 0.311. At 7.5°, the coefficient was 0.706. When the Wrights plugged these values into the lift equation, they determined that the aircraft's velocity in calm air should lie between 23 and 35 mph. The lower velocity corresponded to the steeper angle of attack, so they chose that as the minimum velocity that would allow them to fly within the desired range of angles.

$$L = 0.0033\ V^2\ SC_l$$

$$625 \text{ pounds} = 0.0033 \times V^2 \times 500 \text{ ft}^2 \times 0.706$$

$$V^2 = \frac{625}{1.165}$$

$$= 536.5$$

$$V = 23.2 \text{ mph}$$

Knowing that thrust must equal drag in steady level flight, the Wright brothers could calculate the needed thrust by using a drag equation similar to the lift equation. Drag, in this case, meant total drag, the sum of induced drag, D_i, and parasitic drag, D_p. The only difference between the two equations is a substitution of a drag coefficient (C_d) for the lift coefficient. But the Wrights did not determine a drag coefficient directly in their wind tunnel experiments; they determined the ratio of drag to lift instead. Multiplying this by the lift coefficient yielded the drag coefficient:

$$D_i = kV^2 SC_l (C_d/C_l)$$

where k is Smeaton's coefficient (evaluated at 0.0033), V is velocity, S is the wing surface area, C_l is the lift coefficient, and C_d/C_l is the drag-to-lift ratio.

The Wrights calculated parasitic drag, the drag due to the frontal area of the aircraft plus its pilot, using the same equation but substituting the frontal area, S_f, for the total wing surface area, S:

$$D_p = kV^2 S_f C_l (C_d/C_l)$$

This equation shows why the Wrights decided to fly prone. Had they sat upright, the frontal area of the pilot, and hence the parasitic drag, would have been larger, needlessly increasing the difficulty of an already very tough problem. They estimated the frontal area to be 20 square feet. The lift coefficients and drag-to-lift ratios came from their table entries corresponding to airfoil 12. The drag-to-lift ratio was 0.138 at an angle of 2.5° and 0.108 at 7.5°. At the smaller angle of attack, the induced drag is:

$$D_i = 0.0033 \times 35^2 \times 500 \times 0.311 \times 0.138$$

$$= 86.7 \text{ lbs}$$

and the parasitic drag is:

$$D_p = 0.0033 \times 35^2 \times 20 \times 0.311 \times 0.138$$
$$= 3.5 \text{ lbs}$$

The total drag, and therefore the required thrust, at 2.5°, is:

$$D = 86.7 + 3.5$$
$$= 90.2 \text{ lbs}$$

At a 7° angle of attack, only 71 pounds of thrust were needed, but the brothers decided to play it safe and design their aircraft according to the larger number.

The power that the engine must produce is the rate at which work is done in pushing the craft through the air. The Wrights calculated the power, P, as the product of the total drag, D, and the velocity, V

$$P = DV$$

(The units for power, foot-pounds per second, are the same as used in describing the power available to a hydroelectric plant in chapter 3.) Again planning for a worst case scenario, the Wrights decided to calculate the power necessary to fly the plane at the highest velocity and greatest amount of drag encountered within the preferred range of angles

$$P = 90 \text{ pounds} \times 35 \text{ mph}$$
$$= 3{,}150 \text{ mile-pounds/hour}$$
$$= 8.4 \text{ HP } (1 \text{ HP} = 33{,}000 \text{ ft-lbs/min}).$$

These calculations suggest the extent to which the Wright brothers had become engineers. Not only did they design the propulsion system of their aircraft according to the best available scientifically collected data but their performance specifications also left a reasonable margin of safety, or so they believed. Their refusal to overdesign the flying machine made good engineering practice and also reduced the amount of material needed to build the aircraft and thus increased the chances for its success.

In the remarkably short period of six weeks, Orville Wright and Charles Taylor designed and built a four-cylinder engine that exceeded their previous specifications by generating 12 horsepower and weighing 180 pounds. It had no carburetor and used no spark plugs. Sparks from a magneto (a small electrical generator) ignited the gasoline that flowed under the force of gravity into the cylinders and mixed with air as it passed over the hot metal. The velocity of the Wright-Taylor engine could not be changed in flight because there was no carburetor and no accelerator.

The Wrights expected the propeller problem to be solved much more easily than the engine problem. After all, more than a hundred years of experience with propeller-driven ships must have produced a wealth of information on propeller design. Wilbur found none. Refusing again to quit, the two brothers decided to design and build their own. Tackling this problem by trial and error would have cost them months, given all the possible propeller designs. They needed a theory, a predictive equation, to guide their design choices. Wilbur was the first person ever to recognize that a propeller is nothing but a "twisted wing," with the lift forward, not up. As such, its behavior ought to be predictable according to equations similar to those used to calculate a wing's lift.

Concentrating on a wing shape that performed particularly well in the wind tunnel, the Wrights learned that the propeller's rotation rate, surface area, degree of twist, and width all played important roles in determining thrust. They used their theory to design prototypes and then revised their developing theory according to the performance of the prototypes. The propeller theory they created proved so effective that they were able to predict the thrust produced by their propellers to within 1 percent of the actual value. Orville and Wilbur finally decided that their powered aircraft should carry two counterrotating propellers, each 8.5 feet long rotating at a rate of 330 RPM to give 90 pounds of thrust. It is interesting to note that they used propellers in the same way as ships—they placed them behind the vehicle to push it forward. They completed the propulsion system by using gears and bicycle chains to connect the propellers with the engine.

"It Seems to Be Our Turn to Throw Now"

Upon hearing of Langley's first failure with the Great Aerodrome, Wilbur remarked, "I see that Langley has had his fling and failed. It seems to be our turn to throw now, and I wonder what our luck will be" (Combs 1979). Luck eventually had something to do with their success but not much.

Figure 4.16 The *Flyer* before it was wrecked on December 17, 1903. The rudder in the rear of the aircraft (to the right) is now free to move under command from the pilot who lay alongside the engine. (Photograph courtesy of the Library of Congress.)

Table 4.1
Comparison of the Wright's Flying Machines

Year	Wt★ (lbs)	Span (ft)	Chord (ft)	Area (ft^2)	Camber
1900	190	17	5	165	1/23
1901	240	22	7	290	1/12
1902	250	32.1	5	305	1/24
1903	605	40.3	6.5	510	1/20

★ With Orville as pilot
Data from McFarland 1953

The Wright brothers left Dayton once again for Kitty Hawk on September 23, 1903. Once more, they assembled their flying machine, called the *Wright Flyer*, on the sandy flats below the Kill Devil Hills (figure 4.16). Its total wingspan stretched further than any of their previous aircraft, 40 feet 4 inches (table 4.1). This oddly precise length had a purpose. The engine weighed 35 pounds more than the pilot and was anchored to the wing just right of center, so the right wing was made four inches longer than the left to provide the extra lift that would be needed to keep the plane from rolling to the right. The large wingspan contributed to a total wing surface area of 510 square feet, more than 200 square feet larger than their 1902 glider. The *Flyer* carried movable twin rudders linked to the wing-warping control system.

Ground tests of the propulsion system gave the Wrights well-deserved good news as its performance exceeded all expectations. Only 2.5 horsepower was lost as energy traveled from the engine through the gears and chains on its way to the propellers. This left 9.5 horsepower to drive the propellers, more than had been planned. The propellers proved to be 66 percent efficient in converting the power they received into thrust such that they produced 132 pounds of thrust, 42 pounds more than anticipated. These were the most efficient propellers yet devised. Orville and Wilbur knew their craft would fly. Now all they needed were the right conditions for the chance to prove it.

Perhaps pressured by Langley's efforts, they planned their first powered flight for November 4, 1903, before Langley's second attempt to fly the Great Aerodrome. Problems with the engine and drive shafts scuttled this plan. The repairs were made, and they waited out a stretch of bad weather, hoping to fly on November 28. This time a crack in one of the propeller shafts forced a delay and Orville left for Dayton to get a new one. By the time Orville returned to Kitty Hawk, the Great Aerodrome had crashed for the second and last time.

December 13 dawned with perfect weather for flying; a clear day and moderate steady breezes. But December 13 was a Sunday, and the boys had promised their father, a bishop, that they would not work on the Sabbath. The first powered flight was put off for the third time. On December 14, they were ready to go. Wilbur won a coin toss to decide who would pilot the *Flyer* on its maiden voyage. The engine's crankshaft turned over and drove the propellers, creating what must have been an almost unbearable racket. Wilbur rode the *Flyer* as it accelerated down the narrow wooden runway they had laid out on the sand. Finally, the craft lifted into the air. As the *Flyer* took off, Wilbur overcontrolled the front elevator, putting the nose up at a dangerously steep angle. The *Flyer* stalled and crashed 60 feet past the end of the runway. Wilbur blamed himself, the first case of pilot error causing an airplane crash. This first attempt did not meet the criteria for successful, powered, piloted flight, and the Wrights' considered it a failure because Wilbur had not sustained the craft aloft and had failed to produce a controlled landing. Fierce winds delayed the next trial three days.

Conditions again became favorable for flying on the morning of December 17, 1903. The wind was blowing off the ocean at velocities between 25 and 30 mph. This time it was Orville's turn to be the pilot. Only a handful of people were on hand to watch him lie down on the bottom wing of the *Flyer* as it sat on the runway. Three instruments had been mounted on board to measure the flight: an anemometer to determine relative wind velocity, a stopwatch to measure the flight's duration, and a device that recorded the number of times the propellers turned so they could calculate horsepower.

Under Orville's control, the *Flyer* sped down the runway and once again lifted into the air. The flight was erratic and looping as Orville struggled with the delicate pitch control mechanism, but 12 seconds later the *Flyer* touched down under control and the first piloted, powered flight was over. The *Flyer* had covered 120 feet in the air at an average ground speed of about 7 mph. The most famous photograph in aviation history shows the moment of liftoff (figure 4.2). Orville reiterated the criteria for a successful flight in observing that this was "the first in the history of the world in which a machine carrying a man had raised itself by its own power into the air in full flight, and sailed forward without reduction of velocity and had finally landed at a point as high as that from which it started."

Three more flights followed that day with Orville and Wilbur taking turns at the controls. Wilbur piloted the *Flyer* on its longest flight: 852 feet in 59 seconds. He appeared headed toward Kitty Hawk, four miles away, when air turbulence over a sand hummock forced him to make a rough landing that slightly damaged the aircraft. The historic plane never flew again. As the brothers and a few helpers moved the *Flyer* back to the hangar for repairs, a gust of wind flipped it over and damaged it badly. The parts were shipped back to Dayton, later reassembled, and after a prolonged feud with the Smithsonian over who should be credited with the first powered flight (the Smithsonian claimed Langley), the *Flyer* was placed in the Smithsonian Institution where it remains one of the most popular exhibits.

Looking back on the events of December 17, Orville could hardly believe they had mustered the courage to attempt powered flight. The brothers together had amassed only two hours of flying experience before piloting the *Flyer*. Furthermore, they had abandoned their prudent custom of flying a new aircraft as a glider before attempting full-scale tests. They may have felt that they needed to rush the *Flyer's* trials because Langley appeared so close to solving the flying machine problem, or they may have simply become extremely confident in their abilities as engineers and pilots.

Predictions Come True, Almost

The ability of Charlie Taylor and Orville Wright to milk their engine for 12 horsepower, rather than 8, allowed the Wright brothers to build a slightly larger, sturdier aircraft than originally planned (table 4.2). The difference between 8 horsepower and 12 horsepower allowed the *Flyer* to attain greater velocities and, in combination with a larger wing, greater lift than anticipated. The extra lifting capability was needed to accommodate the added weight. With more lift, however, comes more induced drag. The more powerful engine coped with this potential problem by balancing the greater drag with more thrust.

From Experiments to Practical Application

With their successful test flights of December 17, 1903, Wilbur and Orville Wright had solved the three major components of the flying machine problem: lift, propulsion, and control. A dream shared by people since the ancient Greeks had finally become reality. Although

Table 4.2
Comparison of Planned with Actual Wright Flyer

Dimension	Planned	Actual
weight	625 lbs	750 lbs
wing area	500 ft^2	510 ft^2
power	8 hp	12 hp
relative speed	23 mph	34 mph
thrust	90 lbs	132 lbs

Data from McFarland, 1953

the major period of experimentation was behind them, the Wright brothers knew that substantial work lay ahead to convert their successes into a truly useful, practical airplane. Military uses came to mind. They thought the military would want to use aircraft for surveillance during wartime, a use to which balloons had been put forty years before in the Civil War. Indeed, the first serious attempts to purchase the new flying machines were made by representatives of military organizations.

The Wrights set about to improve the *Flyer* shortly after returning to Dayton. They arranged to conduct flights from a pasture a few miles outside of the city in order to avoid the expense of traveling to Kitty Hawk. Over the next two years, they made hundreds of flights with several versions of the 1903 *Flyer* as they struggled to increase the engine's power and improve the control systems. Pitch control posed an especially troublesome problem. Their aircraft had a dangerous tendency to rise and swoop as Wilbur or Orville manipulated the sensitive elevator. Both of the brothers suffered through several crashes until they moved the elevator even further in front of the wings to increase its leverage.

The 1905 *Flyer* produced by these efforts eventually gained acclaim as the first truly practical "aeroplane," as such machines were then called. A new engine had been built to deliver 25 horsepower, more than twice the amount produced by the engine on the 1903 *Flyer*. New propellers incorporated design changes that improved the thrust. Although the Wrights reduced slightly the total surface area of the wing compared to the 1903 *Flyer* (503 square feet vs. 510 square feet), the lift was increased in the 1905 *Flyer* to accommodate its greater weight. The aircraft's faster velocity and an improved design for the wings' leading edges no doubt accounted for the improved lift. For the first time, they gave the pilot independent control of each of the three control systems. While this had the potential for reducing stability in flight, it permitted complete control over the craft, something no other aeroplane of the time offered or would offer for several years. On October 5, 1905, Wilbur flew the new *Flyer* a distance of 24.2 miles and kept it in the air for 39 minutes 23.8 seconds, shattering records for duration and distance. This single flight lasted longer than all their previous powered flights combined.

The Business of Flying

The Wright brothers continued to improve their aeroplanes over the next few years by adding seats for a pilot and a passenger, for example, but the days of major discovery and innovation were over. The brothers' attention was increasingly diverted to the business of flying. Now that they had built a practical aeroplane, they sought the financial reward they believed they had earned with their hard work and intelligence. The brothers decided to go into the business of manufacturing and selling aeroplanes as soon as they obtained patents for their inventions.

The Wright brothers' determined struggle to gain patent protection drove them to shroud their work in secrecy so that unscrupulous industrial spies would not steal their designs and undermine their future in commercial aviation. While the brothers eventually emerged victorious from a dozen lawsuits over patent infringement, they paid a price in lost friendships, tarnished reputation, and, ultimately, loss of supremacy in the new field of powered flight.

Orville and Wilbur flew little during the interval between 1905 and 1908, but when they reemerged they swept away all doubts about their claims. In December of 1908, for example, Wilbur set a record by flying for 2.3 hours and covering 77 miles. This success, along with the highly publicized demonstration flights made by Wilbur that same year in France, helped convince the Signal Corps of the United States Army to buy one of the Wrights' aircraft for $30,000. The Signal Corps contract marked a substantial victory for the Wright brothers in their battle to be recognized as the leaders in aviation and to reap the financial reward they deserved. As early as 1905, they had written to their congressman, offering to sell flying machines to the U.S. government or to furnish the government the information and patent licenses it would need to build its own aircraft. Representatives of the War Department's Board of Ordnance and Fortifications rejected the offer saying they would need proof of the *Flyers'* "practical state" before seriously considering such a proposal. Determined to give their government first crack at the new technology, the Wrights resubmitted their offer to the War Department, only to be rebuffed again with the comment that it would not be considered until they had shown their machine capable of carrying a pilot in level flight. By this time, the Wrights had flown one of their planes over two miles and gained sufficient control to fly it in circles.

The War Department's total lack of awareness regarding the Wright brother's achievements is hard to imagine in these days of television, investigative reporting, and the capacity for nearly instantaneous communication worldwide. At the very least, the government could have sent an observer to Dayton to see what Wilbur and Orville were up to, but the government must have been inundated with requests for support from flight enthusiasts all over the country who thought they had solved the flying machine problem. Besides, the military had just been heavily criticized by Congress for sinking $50,000 into Langley's ill-fated project, and bureaucrats at all levels must have been wary of repeating the same mistake.

The Wright brothers' secrecy certainly did not help. As a consequence of their tight security measures, the newspapers either failed to report the Wrights' work or published hopelessly inaccurate accounts. The failure to report the flights is particularly revealing. People who lived around the pasture where the Wright brothers were flying almost daily asked the publisher of the local newspaper why they had seen nothing written about it. The publisher replied that he had heard the rumors but simply did not believe them. Many people remained skeptical that piloted flight was possible. Even Wilbur Wright, just a few years before, had held up human flight as the "standard of impossibility."

Shaken by the U.S. government's rejection of their work, they approached the governments of England, France, and Germany, all of which expressed interest but were slow to offer contracts. The eventual accumulation of several contracts in addition to the Signal Corps contract gave the Wrights the financial resources they needed to start their own aircraft manufacturing company. The Wright Company was organized in 1909 with Wilbur as president and Orville as vice-president. By this time, however, the Wrights had competition. Glenn Hammond Curtiss, a builder of motorcycle engines and holder of the land speed record (137 mph on a motorcycle), had successfully designed and flown the *June Bug* in 1908 with support provided by Alexander Graham Bell, the revered inventor of the telephone. The fact that Curtiss won *Scientific American's* prize for being the first American to fly *publicly* one kilometer in a straight line testifies to the closeness with which the Wrights guarded their work. Curtiss and the Wrights became bitter rivals in a lawsuit over the Wrights' contention that Curtiss had infringed on their patents, a suit the Wrights eventually won. The Wrights' company went through several reorganizations before becoming the Dayton-Wright Company in 1917. This company became one of the major suppliers of airplanes for the war effort. Unfortunately, Wilbur did not survive to see this success. He died of typhoid on May 30, 1912. Orville sold his share of the company in 1915 in order to return to Dayton and the process of discovery that he loved.

Despite their pioneering success, the Wright brothers were slow to abandon their early designs for more modern forms. Other innovators in the United States and Europe introduced the modifications that created the modern airplane including: replacement of the wing warping method of roll control with ailerons; adoption of the single, self-supporting wing; replacement of the *Flyer's* stubby open frame with a longer, entirely enclosed fuselage that reduced drag; transferring the elevator from the front of the plane to the tail for improved stability and control; and moving the propellers to the front of the plane where they work to pull the plane forward rather than push it from behind. The Wright companies eventually designed planes with these features, but they were not the primary innovators. This should not detract from the Wrights' achievements, however, because their contributions were much more fundamental and opened the door for the rapid innovation that followed.

The business of airplanes, initially, was almost entirely military. The advent of World War I in 1914 no doubt provided great impetus for the rapid improvements in flying machines that followed the Wrights' aircraft of 1908. Only 43 airplanes were manufactured in the United States in 1913, but five years later the Dayton-Wright Company alone produced 3,500 aircraft. Among the first customers for airplanes after the war was the U.S. Postal Service. Airmail became commonplace in the United States by 1920. (Charles Lindbergh began his flying career as an airmail pilot.) Commercial passenger service followed close behind. Although "airliner" routes appeared first in Europe, air travel became so popular in the United States that the number of airplane passengers in the United States exceeded all those in the rest of the world by 1930. The rapid improvement of airplane technology can be seen also in the pace at which cruising

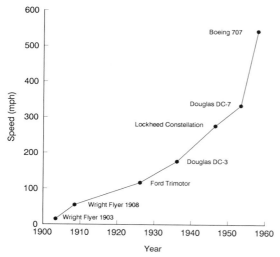

Figure 4.17 Progress in the cruising speed of commercial aircraft produced after the original Wright *Flyer*.

speed increased over the years (figure 4.17). Orville Wright survived to see these dramatic changes, including Chuck Yeager's historic flight in a jet aircraft to break the sound barrier, thirty-three years after Wilbur's record-setting flight with the 1905 *Flyer*. During his last flight, Orville took the controls of Lockheed's famous *Constellation*. He died five years later in January, 1948 at the age of seventy-seven.

The Wright Stuff

A great many people, some with much more formal training than the Wright brothers, had broken their picks on the flying machine problem. Yet only four years after beginning to study the problem and a mere three years after their first gliding flights, Orville and Wilbur were making controlled powered flights. This is truly remarkable considering that they had to reestablish basic data on lift and drag, design a craft that would endure the stresses of flight, determine what control systems were necessary, design those control systems, learn to be pilots, develop a theory for predicting propeller performance, and design and build their own propulsion system including an engine and propellers.

The Flying Machine Problem

Why were the Wright brothers so successful? Aspects of their characters, especially Wilbur's attention to detail, no doubt played major roles in separating the Wrights from the rest of the pack. But there are general lessons that emerge from the way in which they went about solving this problem. Contrary to popular accounts, the Wright brothers were not bird-watching tinkerers working from a blank slate. Their notebooks and personal letters reveal a sound quantitative understanding of basic physics and an unflinching demand for careful, logical designs. Looking back on their work, Charles Taylor commented that the Wright brothers "sure knew their physics. I guess that's why they always knew what they were doing and hardly ever guessed at anything." When the Wrights observed that the existing coefficients for lift and drag could not be used to predict accurately the behavior of their gliders, they undertook careful scientific studies to increase the accuracy of these coefficients. Better data, combined with systematic field tests of their designs, gave the Wright brothers a distinct advantage over their competitors and no doubt contributed to their rapid progress. The interplay between scientific investigation and engineering design that characterized the Wright's work thus anticipated many of the successes in modern technology.

Theory Catches Practice

As Orville and Wilbur Wright tackled the practical problems of making a flying machine in the United States, engineers, mathematicians, and physicists in Europe struggled to translate Newton's and Bernoulli's ideas into detailed theories of lift and drag. Actually, a crucial development in the theory of aerodynamics occurred in 1894, well before the Wrights' most historic flight, when an English engineer and automobile manufacturer named Frederick Lanchester established a link between aerodynamic lift and a concept called circulation. Unfortunately, Lanchester's description of his ideas proved so difficult to understand that they had little impact until 1907, well after the Wrights' major successes.

The value of Lanchester's contribution arose from the difficulty faced by aeronautical engineers in accurately determining the aerodynamic lift that a particular wing shape will produce. Bernoulli's principle provided one general explanation of lift but failed to yield accurate values for lift especially in the absence of detailed information about airflow velocity over a wing. Lift might have been determined by precisely measuring the air pressure at a large number of points over the surface of a wing in an air flow, but this proved difficult to accomplish. Lanchester's "circulation" represented a mathematical quantity directly related to lift that was, and remains, much easier to determine than actual pressures. *Circulation* exists when the air flowing over the top of the wing does not move at the same velocity as the air flowing over the bottom of the wing. Circulation incorrectly implies that air particles are somehow circulating around the wing. Rather, imagine holding a ping-pong paddle at arm's length with the face of the paddle perpendicular to the direction of a stiff breeze. If the wind strikes the upper and lower halves of the paddle at the same velocity, then the amount of circulation is zero and the paddle does not rotate. But if the wind striking the upper half of the paddle is moving with a greater velocity than the wind hitting the lower half, the circulation exists in an amount proportional to the difference in velocities and the paddle rotates in your hand. As Bernoulli recognized, velocity differences in fluid flow imply corresponding pressure differences. Pressure differences between the upper and lower surfaces of a wing produce lift. Because circulation is related to velocity, then it is also related to pressure and lift.

The conversion of Lanchester's idea of circulation into a useful tool for predicting lift required the efforts of two other people. A German mathematician named M. Wilhelm Kutta was stimulated to tackle the problem of lift by Lilienthal's dramatic exploits with gliders. In trying to calculate the lift generated by wings such as the ones used by Lilienthal, Kutta discovered a method for calculating the velocity of airflow over a wing. He also developed qualitative arguments for circulation quite independently of Lanchester. He published his ideas in 1902, the same year in which a Russian mathematician, Nikolai Joukowski, built the first wind tunnel in Russia. Four years later, Joukowski converted Lanchester's and Kutta's concepts into a simple mathematical expression relating lift, velocity, and circulation that came to be known as the Kutta-Joukowski theorem. Once the velocity of airflow over a wing was calculated, the circulation and lift could be determined in turn. The Kutta-

Joukowski theorem thus provided the first practical method for calculating lift precisely.

Lift depends on the interaction between a wing and the air flowing around it, but the significance of this mysterious boundary remained unknown to Kutta and Joukowski despite their successes in developing a method for determining lift. Ludwig Prandtl, a German physicist, began to unravel this mystery in the first decade of the twentieth century. Prandtl guessed that friction retards the flow of air in a thin "boundary layer" immediately adjacent to the wing's surface. The effect is so great right at the wing's surface that air particles there have no velocity. The concept of a boundary layer provided invaluable new insights into the phenomenon of aerodynamic drag. Drag inevitably accompanies lift, and a detailed theoretical understanding of both gave a tremendous boost to engineers who sought better designs for wings.

Prandtl also made important contributions to our modern understanding of lift. According to Newton's third law, Prandtl reasoned, lift must be accompanied by a downwardly directed airflow behind the aircraft. He traced the source of this downward force to vortices of air that developed at the tip of each wing when the aircraft was in flight. The theory that Prandtl developed to relate these vortices to lift supported methods for determining how lift was distributed over a wing's surface.

These scientific studies, Prandtl's work especially, laid the groundwork for predictive wing design. By 1919, Prandtl had produced a complete theory to explain the action of wing shapes. Rather than laboriously testing wing designs in wind tunnels, it became possible to design wings from theory and predict their behavior before conducting wind tunnel experiments. The rate at which the practice of flying progressed after Prandtl's work became widely known in the 1920s is truly astonishing. The Wrights' success in overcoming practical problems of flight joined with the theoretician's success in understanding the physics of flight provided a powerful combination leading to advances in aeronautics. The fact that the Wright brothers were able to build a flying machine without benefit of a detailed physical theory to guide them makes their accomplishment all the more remarkable.

The development of the airplane thus illustrates an interaction between scientific discovery and engineering innovation that differs in an interesting way from the interplay that led to the telegraph or hydroelectric power. The Wright brothers took a scientific approach to an engineering problem, but they used little or no scientific theory in their work largely because detailed expressions of that theory were only just coming into existence. The cases presented in chapters 2 and 3, however, reveal that a scientific understanding of electromagnetism preceded and directly stimulated development of the telegraph and the generator. The joining of theory with practice after the Wrights' triumph greatly accelerated development of a technology that has influenced our society so broadly it seems impossible to calculate its effects.

Exercises

1. Invoke Bernoulli's principle to help explain the following:
   (a) An umbrella feels lighter on a windy day.
   (b) When boats or cars pass each other moving in the same direction, they tend to be drawn together.
   (c) When you turn on the shower, the shower curtain moves in and wraps around your leg.
   (d) The convertible top of a moving vehicle bulges upward.
   (e) A roof is blown off by a tornado. Does it make sense to open the windows of your house to prevent this?
   (f) Windows can be blown out of tall buildings on a windy day.
   (g) Papers often fly out of open car windows.

2. The Wright engine was close to 12 horsepower, but the transmission efficiency was about 95 percent, and the propeller efficiency was approximately 66 percent. As a result, what was the actual power available to the Flyer?

3. Modern aeronautical engineers use a slightly different equation for calculating lift than the Wright brothers used:

$$L = \frac{\text{air density}}{2} SV^2 C_l$$

   L = lift in pounds; S = surface area in square feet; V = relative velocity in *feet per second* (not mph as before); and C_l = lift coefficient.
   (a) Use this equation to determine whether the 1903 *Flyer, as planned* (see table 4.2), could have generated enough lift at the steepest desirable angle of attack (7.5°) at 23 mph. The lift coefficient corresponding to this angle of attack is 0.706, and the air density at Kitty Hawk is 0.00238 slugs/ft³ (don't worry about this strange English unit for density— engineers don't always use the same units as scientists).
   (b) What is the minimum relative velocity (in miles per hour) that the *actual* 1903 *Flyer* had to sustain in order to stay aloft at the steepest

desirable angle of attack (7.5°)? Compare your answer to the craft's relative velocity on its inaugural flight (34 mph). Given that the Flyer's pilot could not alter engine speed once the craft was airborne, how might the pilot have adjusted lift according to slight changes in wind velocity?

(c) Estimate the minimum relative velocity that must be maintained by a Boeing 747 jet airliner to maintain the necessary lift at 30,000 feet and at sea level. The relevant specifications for this huge aircraft are as follows:

weight = 800,000 lbs

air density at 30,000 ft = .0009 slug/ft^3

air density at sea level = .00238 slug/ft^3

wing surface area = 5,500 ft^2.

coefficient of lift = 0.47

(d) An efficient flight is achieved when a given load is carried with a minimum of thrust. In steady, level flight, the amount of lift must equal the load (the aircraft's weight) and the amount of thrust provided by the engines must equal the drag. This means that aeronautical engineers try to achieve a large lift-to-drag ratio in order to increase efficiency. The Wright brothers' 1903 *Flyer* had a lift-to-drag ratio of approximately 7. Modern jet liners have a ratio roughly twice that. What changes in aircraft design can you think of that contribute to the greater efficiency of modern aircraft compared to that of the 1903 *Flyer*?

4. What features of the period between about 1890 and 1910 made it a propitious time for the development of flying machines?

Innovation

II

Engineering is not simply applied science. Technological innovations may spring from scientific discoveries or methodologies, as described in the previous three chapters, or they may grow entirely from within the practice of engineering. In the latter case, the engineering achievement may actually precede a complete understanding of the underlying scientific principles, as occurred in the development of the steam engine (chapter 5), or the scientific foundations for a technical innovation may be so broad and remote from the innovation that it can be considered to have arisen largely independently of them, as described in the development of prestressed concrete (chapter 7). Frequently, the interplay between science and engineering is more complex, as in the development of new methods for extracting gasoline from oil (chapter 6). In this example, developments in chemistry and the origin of corporate divisions dedicated to research and development created a climate in which the distinction between scientific discovery and engineering innovation became blurred. This new partnership between science and engineering proved so successful that it established the basic mode by which most large corporations now operate.

Fossil Fuels, Steam Power, and Electricity: Los Angeles Revisited

5

The City Grows: Demand for New Technology

Imagine sun-splashed beaches, rugged mountains, and starkly beautiful deserts combined with uncongested cities and low real estate prices. What would you predict would happen there in short order? This was the situation in the Los Angeles area at the end of the Second World War. It had been rediscovered by hundreds of thousands of servicemen on their way to and from the Pacific and had also become a major defense industry center. Earlier growth of the city had been supported by hydroelectric power, as described in chapter 3, but the influx of new residents stretched that resource beyond its limits. How did the region provide the necessary power for the daily lives of millions of inhabitants?

To fill the projected electricity needs of the postwar boom, the utility companies rapidly returned, albeit on a much larger scale, to the fossil fuel and steam technology that had languished after hydropower reached its maturity in the 1920s. Steam power ascended to dominant status, replacing hydropower as the major source of electricity for Los Angeles at an astonishing rate (figure 5.1). This remarkable transformation probably could not have been supported by the old workhorse of the First Industrial Revolution, the steam piston engine run by a coal-fired boiler. The task no longer involved twenty streetlights run by a 300-kilowatt (kW) generator nor could it be satisfied by 10,000-kW steam plants.

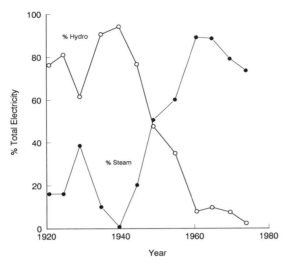

Figure 5.1 Electrical energy from hydropower (open circle) and steam power (filled circle). (Data provided by the Los Angeles Department of Water and Power.)

Instead, the new era of electric power turned to the steam engine's modern cousin, the steam turbine. This "prime mover" was developed in the late nineteenth century and was as important to steam electrical generation as the water turbine was to the tapping of hydroelectric power. The steam turbine did not act alone, however, to reverse the fortunes of steam technology. It had existed for over fifty years before overtaking hydropower in Los Angeles.

The late 1800s also witnessed the discovery of petroleum and the advances in chemical engineering which made refining it into many useful products possible. This was not only an important step leading to the rise of the automobile and the airplane as modern forms of transportation but it also gave us another major source of fuels for producing electricity. The birth of Southern California's own oil industry in the 1890s and its maturation in the 1920s also fostered the use of steam plants for electric power generation in the area.

In order to appreciate the importance of the steam turbine's contribution to electrification, it is worth tracing the early development of steam power from its humble origins in the early eighteenth century to its use as the prime mover of industry and urban civilization.

Origins of the Steam Engine

As discussed in chapter 3, electricity may be provided by harnessing the energy of falling water to drive an electric generator. The use of running water as a source of power for human needs dates back to the Middle Ages. The giant step forward for the Industrial Revolution was the extraction of the chemical energy stored in fuels like wood and coal to produce steam power to run pumps, looms, mills, and other machines (see box 5.1). As with hydropower, this inevitably culminated in the linking of the steam engine to the electric generator.

Steam plants must convert, as efficiently as possible, the chemical energy of fuel to the thermal energy of steam to the mechanical energy of a steam engine to electrical energy of a generator (figure 5.2). The technology for accomplishing these energy conversions has diverse roots in the invention of the steam engine in the eighteenth century and the development of the electric generator in the nineteenth century.

The ancestry of the modern steam engine goes back to the early 1700s when inventors such as Thomas Newcomen, a British iron maker, designed machines that used steam to pump water out of mines and wells. In these early engines, the piston was pulled to the top of a vertically mounted cylinder by the weight of a lever arm while hot steam was allowed to enter the cylinder below the rising piston. When the cylinder was cooled with cold water, the steam condensed creating a partial vacuum within the cylinder, and atmospheric pressure forced the piston downward in the return stroke. Raising the piston and injecting more steam into the cylinder reheated it and started the next cycle. Because the piston was not actually driven by steam pressure but by the weight of the external air, some experts on technology prefer to call such devices *atmospheric engines*. The simple reciprocating motion of the piston in atmospheric engines suited the tasks of pumping, hammering, and sawing, but the great utility of the steam engine was not fully realized until after 1769 when James Watt revolutionized its design.

Watt, a Scottish instrument maker and inventor (figure 5.3) introduced several ingenious modifications that dramatically improved the efficiency and the utility of the steam engine (figure 5.4). His engines burned only one third of the fuel required by

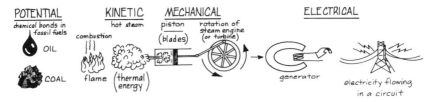

Figure 5.2 Energy conversion: steps in generation of electricity by fossil fuel steam plants.

Figure 5.3 James Watt. (From *A History of the Growth of the Steam Engine*, by Robert H. Thurston, D. Appleton and Co., New York, 1902.)

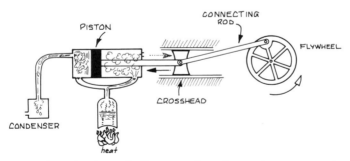

Figure 5.4 A schematic diagram of a reciprocating steam engine.

Coal, Oil, and Electricity: Energy vs. Power

The generation of electricity at a facility such as the Banning-Alameda steam plant begins by burning fuel to heat water. Fossil fuels, such as coal, oil, and natural gas, serve this purpose better than wood or many other combustible materials because they contain more energy that can be tapped to do work. The two most commonly used fuels are coal and fuel oil derived from the refining of petroleum. Various grades of these fuels depend, in part, on how much heat or thermal energy each provides when burned. This is the chemical energy of the fuel "stored" in bonds between atoms in the materials. Coal is largely graphite (composed of carbon atoms) and some impurities including sulfur. Oil consists of molecules (hydrocarbons) containing carbon-hydrogen bonds and some impurities, which often contain sulfur. A pound of coal "contains" or produces upon combustion about 11,400 Btu of thermal energy, almost 23 million Btu per ton, which is transferred as heat. For comparison, a gallon of oil contains 140,000 Btu, or about 5.9 million Btu per 42-gallon barrel. On a per weight basis, oil has 1.7 times the amount of energy as coal.

The Btu, or British thermal unit, is a respectable measure of energy, but it is more useful for us to relate these values to electricity. Electrical energy is typically measured in kilowatt-hours (kWh). The kilowatt itself is a unit of power equal to 1000 watts and refers to the rate of energy use, that is, how much energy is expended per unit time. The kilowatt-hour describes the total amount of energy used over a given period of time. The energy available in one ton of coal is:

$$23{,}000{,}000 \text{ Btu} \times \frac{0.000293 \text{ kWh}}{1 \text{ Btu}} = 6740 \text{ kWh}$$

whereas one ton of fuel oil would yield 11,500 kWh. For one barrel of oil, the energy content is:

$$5{,}900{,}000 \text{ Btu} \times \frac{0.000293 \text{ kWh}}{1 \text{ Bu}} = 1740 \text{ kWh}$$

The above values represent energy. Power depends upon the time over which energy is expended. For example, when one ton of coal burns over two hours, an average of 3370 kW of power is expended:

$$\frac{6740 \text{ kWh}}{2 \text{ h}} = 3370 \text{ kWh}$$

But if the coal were burned in only one half hour, a more powerful process involving 13,480 kW would result.

Box 5.1

atmospheric engines to perform the same amount of work. Watt achieved this remarkable advantage by using an external condenser, rather than the cylinder itself, to cool the steam to water, and his engines thus avoided expending energy to reheat the cylinder in every cycle. He and his business partner, Matthew Boulton, grew so confident of this efficiency advantage that in 1776 they encouraged miners to buy their engines by accepting as the purchase price one third the value of the fuel saved by not using an atmospheric engine. In 1782, the firm of Boulton and Watt produced the first "double acting reciprocating engine" whereby steam pressure was used to drive the piston first one way and then back to its original position rather than relying on atmospheric pressure to drive the return stroke. Watt also designed a system of gears and rods that converted the back-and-forth motion of the piston into circular motion of a wheel. Energy in the rotating wheel could be used directly or stored in a large, freely rotating wheel called a *flywheel*. Systems of belts and gears transmitted this energy to a great variety of other devices such as looms. These improvements in the design of the steam engine opened the way for steam power to come into common use and to catalyze expansion of the First Industrial Revolution. They also made Watt and Boulton rich men.

Steam Power

The steam engine's performance depended upon the amount of steam pressure, the design of the piston, and the speed of the piston's movement. Watt developed a rather simple equation to predict the output of his engine in terms of horsepower (HP):

$$HP = \frac{PLAN}{33,000}$$

The symbols refer to steam pressure (P, in pounds per square inch or psi), the distance the piston travels in the cylinder (L, in feet), the area of the piston head (A, in square inches), and the number of strokes per minute of the piston (N). The numerical term arises from Watt's definition of one horsepower being equal to 33,000 foot-pounds per minute (also equal to 0.746 kW). Imagine a horse (with the aid of a rope over a pulley) lifting at a constant rate a 1000-pound load 33 feet off the ground in a period of one minute.

We can use this formula, which accounts for losses due to poor heat transfer and friction, to calculate the output of a steam engine similar to the one installed in the first Los Angeles steam plant at Banning and Alameda streets. Typical values for the operation might be: steam pressure of 75 psi, a piston with a length of 4 feet and area of 200 square inches, and 120 strokes per minute:

$$\text{power} = \frac{75 \times 4 \times 200 \times 120}{33,000} = 220 \text{ HP}$$

The piston area corresponds to a circle with a diameter of 16 inches, and the steam pressure is about five times atmospheric pressure. In terms of kilowatts:

$$\text{power} = 220 \text{ HP} \times \frac{0.75 \text{ kW}}{1 \text{ HP}} = 165 \text{ kW}$$

Los Angeles' first electric station, the Banning-Alameda plant, probably used two generators of approximately this size to run its electric generators.

New Technology: the Steam Turbine and Turbogenerator

Prior to 1904, all steam plants in Los Angeles used reciprocating steam engines to drive their generators. Drawbacks associated with this use of these engines (see discussion of efficiency that follows), however, encouraged engineers to search for other "prime movers" to convert thermal energy into rotational energy. It has been said that the steam turbine "is the most fundamental change in the steam engine since Newcomen's day" (van Riemsdijk and Brown 1980), a claim that probably does a disservice to James Watt.

Development of the steam turbine awaited the pioneering efforts of Gustav Laval, a Swedish engineer, who in 1882 was trying to build a better cream separator. His first efforts produced turbines that suffered mechanical problems because they rotated too rapidly (40,000 RPM), although he eventually succeeded in designing turbines to generate electric power.

The concept of using steam turbines to drive electric generators is credited to Sir Charles Parsons, the noted British engineer (figure 5.5). He modified Laval's design and built a turbogenerator in 1884 that rotated at the safer, more sedate rate of several thousand revolu-

Figure 5.5 Sir Charles Parsons. (Photograph from *A Short History of the Steam Engine* by H. W. Dickenson, 1963. Reprinted by permission of Frank Cass & Co Ltd, London.)

tions per minute (RPM). The problem had been to find a method of forcing steam onto the turbine's blades without producing excessively fast rotation. Parsons solved this problem by "pressure staging" in which turbine blades were arranged in a series along the rotor shaft. As the steam "writhed through the turbine" (Hodgins and Magoun 1938) past the turning blades, its pressure dropped gradually so that the speed could be controlled more precisely.

Operation of the steam turbine depends upon the kinetic energy of hot steam turning the blades on a shaft as rapidly and efficiently as possible. As with water-powered turbines, there are two different mechanisms (impulse and reaction) that can be used, but both involve "axial-flow" with steam moving parallel to the rotating shaft. The steam that turns the rotor blades is injected at high temperature and pressure (e.g., 1000°F and 2000 psi). The turbine blades fit snugly in successive compartments so that steam moves along the rotor at hundreds of feet per second, alternately moving past fixed blades and rotating blades until it reaches the condenser. The fixed blades serve as nozzles to direct the steam toward the moving blades at the optimal angle (figure 5.6). Parsons built reaction turbines with the blades shaped like airfoils, as in the Francis turbine, and the steam flowed over them to produce "lift." Impulse

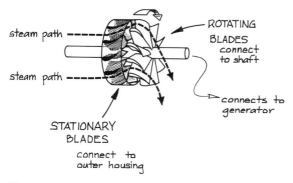

Figure 5.6 A schematic diagram of a steam turbine.

turbines work like the Pelton wheels in that steam is shot directly at bucket-shaped blades.

The steam turbine, like the steam engine, utilizes a separate condenser to help draw the steam through the blades and ultimately form water that can be recycled through the turbine. Condensing the steam leads to a large drop in pressure as the steam goes from the front to the end of the turbine; this effect adds to the power of the turbine by increasing the steam's flow rate. A large turbine requires a considerable volume of water to cool the steam, thus utility companies site steam plants near large sources of water.

In contrast to the reciprocating engine, the steam turbine suited the linkage with the electric generator almost perfectly because both operated by rapid rotation around a central axle. The process of converting the reciprocating motion of the "Watt engine" into rotary motion required intervening mechanical devices such as connecting rods, eccentric gears, and belt dives that also acted as sources of friction reducing the efficiency of the energy transfer. The reciprocating engine was also encumbered by the inertia of the piston, which changed direction rapidly and completely. Although the steam turbine blades must move, they rotate continually in the same direction, making the turbine inherently better suited than reciprocating engines to turn the rotor of a generator, as well as allowing them to run more quietly. The turbine could easily develop rotation rates of thousands of revolutions per minute (1800 and 3600 RPM being common) compared to hundreds of revolutions per minute for reciprocating steam engines. Operating the electric generators at higher rotation rates led to higher voltages and more efficient

Fossil Fuels, Steam Power, and Electricity

transmission of power. The rotation rate of steam turbines also proved easier to control than that of reciprocating engines.

In addition to these technical advantages, several nontechnical factors favored steam turbines. In the first two decades of this century, the increasing demand for electricity was met in part by building ever larger reciprocating steam engines to develop greater power. These engines took up correspondingly larger spaces. Equal amounts of power could be produced by relatively smaller steam turbines, a feature that made them well suited to power plants in urban areas where land was commanding high prices. Over the years, however, turbines have grown larger too, so that a modern version is about 90 feet long and weighs 100 tons.

In contrast to the turbines employed in hydroelectric plants, steam turbines are usually constructed so that the rotor shaft is horizontal. Steam turbogenerators also rotate much faster than turbines in hydroelectric plants, which typically rotate at less than 300 RPM. Turbogenerators are also used in nuclear power plants, which use the heat from radioactive decay to boil water to produce steam.

Postwar Resurgence of Steam Power Plants

It took nearly fifty years for steam power to overtake hydroelectricity, as shown in figure 5.1, although it began to be more competitive in the late 1800s and early twentieth century. At the close of World War II, the increased demand for electricity could not be met by hydropower alone because most of the major rivers had been tapped. Utility companies in the Los Angeles basin were forced to reconsider fossil fuels as their primary energy source.

One factor contributing to the renewed appeal of steam power was the development of the oil industry in Southern California. Petroleum deposits had been evident in the area well before the Spaniards arrived: Native Americans used the viscous "brea" (tar) for caulking boats and waterproofing baskets. Various attempts to exploit this resource had been made in the mid-1860s, but the oil was not suited to the refining techniques then available. The petroleum industry began in earnest in Los Angeles in 1892 when Edward Doheny, a local civic leader, had a well dug near the center of the city. New refining techniques allowed the oil to be converted to more moderately priced fuels and other products. Over two thousand wells were drilled in the next five years in Los Angeles, and the expertise gained in drilling and refining benefitted the subsequent exploration and development of the large coastal oil fields discovered some twenty years later, marking the maturation of the Southern California oil industry.

Despite this poor efficiency relative to hydropower installations, fossil fuel steam plants offered a distinct advantage in the early days of electrification in Los Angeles because they could be located close to sites where power was needed. In those days, transmission of electricity over long distances posed difficult technical problems, so that steam plants offered a reasonable alternative. However, they could not continue to offer an economically viable solution to the rapidly increasing demand for electricity in the late nineteenth century. Duplication of small steam plants, like the Banning-Alameda plant, presented none of the economies of scale offered by large centralized power plants developed later on. Coal-fired power plants required the costly importation of fuel because local deposits were inadequate to meet the need for energy. Oil had been discovered in the Los Angeles basin, but until the mid-1890s there was no capacity for refining it into useful fuel, and it was not until the 1920s that a full-fledged petroleum industry developed. Given these constraints, steam power took a back seat to hydropower in Los Angeles until after World War II. By then the steam turbogenerator had become the engine of choice for generating electricity from fossil fuels.

Development of a local, abundant source of fossil fuel reduced the need for costly imported fuels and thus removed one of the impediments hindering the development of steam plants in Los Angeles. These petroleum deposits supplied fuel oil to run generators and during World War II also provided ample natural gas to substitute for the strategically important oil. Most steam plants now use oil as their primary energy source, but they can switch to gas when required, for example, to meet air quality standards.

Among the first steam plants to be built for the Los Angeles market after World War II was Southern California Edison's (SCE) Redondo Beach Plant, which began producing electricity in 1948. The company chose to locate the new facility on a site southwest of the city previously occupied by the main power station for Henry

Huntington's Pacific Electric Railway. The urgent need for electricity forced the utility to complete this new plant in less than two years. Sacrificing power for speedy construction, SCE equipped the Redondo plant with smaller generators than those that were technically feasible at the time. Its generating capacity was about half that of facilities built a few years later, but the four generators of the Redondo Beach Plant had a total capacity of 270,000 kW, or about one-fifth that of Hoover Dam. Southern California Edison constructed the Redondo plant as the first step in an ambitious expansion program that placed ten successively larger steam plants into service by the mid 1970s. The resurgence of steam power was underway.

Widespread utilization of the steam turbine was essential to the proliferation of steam plants to meet the city's growing demand for electricity. In 1904, steam turbines had been introduced in Los Angeles at the Pacific Electric Company's Redondo Beach Plant and had gradually replaced reciprocating engines in steam power plants while hydroelectric facilities provided the majority of power. The rapidly expanding post–World War II demand for electricity, however, prompted their increased utilization. Without this technology, it is unlikely that the power companies could have met the challenge.

Fossil Fuels to Electricity: Efficiency

The purpose of a fossil fuel steam plant is to transform as efficiently as possible the chemical potential energy in the fuels into electrical energy at an appropriate rate to provide adequate power. For the 300-kW Banning-Alameda plant of 1882, this included the efficiency of the boiler and steam engine in converting chemical energy of the fuel first into thermal energy and then into mechanical (rotational) energy and the efficiency of the Brush generators in converting rotational energy into electrical energy. Unfortunately, data for determining its overall efficiency are no longer available for that long-defunct facility. The plant's operating load and fuel consumption can be estimated, however, by making reasonable assumptions based on standard equipment in use at the time.

Assuming that the twenty-one arc lamps operating in Los Angeles in 1882 were rated at 4000 W each and remained lit five hours per day, then the amount of electrical energy needed each day can be estimated as follows:

energy needed = 21 lamps × 4000 W/lamp × 5 h

= 420,000 Wh or 420 kWh

Given that one ton of coal contains 8,500 kWh of potentially useful energy (box 5.1), it appears that the Banning-Alameda plant could have produced the needed electricity by burning much less than one ton of coal during the time the lamps remained lit.

No process operates with complete efficiency according to the second law of thermodynamics (box 5.2), so that all of the energy in the fuel cannot be converted to electrical energy. The type of steam engine likely to have been employed at the Banning-Alameda plant probably never operated at an overall efficiency greater than 15 percent going from the coal's energy to the electrical energy produced. That means that at least 85 percent of the energy liberated from the coal was lost and never contributed to electricity production. Most of this waste should be blamed on the steam engine and boiler, not the generator. Steam engines lose a great deal of heat in condensing the steam back to water and in the friction between moving parts in the engine and generator.

The amount of electrical energy obtained from 1 ton of coal is only 15 percent of 8,500 kWh or 1,275 kWh. The Banning-Alameda plant must have burned approximately one-third of a ton of coal to run the arc lamps:

$$\frac{420 \text{ kWh}}{1275 \text{ kWh/ton}} = 0.33 \text{ tons}$$

This is about 660 pounds of coal per day or 241,000 pounds (120 tons) per year. Using lighting at this rate requires a power output of:

power = 21 lamps × 4000 W/lamp = 84,000 W or 84 kW

which is easily within the capacity of a 300-kW steam plant even allowing for losses in transmission and a cushion of excess capacity.

The efficiency of the postwar Redondo Beach Steam Plant can be readily calculated from company data. In 1948, the plant's first year of operation, fuel oil with a heat rating of 8,109,595 million Btu was burned to generate 716.4 million kWh of electricity. This

Thermodynamics

In 1798 Count Rumford, who was actually an American named Benjamin Thompson, observed when boring cannons that the metal cylinders and boring tools became very hot, and he concluded that some of the work was being converted to heat. Forty-five years later the English scientist, James Joule, showed in careful experiments that a stirrer powered by a falling weight raised the temperature of a water bath by a measurable amount. Knowing the mass of the weight, the distance it fell, the mass of the water, and the temperature rise enabled him to calculate the relation between work and heat. In the 1820s the French physicist, Sadi Carnot, had carried out theoretical investigations of "heat engines" to determine what factors affected their efficiency in order to design better steam engines. His results showed that no one could ever build a perfectly efficient engine.

These studies along with work by two other great physicists, Rudolph Clausius of Germany and Lord Kelvin (William Thomson) of Great Britain, resulted in interpretations called the three laws of thermodynamics, which can be stated in gambling parlance as: (1) you can't win; (2) you can't break even; (3) you can't quit.

We will concern ourselves here with the first two laws. They prohibit the existence of "perpetual motion machines," which purportedly generate more energy than is put into them initially. (The third law states that you cannot reach the temperature of "absolute zero.")

The first law is also phrased as "conservation of energy" for ordinary chemical and physical processes (as opposed to nuclear reactions). It basically says that in these processes, energy cannot be created ("you can't get something for nothing") or destroyed. When a lump of coal burns, the carbon forms new chemical bonds with oxygen and heat is also released. The total energy in the chemical bonds of the reactants (carbon and oxygen) equals the chemical bond energy of the product (carbon dioxide) plus the thermal energy given off. Thus, the energy balance sheet is satisfied.

The second law states that no conversion of heat to work is 100 percent successful because some heat is lost as thermal energy to the surroundings. An obvious example is the automobile engine in which most of the chemical energy in the gasoline is given off as heat rather than converted to the kinetic energy of the moving vehicle.

Remarkably, these laws, which now figure so prominently in the operation of machines, were not understood until well after Watt improved the steam engine in 1769. In this case, scientific understanding followed technical advancement and then helped guide further improvements in engine design.

Box 5.2

fuel energy can be restated as 2,370 million kWh. Therefore, most of the heat from burning the oil is lost, and the efficiency is given by:

$$\% \text{ efficiency} = \frac{\text{energy produced}}{\text{energy available}} \times 100$$

$$= \frac{716{,}000{,}000}{2{,}370{,}000{,}000} \times 100 = 30.2\%$$

Although this does not seem to be spectacular, recall that the best reciprocating steam engines were only 15 percent efficient.

Modern plants that use steam turbines in generating electricity have an overall efficiency in the 35 to 40 percent range and produce up to 1.3 million kW of power or 1300 MW. Such steam plants may seem inefficient compared to hydroelectric plants, but if the energy expended in heating the water to produce steam is not counted, then the efficiency of a steam plant rises to 80 to 90 percent. This energy expenditure is analogous to the energy invested in raising water to a high level above a water wheel. That number is obviously ignored in calculating the efficiency of a hydroelectric plant because the sun provides that energy free of charge.

This significant efficiency advantage of turbines over reciprocating steam engines was extremely important to electric utilities in Los Angeles during the early 1950s. It meant that they could provide two or three times as much electricity by burning the same amount of fuel. This achievement led to lower prices for electricity, which in turn stimulated more demand. By using steam turbines, utilities could provide more power at lower unit cost and, presumably, greater profit, something Watt and the other leaders of the First Industrial Revolution would have keenly appreciated.

Panacea for Electricity Needs?

The generating capacity of steam plants grew quite large after the Redondo Beach plant. The Mojave Generating Station, built near the Colorado River in the late 1960s, compares favorably with the Hoover Dam in producing a total of 1,500,000 kW or 1,500 MW from its two turbogenerators. The success of the turbogenerator has helped us attain a higher standard of living through the use of relatively cheap electric energy to run home appliances like refrigerators, washers and dryers, televisions, air conditioners, scientific and medical instruments, computers, robots, and industrial machinery.

A price is paid for abundant electrical power, however, in terms of environmental problems like air pollution, acid rain, and the potential of the greenhouse effect. Many people now look to alternative energy sources to satisfy the continually expanding thirst for electricity in Southern California and elsewhere. Energy conservation would help reduce the immediacy of these problems, but as the population continues to swell, the demand for electricity is certain to increase as well.

Exercises

1. Describe the energy conversions involved in going from coal to electricity. Name each type of energy and the process that converts it into another form. Give three examples of energy conversion that occur after the electricity reaches the consumer.

2. Using data from the text, show that oil has about 1.7 times the energy content of coal.

3. A generator requires 36 horsepower from a reciprocating steam engine to operate a lighting system. The engine runs at 75 cycles per minute, but an arrangement of belts and gears is needed to turn the generator to 750 RPM. The piston itself is one foot in diameter and has a stroke of two feet.
   (a) Calculate the steam pressure needed to run the generating system.
   (b) What is the maximum number of 700 W arc lamps that the above system can illuminate?
   (c) Why is your answer for (b) likely to be overly optimistic? (Consider the processes involved in the transmission of electricity.)

4. Compare and contrast the steam engine and the steam turbine. List two similarities and two differences.

5. In 1948, the Redondo Beach Station generated about 716 million kWh of electricity. How many barrels of fuel oil would be needed to accomplish this output? (Use the efficiency value and energy rating for oil given in this chapter.)

6. The United States used a total of 2.6 billion kWh of electricity per day in 1987; it also produced approximately 8.7 million barrels of crude oil per day. Could the United States supply all of its electricity from domestic oil? (Assume that a gallon of crude oil can be refined to yield 0.30 gal fuel oil.)

7. Major league baseball introduced night games in the 1930s to the displeasure of many baseball purists, only to later adopt other technological heresies such as artificial turf and multipurpose domed stadiums (as well as that ultimate atrocity, the designated hitter). At a night game, fans cannot

fail to marvel at the apparent power of the lights arrayed around the field and also wonder how much it costs. There are 999 1000-W lights on the towers in L.A.'s Dodger Stadium, and a game requires about four hours of lighting on average. The Dodgers play sixty night games each season and draw three million fans for all eighty-one home games at about $6 per ticket. Electricity costs about $0.10 per kWh.

(a) Calculate the number of barrels of oil needed to provide energy for the lights from turbogenerators. (Write down a guess before you attempt to do the calculation.)

(b) Is the cost of lighting the field a significant part of the ticket price?

Gasoline: From Waste Product to Fuel

6

A New Technology Demands a New Energy Source

When we look at late ninteenth-century photographs of America and many other countries, the difference from the present is startling. Aside from the smaller size and number of skyscrapers in cities, we note the absence of the complex network of abundant paved roads, especially freeways, that are now commonplace. In the countryside, the vast system of highways did not exist. What caused this monumental reworking of the land?

The automobile powered by the internal combustion engine was, of course, the major influence. But the mere invention of the machine did not guarantee our dependence on it. A readily available source of energy to run the engine was essential in unleashing its impact. The history of transportation would certainly be very different if cars and trucks had to rely on coal, wood, or electric batteries rather than a liquid fuel with a high energy content. Gasoline is one of the most important chemical mixtures produced today. In a relatively brief period of time, it has had pronounced—perhaps irreversible—effects on the economics and politics of every country in the world. It has changed the way in which most inhabitants of this globe live. The development of the technology of refining crude oil (petroleum) into a sophisticated chemical fuel stands as one of the most important technological stories in modern history. Over the last century, the production of oil and gas has changed the sources of energy to do work. Our dependence on the labor of horses and humans has declined markedly as fuels derived from petroleum powered more and more machines in the period 1850 to 1959 (figure 6.1). Consequently, as figure 6.2 demonstrates, oil production increased nearly exponentially during the period 1910 to 1967.

About 125 years ago, however, there was essentially no market for the gasoline produced in the early refineries. Naturally occurring, or "straight-run," gasoline was a by-product in the refining of crude petroleum to produce the two major desired products: kerosene as a fuel for lamps and lubricants for machinery. Early refiners, such as John D. Rockefeller, faced a major disposal problem in connection with this nonmarketable product. The gasoline fraction from their petroleum distillation was a low boiling, volatile, flammable hydrocarbon mixture and therefore a serious fire hazard at these first petroleum refineries. The typical solution for this problem was to dump the gasoline into adjacent rivers and hope it would evaporate before the river caught on fire!

A social cost for the development of the industry was the waste in production as well as the risk of accidents. Only a few geologists urged more responsibility in the placement of wells, in the control of well pressure, in dealing with the dangers of flooding and fires, and in the use of methods to prevent premature depletion of the wells. These concerns were essentially ignored by oil producers. Many of the explosions and fires from kerosene lamps were due to refiners who did not monitor carefully their methods of separating out the more volatile, and more flammable, fractions. Only when these components proved to be more valuable as by-products did excessive dangers due to inferior grades of kerosene disappear in the late 1880s. In addition, consumers certainly benefitted from the decline in the cost of kerosene—from \$0.45 per gallon in 1863 to \$0.06 in the mid-1890s. The reason for this dramatic decline was twofold: more sources of crude petroleum were discovered and, equally important, more efficient methods of refining petroleum into salable products were developed.

In 1876 Otto built the first practical internal combustion engine. Gasoline proved to be a good fuel for this new engine. Motor vehicles using these internal combustion engines appeared first in the late 1880s. At the beginning of the twentieth century, the great industrialists, Henry Ford and John D. Rockefeller, won the com-

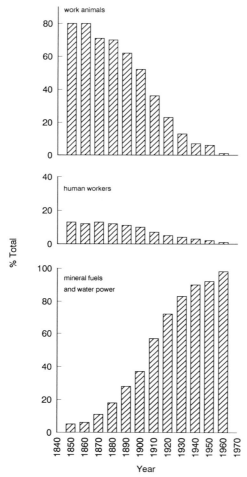

Figure 6.1 The relative changes in the major energy sources from the mid-nineteenth to mid-twentieth centuries. (Adapted from *Aramco Handbook*, 1968 edition. Arabian American Oil Company, Dhahran, Saudi Arabia.)

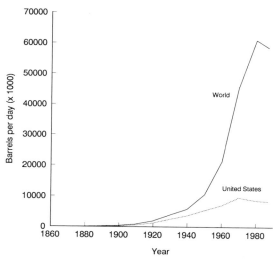

Figure 6.2 World and U.S. oil production, 1860 to 1987. (Data from *Oil Economics Handbook*, 5th ed., by Gilbert Jenkins, Elsevier Science, New York, 1989.)

petition among the gasoline-powered internal combustion engines, the steam engine, and the electric battery-powered engines as the standard of choice for the emerging automobile. By 1910 it was clear to refinery owners that the demand for gasoline to power these new automobiles would quickly exceed the gasoline available from the simple distillation of crude oil, that the demand for kerosene would continue to drop dramatically as the illumination of cities by Edison's electric light bulbs continued to spread, and that huge profits could be made by refiners that developed processes to take advantage of this shift in the market. The next fifty years brought forth significant changes in the processing of crude oil so that the yield of gasoline manufactured from a barrel of crude oil could be increased from 10 percent to as much as 50 percent. Dramatic advances occurred as well in the manufacturing of heating oil and petrochemicals such as plastics. In 1930, 18.5 billion gallons of this "waste product" were sold; by the 1980s gasoline production averaged around 110 billion gallons per year.

Today, the energy consumed by American vehicles is fifty times greater than the combined energy output of all the electric power companies in the United States. Each American uses the equivalent of over 450 gallons of gasoline per year or twice as much gasoline as each Australian, four times as much as each European, and ten times as much as each Asian. American autos travel about half of the over two trillion miles traveled by all autos worldwide, and about three of every four Americans commute to work in their private automobiles and trucks. Almost every item in an American's daily life—food, clothes, shelter, and so forth—is at some point moved by trucks in the process of converting natural resources into manufactured products and in delivering those products to the consumer. Each year American automobiles and trucks consume between 100 and 120 billion gallons of gasoline to travel over one trillion miles at a cost in excess of $130 billion. Although trucks account for about one-fifth of the motor vehicles on the road, each truck travels 65 percent more miles than the average auto does.

The goal of this chapter is to show how we got to this point by describing some of the key developments in the history of manufacturing gasoline. In particular, we want to emphasize the role of refining crude oil into useful products and the technical innovations that helped us to accomplish this.

The Automobile: Otto and Ford

At the end of the nineteenth century the petroleum industry suffered from the widespread use of Edison's electric lights, which helped eliminate the need for kerosene. However, Henry Ford's development of the first mass-produced automobile created new markets for petroleum products, especially gasoline. The automobile, like most new technology, depended upon prior developments, and was, in fact, a late nineteenth century invention.

In 1866, Nicolaus A. Otto and his partner obtained a German patent for a two-stroke cycle atmospheric engine. This engine was the first successful internal combustion engine, and they sold about 5,000. At the Philadelphia Centennial in 1876 (and elsewhere that year), Otto exhibited a practical engine that used four strokes to achieve one power cycle—a major improvement over earlier engines. The four strokes are as follows: first, the piston goes down and draws in the fuel and air mixture to the cylinder; second, the piston moves upward to compress the mixture of fuel and air; third, the ignition (by a spark plug in today's engine) causes the compressed mixture to ignite and drives the piston down in its power stroke; and fourth, the piston moves upward and pushes out (exhausts) the burned fuel mixture.

The 1876 Otto engine was a great commercial success because it proved that internal combustion (i.e., combustion within the cylinder itself) could serve as a fixed power plant in place of the steam engines. Furnaces and boilers were no longer needed to generate the steam to power the pistons. The "steamers" had lower thermal efficiency, needed a dependable supply of soft water, and could not match the increasing horsepower of the gasoline powered machines. This utility of the gasoline engine would lead to the automobile and the airplane as common forms of travel.

Henry Ford grew up on a farm outside Detroit and became interested in things mechanical. As a young boy he was considered a wizard at fixing watches. Ford gradually gained a solid reputation as a skilled machinist who could repair steam engines and who had repaired an Otto engine that defied other mechanics. Ford later told how he had built for himself an engine modeled after the Otto just to be sure he understood its principles. This early curiosity was to pay off handsomely years later.

As a young man, Ford worked for the Edison Illuminating Company in Detroit, between 1891 and 1899, rising to become its chief engineer. All that time, he was thinking about building a horseless carriage using the internal combustion engine. He set up a shop in a small building behind his house in Detroit and in June 1896 successfully operated his first car, which he called a quadracycle (figure 6.3). After two failures at forming successful car companies, Ford realized in 1901 that to gain financial support he would need to produce a racing car and thereby catch the imagination of the public. In October 1901 he won a well-publicized race in a car he built especially for speed, and the next year he convinced the famous racer, Barney Oldfield, to race in his new "999" car. The victory of the "999" car led directly to the founding of the Ford Motor Company in June 1903, seven years after demonstrating his first car.

Fortunately for Ford, by 1904 the country was in a period of rapid growth and high prosperity, and he immediately found a good market for his cars. His idea was to produce a low-priced family car that would sell to a wide market. However, he had trou-

Figure 6.3 Henry Ford in his first automobile in 1896. (From the collections of Henry Ford Museum and Greenfield Village.)

ble convincing his backers of the wisdom of this idea, so in 1907 he bought them out. In 1908 Ford introduced his Model T and began to revolutionize automobile manufacturing. Production reached nearly 6000 Model T Fords in 1908, each selling for $850. By 1916 he was producing almost 600,000 per year and selling them for only $360 each. This remarkable drop in price came about because of Ford's decision to automate his factory by the use of a moving assembly line. Soon his procedures became known as the method of *mass production*. In January 1910 Ford began operation in a new plant at Highland Park and by April 1913 he had installed his first moving assembly line there. Not only had Ford succeeded in designing an inexpensive and simple car but he had also achieved his goal of designing an inexpensive and rapid means of producing his cars. In doing so, he designed both a new machine and a new manufacturing process.

Obtaining Crude Oil: Drake and the First Oil Well in the United States

The initial problem faced by petroleum producers was the very basic one of finding it. Until the beginning of the nineteenth century, illumination in America remained almost as primitive as it had been in ancient Greece or Rome. In the 1800s, tallow (animal fat) candles or lamps fed by whale oil were the principal sources of lighting. Whaling was a major industry both for the U.S. home market and for the very lucrative export market. However, intensive hunting began to deplete the number of whales. At the same time, a relatively small amount of lamp oil was being produced by skimming crude oil from ponds and subjecting it to primitive processing methods to make a crude version of kerosene. A growing demand for this lamp oil created strong incentives for entrepreneurs to produce more, but it was not until the 1850s that anyone thought of drilling for oil to find the sources of the seepages.

Dr. Francis Brewer, a medical graduate of Dartmouth College, had an interest in petroleum as a medicine. His family ran a lumber firm in northwestern Pennsylvania, in Titusville. Oil had been reported in this area in the mid-1700s, and George Washington had owned land there that contained a "bituminous spring." Brewer spent considerable time examining an oil spring on his family's property. In mid-1853, the firm leased the spring to a Titusville man for five years to gather the oil. In the fall, while visiting friends and relatives in New Hampshire, Brewer took along a bottle of his oil to show his uncle, a professor of surgery and obstetrics at Dartmouth. A professor of chemistry also examined the oil and predicted that it was valuable but would never become an article of commerce because of limited quantities!

However, a few weeks later the bottle came to the attention of George Bissell, a young New York lawyer, who was visiting the campus of his alma mater. Struck by the similarities between "coal oil" (a liquid produced by heating coal) and this petroleum, he guessed that petroleum might yield a good illuminant. Bissell contracted with Dr. Brewer to lease oil rights on 12,000 acres of timber land for $5,000 and founded the Pennsylvania Rock Oil Company in December 1854. The new company engaged Professor Benjamin Silliman of the Yale chemistry department to analyze the oil and prepare a report. Upon receiving the report, the company's financiers acted quickly because of Silliman's reputation and his optimistic conclusions that 50 percent of the crude oil could be distilled into lamp fuel and 90 percent of the crude would yield commercial products.

They wanted someone to go to Titusville to begin the explorations for oil. Edwin L. Drake of New Haven was recovering from an illness that forced him to retire as a conductor on the New York and New Haven Railroad after eight years. Drake had no experience in the oil business, but he was available and would be relatively inexpensive to employ since he had a railroad pass. In December 1857, Drake made the trip to Titusville. Bissell referred to Drake as "Colonel" Drake to impress the local folks in Titusville, and the title stuck for the rest of his life. Because of favorable reports from Drake, appropriate leases were obtained, and the Seneca Oil Company was formed. Drake was appointed general agent, and he went to Titusville in May, 1858, to begin oil recovery operations. At first Drake reactivated the old machinery and tried the old methods of accumulating the oil by enlarging the pits and gathering it as it seeped up. His failure to obtain much oil was so discouraging that he finally decided to try drilling.

The most experienced drillers at that time were those who drilled salt water, or brine, wells to obtain salt for preserving meat and hides. In the early spring of 1859, Drake found an experienced man, William Smith, who was also a blacksmith with experience making drilling tools. Drake had designed and constructed an engine house for a six-horsepower stationary engine with a tubular boiler to power the drilling. He designed and built a 30-foot wooden derrick in which to swing the drilling tools. (Derrick, incidentally, was the name of a famous seventeenth-century London hangman.) Drake's men began digging a hole and cribbing it with timber, the standard method for creating salt wells. Water kept flowing into the hole faster than a pump devised by Smith could carry it out. So Drake proposed that they stop the cribbing process at the 16-foot

Figure 6.4 The Drake oil well with Edwin Drake, *center*, 1861. (Photograph from the collection of the Drake Well Museum, Titusville, Pennsylvania.)

level and drive an iron pipe through the clay to rock. This method was successful to a depth of 32 feet, at which point Smith shifted to steam powered drilling at a rate of three feet per day. By late August, after working seven days a week, he had only a mud hole to show for his summer's work.

On Sunday, August 28, 1859, he went out to the well (figure 6.4) for a routine inspection. The workmen had reached a depth of 69 feet the previous night. When Smith looked into the pipe, he noticed a dark film floating on the surface of the water a few feet below the level of the derrick floor. Smith and his son began to collect the oil in whatever buckets they could find. The next morning, Drake arrived and began using an ordinary hand pump to gather the oil. The first oil well in America was producing about eight to ten gallons a day!

Ironically, Drake did not seem to appreciate the significance of this accomplishment. He simply carried out the job he was employed to do with great perseverance but with little skill. While others scrambled to lease drilling lands, Drake worked on improving the equipment for his first well. Several months after the strike, his storage tanks and derrick burned to the ground. He eventually lost all his money and spent the rest of his life in ill health, poverty, and obscurity.

Nevertheless, despite his shortcomings as a driller, businessman, or entrepreneur, Drake was the first to demonstrate the feasibility of drilling for petroleum and obtaining it in substantial quantities. His success with the Titusville well removed the major technical barrier to the development of a new petroleum industry—first as a source of illumination, later as a source of fuels for new forms of transportation, and finally as a source of petrochemicals for plastics and pharmaceuticals.

From Drake's first oil well producing a few barrels in 1859, a total of almost 60 million barrels flowed from all wells in 1899. The total output over the forty-year period reached one billion barrels. Most of this output went to supply the world with the inexpensive illuminant, kerosene, but about two hundred by-products accounted for half of the industry's sales. Naphthas for local anesthetics, solvents for industry, fuel for stoves and the internal combustion engines, wax for pharmaceuticals and candles, oils and lubricants to free machines from friction, heavy oils for the gas industry—this partial list indicates the wide range of petroleum-based products that contributed to the rapid industrial expansion in America and Europe during the last quarter of the nineteenth century.

During the first ten years of the petroleum industry, the basic production techniques were set by adapting methods of salt well boring to give effective methods of drilling and pumping oil and the foundation for refining techniques was also laid. The methods of heating coal, and the refining of the liquid coal oil that was produced, were transferred to petroleum and scaled up efficiently to produce kerosene in large quantities and then a series of by-products. But no American innovation had more effect on the early growth of the petroleum industry than the development of bulk transport of oil to refineries and markets—first by railroads and then by long-distance pipelines.

Making Oil Available: John D. Rockefeller and Standard Oil

The fortune of one of the first great American tycoons was derived from the large-scale refining of crude oil and shipment of the petroleum products. Upon graduation from high school, John D. Rockefeller (figure 6.5) began his business career as an assistant bookkeeper with a Cleveland produce firm. Three years later, he decided to start his own business along with a partner, and they began operations as produce merchants in March 1859. From the beginning they were a financial success, benefiting greatly from the Civil War due to increases in both business and prices, and they became wealthy by the war's end. Late in 1862, they were asked to finance a refinery; they formed a new partnership and began operating their new refinery, Excelsior Works.

Although it started as a sideline investment for Rockefeller, he became increasingly attracted to the refining business as he saw

Figure 6.5 John D. Rockefeller in 1888. (Courtesy of the Rockefeller Archive Center.)

opportunities to use his management skills in a new industry with a good future. He encouraged the use of improved refining methods and cost-cutting measures. By 1865 Excelsior was one of the best equipped and operated refineries in Cleveland. As the Civil War was drawing to a close, Rockefeller decided to get out of the produce business completely and devote his full attention to refining. He was not impressed with his partners, so he proposed that they auction the oil properties among themselves with the buyer taking the refinery and the others getting the commission business. In February 1865 Rockefeller bought the refinery for $72,500; the new firm of Rockefeller and Andrews began business within a few days. In 1866, he hired Andrews's brother to go into the Pennsylvania region to purchase oil. John D. also hired his younger brother, William, to help with another new Cleveland refinery, Rockefeller and Company. Within months William was sent to New York City to set up an office for export sales. The next year John D. persuaded Henry M. Flagler, a grain and produce merchant, to take an active role in the management of a new firm, Rockefeller, Andrews and Flagler, backed by Flagler's father-in-law, who had made a fortune in the California gold rush. The partnerships continued to prosper, and in January 1870, the Standard Oil Company of Ohio was formed to consolidate all the partnerships, including ownership of the Cleveland refinery shown in figure 6.6.

The new company was capitalized at $1 million, and its 10,000 shares were divided among six partners with John D. holding slightly over one-fourth of the shares. In the seven years since he had entered the refinery business, John D. had amassed sixty acres of land in Cleveland; two great refineries; a huge barrel-making plant; lake shore facilities; a fleet of tank cars, sidings, and warehouses in the oil regions; timberlands; and warehouses in New York City. However, in spite of his ability to amass all of these holdings, the great success of the Standard Oil Company was soon to be found in Rockefeller's ability to obtain favorable rates for shipping oil over the railroads and later through pipelines. Part of Rockefeller's business genius was his recognition that the size of his operation was such that the transportation costs could amount to 20 percent of the final product's cost.

By 1880 Rockefeller realized that some new organizational instrument must be found to unite his vast array of companies now

Figure 6.6 The original Cleveland refinery in about 1870. (Courtesy of BP America, formerly Standard Oil of Ohio.)

operating from Cleveland under the banner of Standard Oil of Ohio. In 1873 Standard Oil controlled 10 percent of the refining capacity of 103 U.S. refineries; in 1880 Standard Oil controlled 90 percent! Production of crude oil reached 30 million barrels in 1883; refining capacity grew from 10.5 million barrels in 1873 to 28 million in 1884. Crude quoted at $2.50 a barrel at the wellhead in 1876 fell to $0.78 in 1879. Problems of communication, administration, governmental legislation, and public criticism needed attention. In 1882 the Commonwealth of Pennsylvania sued Standard Oil for $3 million in nonpayment of taxes back to 1868 because it ruled that all capital stock of corporations doing any part of their business in Pennsylvania was fully taxable by Pennsylvania. Even though the courts subsequently ruled that only property located in Pennsylvania was taxable, this got John D.'s attention. To avoid such problems in the future, the Standard Oil Trust Agreement was drawn up and signed in January 1882, at a total value of $70 million. The agreement provided for corporations bearing state names, so Stan-

Refining Crude Oil: Distillation

The early procedures for working with the crude petroleum from wells were based upon the principles of distillation. As a mixture of chemicals is heated in a "still," the lowest boiling components will evaporate first, and the highest boiling components will be vaporized at the highest temperatures. As distillation proceeds, various "fractions" may be collected by cooling the vaporized chemicals. Brandy makers and moonshiners have been making alcoholic spirits for centuries by running their distillates through glass or copper coils that act as "condensers", and are cooled by the air around the coil. More efficient processes use water to cool the condensing apparatus.

In the simple distillation of petroleum, small and easily vaporized hydrocarbons with between one and five carbon atoms per molecule come off first. The next fraction is gasoline, which is a mixture of compounds whose boiling points may go up to 200°C. After this, kerosene (roughly 175°C to 275°C) is distilled off and collected. Then, the next fraction is "gas oil," which boils between 250°C and 400°C. After this are higher boiling compounds that include lubricating oils. In the complex mixtures that make up petroleum, some semisolid residues like asphalt may be left behind in the still. The gasoline produced in this manner is referred to as "straight-run" gasoline, to distinguish it from gasoline produced by cracking processes. Remember that kerosene was the main petroleum product before gasoline demand expanded.

Later on it was discovered that the "gas oil" fraction, which is part of the heavy petroleum fraction, was an excellent feedstock for the thermal cracking that greatly increased the yield of gasoline from each barrel of crude oil and reduced the amount of coke formed in the reaction vessel (still).

Box 6.1

dard Oil of New Jersey and of New York were formed in 1882, and companies in Indiana, Iowa, Kentucky, and California followed.

The Molecular Architecture of Gasoline

The problem the petroleum industry soon faced was how to supply enough fuel for that popular new mode of transportation, the automobile. As early as 1908, a few inventors recognized the need for new chemical processes that would provide substantial increases in the availability of gasoline. Exploration techniques were not well developed, and the surest approach to increasing supply was to refine the available crude oil more efficiently. The portion of the crude petroleum boiling above 250°C, called "heavy gas oil" and recognized by Benjamin Silliman in 1855 as a source of gasoline, was now available in large quantities due to a weakening demand for kerosene. Before we examine the early twentieth-century breakthroughs in the technology of refining crude oil, let us first look at some of the physical and chemical properties of gasoline.

Gasoline is a liquid fuel designed to give off vapors that can be ignited by the spark generated within each cylinder of an internal combustion engine. Modern gasoline is a complex chemical mixture, frequently consisting of over one hundred different chemical substances. To achieve maximum efficiency from an internal combustion engine working under a given set of performance and environmental requirements, a refiner will blend appropriate combinations of chemicals. With a density of about 0.75 grams per milliliter (g/ml) at 15.6°C, gasoline is considerably less dense than water (1.00 g/ml). Its *heat of combustion*—the amount of thermal energy released when it is burned—is 115,000 Btu/gal (or 32,000 kJ/L). Its *flash point*, the temperature at which liquid gasoline gives off sufficient vapors to ignite with the air near the surface of the liquid, is very low—43°C.

The chemical compounds in gasoline are composed primarily of two elements: carbon and hydrogen. Thus, most of the chemical components of gasoline are *hydrocarbons* and their derivatives, that is, various combinations of carbon and hydrogen that may also contain other elements. One of the most important features of carbon is its ability to *catenate*, or to form chains composed of a backbone of carbon atoms. Sometimes these chains can link at the ends to form ring molecules. For a premium gasoline, each of the major hydrocarbon chains contains between four and twelve carbons, and most of these compounds vaporize within the temperature range of 30°C to 210°C. As is generally the case, the smaller or lighter compounds vaporize at lower temperatures.

Hydrocarbons are divided into various categories and subcategories according to their chemical composition and molecular structure. Although the proportion of each of these types of compounds has been known since the turn of the century, detailed information about these compounds is now available due to recent modern analytical techniques. The following examples include some of the more important chemical components of gasoline.

Alkanes, also called paraffins, are hydrocarbons that contain only single bonds between any two carbons or carbon and hydrogen atoms forming the molecule. A single chemical bond is formed by two elements sharing a pair of electrons. Since carbon is a tetravalent (forms four bonds) element and hydrogen is a monovalent (one bond) element, notice that each of the following formulas contain four single bonds to each carbon atom and one single bond to each hydrogen atom. Such hydrocarbons are said to be "saturated" since they contain the maximum number of single bonds.

Butane is the smallest alkane found in gasoline, and thus it has the lowest boiling point. Its general formula is C_4H_{10}, and it can exist in two forms or *isomers* (box 6.2). Only n-butane, the form with all four carbons in a straight chain, is present in gasoline. The typical concentration of n-butane is 3–4 percent. The chemical structure is often depicted as:

$$CH_3-CH_2-CH_2-CH_3,$$

n-butane, with a boiling point of $-0.5°C$. The other isomer is:

$$CH_3-CH-CH_3,$$
$$|$$
$$CH_3$$

which boils at $-11.6°C$. Both are gases at ordinary room temperatures (20°C–25°C) that are found dissolved in liquid hydrocarbons.

The next hydrocarbons in the series are called *pentanes* and have the general formula C_5H_{12}. A typical premium gasoline with

Isomers: Varying the Molecular Structure

The rules of hydrocarbon chemistry are fairly simple, especially for the class of compounds called alkanes in which there are roughly two hydrogen atoms for every carbon atom (actually twice the number of carbons plus two). Each hydrogen atom must be connected to a carbon atom, and each carbon atom must have four connections total to carbons and hydrogens. Thus, a carbon can attach itself to two hydrogens and two carbons, or to three hydrogens and one carbon, and so on. Each carbon also must be attached to at least one other carbon, the chain-forming ability of carbon that makes it such a fascinating element.

These guidelines open up a world of myriad shapes and arrangements for the area known as *organic chemistry*. As some of the examples in the text have shown, there are several ways to arrange the atoms to get quite different structures for exactly the same chemical composition. These molecules are called *structural isomers* and are a source of great interest to organic chemists. One arrangement may have just the right properties to be a very useful material, whereas a slightly different ordering may produce the opposite effect, which illustrates both the efficiency and perversity of nature. Thus, putting eight carbons in a row yields a fuel that has undesirable properties, but by rearranging the carbons to give iso-octane, you have a fuel that burns very well.

One important point to remember is that molecules are *three-dimensional structures*, although we try to depict them on paper. Normally the actual architecture is more complex than it seems from a simple formula. The spatial orientation of the atoms is crucial to their chemical and physical properties. Two isomers not only may boil at different temperatures but they also will likely undergo chemical transformations in distinct ways.

Box 6.2

a boiling point range of 30°C–200°C contains about 6 percent n-pentane and 3.5 percent isopentane.

$$CH_3-CH_2-CH_2-CH_2-CH_3 \qquad CH_3-CH(CH_3)-CH_2-CH_3$$

n-pentane, b.p. 36°C   isopentane, b.p. 28°C

Several other isomers are possible but are not important for making gasoline.

The next few members of the series are called *hexanes, heptanes,* and *octanes*. Gasoline contains about 12 percent total hexanes but much less of the heptanes. However, when the heptane with seven carbon atoms in a row is used as a fuel, it causes the engine to "knock" very badly and is assigned an "octane rating" of zero (box 6.4).

Although the term octane is commonly associated with gasoline, a typical sample of gasoline may contain only 1–2 percent octanes. The basic "straight chain" molecule is:

$$CH_3-CH_2-CH_2-CH_2-CH_2-CH_2-CH_2-CH_3$$

n-octane, b.p. 126°C

Another important isomer is a branched molecule consisting of a five-carbon chain with two methyl groups (CH_3) on the second carbon and another on the fourth carbon (2,2,4-trimethylpentane, b.p. 90.1°C) and is commonly called *isooctane*. When pure isooctane is used as the fuel in the standard internal combustion engine, the octane number scale is assigned the value of 100 to indicate the absence, or minimum amount, of engine knock.

Even longer alkanes, like decanes (ten carbons) and longer chain alkanes or paraffins, occur in crude oil, but they do not usually occur in gasoline because they are converted into other compounds during the production of gasoline.

Other important compounds include *alkenes*, which contain double bonds between one pair of carbons. One such alkene, butene, can exist in two isomeric forms:

$$CH_2=CH-CH_2-CH_3 \qquad \text{and} \qquad CH_3-C(CH_3)=CH_2$$

1-butene, b.p. −6.3°C   isobutene, b.p. −6.9°C

Ring-like structures, *cycloalkanes*, are found with three or more carbon atoms in them, in which each carbon is linked to two other carbons and to two hydrogens. This yields the same 2:1 ratio of hydrogens to carbons as in the alkenes but very different chemical and physical properties. The most stable and common cycloalkanes have five- or six-membered (counting C-atoms) rings. Cyclopentane is present at the 1 to 3 percent level in gasoline, but cyclohexane is found only in traces because it undergoes breakdown to smaller molecules in the cracking process (discussed in a later section).

Benzene, C_6H_6, is an "aromatic" hydrocarbon and has a characteristic hexagonal structure that gives it special chemical properties. Although gasoline in Europe before the World War II contained 20 to 25 percent benzene, it is now limited to 5 percent. It freezes at 5.3°C, so it cannot be put in gasoline that is used at very low temperatures. It has also been found to be carcinogenic to humans.

Methyl alcohol (methanol) and ethyl alcohol (ethanol) are hydrocarbons that also contain oxygen and are relatively high boiling. Pure methanol is an excellent fuel and is used in some racing cars such as dragsters. In recent years there has been much interest in these chemicals because of the renewed popularity of *gasohol*, mixtures of gasoline and alcohol. One advantage of using gasohol is the decreased dependence on imported petroleum.

CH_3OH CH_3CH_2OH

methanol, b.p. 64.7°C   ethanol, b.p. 78.5°C

Aviation fuel is a different substance than automobile fuel. Only light aircraft with spark ignition engines use normal, unleaded gasoline with a lower boiling range of 30°C–150°C and a lower freezing point range (due to colder weather encountered in flying). Turbo-prop and jet engines use kerosene with a low smoke point, that is, with less than 20 percent aromatics. Military jets usually employ kerosene with a wider boiling point range, 70°C–280°C, and freezing point −50°C to −60°C.

Higher Yield of Gasoline: Burton and the Cracking of Petroleum

The crucial step in providing useful petroleum products for modern consumers is the refining of the crude oil. Today, this is a very complex and technology-intensive process, but it was not always so. At the beginning of the twentieth century, a typical refinery used basic equipment to carry out relatively simple processes but often under very hazardous conditions. Refinery operations were improved because of two major events: (1) large quantities of crude oil were discovered around the world to provide a steady supply of oil to sustain large-scale refinery operations; and (2) the automobile, followed by the airplane, revolutionized transportation. These events stimulated the design of the most important process in the refining of petroleum to produce gasoline: *cracking*. In this process, heat, pressure, and catalysts are used in creative combinations to break large molecules into smaller ones having octane ratings suitable for the synthesis of various blends of gasoline. As we shall see, the history of petroleum in this century is essentially the story of innovations in the methods of cracking oil.

The general challenge facing scientists and engineers was how to convert the myriad of chemical compounds found in crude oil into useful profit-making products, especially gasoline. In order to accomplish this, the oil industry had to recognize its need for the research and development laboratory. This paralleled the experiences of the German dye industry and the American electrical and communications companies.

The person who exemplified this approach that combined the methods of science with technological innovation was William M. Burton (figure 6.7), born near Cleveland two years after John D. Rockefeller financed his first refinery there. As a child, Burton became interested in chemistry, set up a laboratory in his family's barn, and visited frequently in the nearby home of Charles F. Brush, the inventor of the Brush electric generator used to power some early lighting systems. After earning his Ph.D. in chemistry in 1889, Burton went to work for the parent Standard Oil Company for a short period. Then, after a year with Standard Oil of Ohio, he was assigned to the Indiana corporation. In the 1890s, the contributions from a Ph.D. chemist to a petroleum refinery or even the petroleum industry were only partially appreciated. In fact, William Cowan, the vice president in charge of refining, had no idea what he was to do with this newly assigned employee. On their first meeting, he asked Dr. Burton if he had brought his tools along with him. Obviously, the concept of the now familiar industrial research and

Figure 6.7   William M. Burton. (Courtesy of Amoco Corp.)

development laboratory (R&D lab) was a novel idea to petroleum executives of that day.

The parent company directors (without much communication with Cowan) had decided that Dr. Burton was to be the first head of a laboratory they wanted to create at their new refinery in Whiting, Indiana. The laboratory was set up in the second-story bedrooms of a farmhouse on the refinery property near the shore of Lake Michigan. His first experiments, conducted with equipment and instruments that he made or had made in the refinery shops, were designed to determine the chemical and physical characteristics of refinery products: kerosene, waxes, greases, and lubricating oils. This testing led to the development of standardized analytical methods, which could then be run as routine procedures. The work soon became time consuming and not very challenging. So, he turned his attention to ways of improving the chemical processes that produced the products he had been testing. His work involved one essential goal of an industrial research and development laboratory: discovering methods to improve the yield or quality of these products by new approaches to their manufacturing processes.

About 1908, Burton recognized that there was a great need for more light petroleum products and that a handsome profit could be made by supplying them. As an entrepreneur able to see beyond the confines of his own company, he predicted the general acceptance of the automobile and realized that in a few years the demand for gasoline would exceed the amount available from simple distillation of crude oil. As a visionary, he saw that the oil industry would have to adapt its processes and marketing activities to meet these challenges of the automobile industry if the petroleum industry were to continue to prosper in the face of the declining kerosene market.

Burton became general superintendent of the Whiting refinery, and in an age when few refinery supervisory personnel had any technical training, he was outstanding in his educational background and laboratory experience. He had all the authority and facilities to put his training to use in the conception and development of the cracking process. He could orchestrate the laboratory research team, direct his shops in the construction of novel equipment, gather information on potential markets from his sales force, and control the costs through his accounting department. Because refinery superintendents also enjoyed more access and support from the president and directors, a research and development operation could only be carried out through that position in the years around 1910.

Burton and his colleague Dr. Robert Humphreys began their experiments with the knowledge originally included in the report to the Pennsylvania Rock Oil Company in 1855. Those results showed that when a heavy petroleum fraction is heated to high temperatures, it decomposes (or "cracks") into lighter products, which boil in the gasoline range, and into denser products, a viscous liquid and solid petroleum coke. The general term for this process is *thermal cracking*.

On a molecular level, this means that some smaller hydrocarbon molecules are formed by first breaking apart the chemical bonds between atoms in the larger hydrocarbons in petroleum, thus leading to lighter products that boil at relatively lower temperatures. Competing with this process is the combination of molecules to form larger hydrocarbons, which are viscous and high boiling. Another complication is the production of an impure form of solid carbon called coke.

For the raw material, they selected the entire heavy fuel oil fraction composed of the gas oil fraction, the lubricating oil fraction, and the asphaltic residue. This represented all the components that boil above the kerosene range (above 250°C) and is equivalent to about two-thirds of the total crude oil coming into the refinery. The obvious goal of Burton, and other refiners as well, was to crack this entire heavy fuel oil fraction since it was so large compared to the kerosene and gasoline fractions and much less valuable in the marketplace of that time.

However, Burton and Humphreys soon realized that they should work with the gas oil part of the heavy fuel oil rather than with all of the fractions heavier than kerosene. Using the lighter fractions (b.p. 250°C–400°C) as a single feedstock for their new processes meant that more gasoline was produced and less coke formed than when the combined heavy fuel oil fractions were all cracked together. Thus they were able to improve the yield and avoid some of their previous problems.

In the Burton process, a gas oil fraction from the simple distillation of crude oil was passed through tubes in a furnace and heated to the 700°F range for several minutes. Depending upon the pressure, as much as half of the compounds were vaporized and some cracked. From his investigations, Burton discovered that there were many possible reaction pathways (box 6.3) for these chemical processes and that the final product distribution was controlled to some degree by the temperature, pressure, and how long heating was done. The extent of cracking and the amount of gasoline produced increased with the temperature, but so did the formation of very light or gaseous hydrocarbons and the solid residue of coke. With the existing equipment, it was very difficult to purify and improve the quality of the gasoline produced at very high temperatures. Also, it was very difficult to get good heat transfer in the pipes as the coke built up and cleaning the coke from the still and pipes at the end of each batch process was very expensive and time consuming.

During the next two years, Burton and his colleagues learned that subjecting the gas oil fraction to moderate temperatures produced the best cracking. In the laboratory these experiments worked well, but, as often is the case in industrial research and development, the process did not work satisfactorily in the plant under large-scale conditions. The main problem was the evaporation of part of the

Chemistry of Cracking Petroleum

When Burton received the Willard Gibbs Medal from the American Chemical Society on 17 May, 1918, he said:

> Although we know very little about the reactions that occur when petroleum is distilled under pressure, it may be interesting to speculate a little on this subject.
>
> Let us start with the paraffin, $C_{14}H_{30}$ and see what might happen.
>
> $C_{14}H_{30} = C_{13}H_{26} + CH_4$
> $C_{14}H_{30} = C_{12}H_{26} + C_2H_4$
> $C_{14}H_{30} = C_{14}H_{28} + H_2$
> $C_{14}H_{30} = C_{12}H_{26} + C_2H_4$
> $C_{14}H_{30} = 4C_2H_2 + 4CH_4 + C_2H_6$
> $2C_{14}H_{30} = C_8H_{18} + C_{20}H_{42}$
> $C_{20}H_{42} = C_8H_{18} + C_{12}H_{24}$
> $C_{12}H_{24} = C_{10}H_{22} + C_2H_2$
> $2CH_4 + C_2H_2 = C_4H_{10}$
> $3CH_4 + 2C_2H_2 = C_7H_{16}$
> $C_2H_2 + 2H_2 + C_2H_6$
>
> We feel confident that the finished gasoline contains such paraffins as C_8H_{18}, $C_{10}H_{12}$, and $C_{12}H_{26}$, but one of the most interesting results is the formation of free hydrogen as shown in one of the above equations.

From Willard Bibbs Medal Address, "Chemistry in the Petroleum Industry," by William M. Burton. *Journal of Industrial and Engineering Chemistry* 10, 486 (1918).

Box 6.3

gas oil in the heating process before the cracking reactions could reach the desired degree of completion.

Since they had no detailed theory to guide them, they had to rely upon empirical observations to suggest new approaches. They surmised that heating the gas oil longer would lead to more cracking. Therefore, they turned to the use of higher pressures to keep the gas oil in the reaction still for longer periods of time. Since not much was known about the high-pressure behavior of heated hydrocarbons, especially on an industrial scale, allowing for the possibility of explosions was not unreasonable. Furthermore, as the coke layer formed on the bottom of the still, it was found that more and more heat had to be added to maintain the temperature of the cracking process, but the increased heat could burn a hole in the bottom of the still, leading to a fire. It was difficult to build large, high-pressure vessels, which involved riveting the steel plates together. Although Burton believed that he had a safety factor of four in his pressure calculations for the vessels, he actually only had a factor of 0.5 (or 50 percent above the expected danger) due to his overestimation of the strength of the steel at the operating temperature. The still size was limited by the size of the largest steel plate: thirty feet by ten feet. Because the still bottom had to be seamless, the largest Burton still was eight feet in diameter and thirty feet in length. Therefore, before attempting any large-scale production, they wisely decided to try some some controlled bench-scale experiments at higher pressures.

When Burton's supervisor learned of his high-pressure (from 5 to 30 pounds per square inch, or psi) experiments, he ordered them to stop and pursue less hazardous research. At a scientific meeting, Humphreys learned that heavy oils had been heated under a pressure of 50 psi without explosion. Armed with this new knowledge, Humphreys and Burton obtained permission to resume their experiments. They ventured to 75 psi to discover that little of the gas oil vaporized from the still, the cracking proceeded with little coke formation, and that 20–25 percent of the starting material was converted to gasoline. Moreover, the gasoline produced by this method worked well in the internal combustion engine. On the other hand, if the cracking was allowed to run until 40 percent gasoline was produced, the amount of coke deposited was substantial. At a temperature slightly less than 700°F and a pressure of 75 psi, the vapors

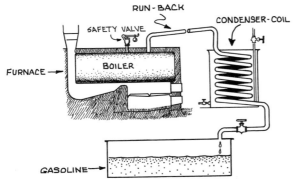

Figure 6.8 Patent illustration of Burton's cracking still. The run-back tube was a long inclined pipe connected to the top of the still and to the condenser. (Adapted from U.S. patent 1,049,667 dated Jan. 7, 1913.)

from the cracking process were about one-fourth gasoline and three-fourths a mixture of kerosene and gas oil.

Separation of this gasoline-kerosene mixture would require several distillations to remove the gasoline, and in the process, the kerosene would become partially cracked and worthless as an illuminant. Therefore, Humphreys invented the *run-back*, a simple but effective distillation pipe running at an upward angle from the still to the condenser (figure 6.8). As the air cooled these twelve-inch diameter pipes, the boiling kerosene would condense and run back into the still, but the gasoline vapors would rise to the condensing coils. By 1911, they knew how to design and operate their process at maximum efficiency and they thought it was ready for full-scale production.

In order to construct a process plant, Burton had to convince management to allocate $1 million to build a plant with six "batteries" of stills (ten pressure stills per battery) and their accessories. He estimated about one year for a return of their construction investment. The directors had little training or experience in the chemistry of petroleum, but they had lived through the age of boiler explosions on trains and steam ships. They feared that the refinery would blow up and result in a major disaster. Furthermore, they were removed from the laboratory and the years of research leading up to Burton's proposal and did not have the confidence that comes from being closely associated with the development of a new idea. The company was also going through a period of uncertainty, due to antitrust law suits, so the directors felt it would be unwise to assume more liabilities for possible explosions even if the potential return on their investment might be quite large.

Antitrust action by the Supreme Court led to the breakup of the parent company in September 1911, and Standard Oil distributed to its stockholders all the equities in its thirty-three subsidiaries. Paradoxically, Burton was in a much more favorable position because he could then seek approval from his fellow directors of the newly independent company, Standard Oil of Indiana. After the breakup, Indiana Standard found itself with excess refining capacity and a great dependence on others for its supply of crude oil. So, it was essential to reduce its dependence on outside suppliers of crude by getting more products from each barrel of oil processed in its refineries. The directors knew Burton well, respected his ability to run the Whiting refinery, and trusted his scientific research results. Approval to build a plant using the new process came swiftly, and construction began at the end of 1911.

The first stills began operation in January 1913, and success was apparent from the beginning. The Burton units performed as expected and doubled the yield of gasoline from crude oil. Within two years Indiana Standard recovered the money spent on the first sixty units. Surprisingly, the price of gasoline did not fall as a result of this doubling of the production capacity. Instead, the cost rose due to the increasing popularity of the automobile. By 1914, another 60 units were installed at the Whiting refinery, with another 120 units at other locations, including Wyoming. During the rest of the decade, half of the company profits ($150 million) came from the Burton process.

Making a Good Thing Even Better

In spite of these successes, there were some difficulties with the Burton process. The slow step in the process was cooling and cleaning out the coke after each batch. The hot residual oil had to be pumped out, and then steam was blown through the tank to remove hydrocarbon vapors. Next, a worker in an asbestos suit and armed with a pick and shovel entered the still to clean out the coke. In order to accommodate the need to return the still to service as quickly as possible while also permitting the cleaner to survive, the worker would wait to enter only when the temperature dropped to 250°F!

The cracked gasoline had a yellow color due to some sulfur contamination. Customers were suspicious of the unusual color, having been stung on other occasions by poor-quality gasoline from refiners with inadequate quality control. Also, in many cars such as the Model T, the gasoline tank was located under the driver's seat. The sulfur odor made it unpleasant to be in the driver's seat if one had spilled a little gasoline during a fill-up. So cracked gasoline was given the name "Motor Spirit" and priced at 10.5 cents per gallon (versus the 13.5 cents for other gasoline in 1913). But the public was not willing to accept the odor and strange yellow color at any price. With a few minor treatments at the refinery, the color and odor problems were corrected and cracked gasoline became indistinguishable from straight-run gasoline.

Improvements to the Burton process itself followed. In 1914, the first 120 stills were equipped with perforated plates to catch some of the coke before it settled on the bottom of the still. This enabled the still to operate longer under more controlled heating conditions, thereby increasing the gasoline produced by about 10 percent. At the same time in California, the principles of distillation were also being applied to the distillation of crude petroleum by Globe Oil and Refining Co. Fifty Burton stills, operating at higher pressures of 90–95 psi, were equipped with "bubble fractionating towers" in 1918. The bottom temperature of the bubble fractionating tower was kept at 595°F (equal to the endpoint of kerosene distillation) so that all kerosene and lighter products (including straight-run gasoline) would be in the overhead fraction. The bubble tower increased not only the yield of gasoline but also the rate of production.

Even as the Burton process matured, other events were having a serious impact on the petroleum industry. In the 1920s, the General Motors Corporation under the direction of Alfred P. Sloan introduced the idea of yearly style changes and began to increase its share of the automobile market. In 1927 Ford ceased production of the Model T and reluctantly began to meet the competition by adding new versions. By 1927 the automobile was America's leading industry in value of product. Fifty-five percent of American families owned automobiles—one automobile for every 4.5 persons. The demand for gasoline in the 1920s exceeded the available supply of gasoline produced by noncracking methods. Also, the improved internal combustion engines in such cars as the Cadillac compressed the gasoline-air mixture in the cylinder much more than the older engines. To avoid knocking, these engines with higher compression ratios demanded gasoline with higher octane ratings (box 6.4). Even so, in the late 1920s there was a fear of a petroleum shortage because the demand for gasoline was so great as a result of the successes of the automobile and also the new aviation industry.

The Burton process, with its subsequent modifications, brought Indiana Standard the greatest windfall of profits relative to initial investment in the history of the petroleum industry. Approximately one-half of their profits of $366 million during the period 1913–1922 came from the Burton process (R&D costs were $236,000). The process had succeeded in doubling the amount of gasoline obtained from a barrel of crude oil. Indiana Standard believed that it was the best possible way of producing gasoline from the inexpensive gas oil and that their patent position was strong and secure; they seemed to have no fear that any competitive processes would threaten them. This attitude, plus the problems of World War I, caused Indiana Standard to lose its leadership role in research as early as 1918.

Burton and his colleagues were all chemists, scientists trained in the *batch* process of the lab bench. Their attention was focused more on the starting material and the final product than on the processes that led from start to finish. It remained for chemical engineers who were more interested in the mechanics of the process—as a topic of significance and possible innovation—to develop efficient *continuous* processes. Furthermore, Burton and his colleagues were promoted into management positions as rewards for their initial successes in their research positions. As a result, the Burton process reigned supreme only from 1913 to 1919, but it produced essentially 100 percent of the cracked gasoline. Then it began to lose its market position so that by 1925 only half of the cracked gasoline was produced by the Burton process and by 1929 only 10 percent. The bold competitor ended as a sedate giant.

Knocking, Gasoline Rating, and Octane Number

To the motorist, engine knocking can be annoying as well as fuel wasting and damaging to the car's engine. A good-quality gasoline must have a high antiknock value, or "octane number," to prevent engine knock at all speeds and loads. Ideally, the spark plug should ignite the vapors of the fuel-air mixture, and the combustion should proceed smoothly and evenly down the cylinder. Knock is thought to be the result of chemical reactions that begin in the fuel-air mixture as soon as it is drawn into the cylinder during the intake stroke. As the mixture is progressively heated by the cylinder walls, by compression prior to the power stroke and by burning gases after the spark plug fires, some of the hydrocarbons undergo a series of cracking and oxidation reactions which produce unstable compounds. The spreading flame front sweeps across the combustion chamber heating and compressing the unburned portion of the fuel so that these unstable compounds autoignite and detonate throughout the cylinder instantaneously. Rather than being pushed down in a smooth power stroke, the piston is given a sharp, hard rap to which it cannot respond because it is physically connected to the crankshaft. Throughout the combustion chamber, this autoignition causes high-frequency pressure fluctuations in the audible range. The human ear hears the sharp, pinging metallic sound called knock or spark knock. Fuel energy is wasted in the form of these pressure waves and in increased heat radiated to the surrounding engine parts and the cooling system. Prolonged knocking overheats valves, spark plugs, and pistons and contributes to their wear.

In 1919, Charles F. Kettering and Thomas Midgley, Jr. began a series of experiments to determine the cause of knock and how to prevent it. Their work led to the discovery of a significant antiknock compound, TEL or tetraethyllead. (Think of four CH_3CH_2-groups attached to a lead atom.) In 1926, Graham Edgar at the Ethyl Corporation developed the octane scale to measure the antiknock characteristics of various gasolines. When the alkane, n-heptane, is burned in a standard internal combustion test engine, it produces a large amount of knocking. So n-heptane was selected as the fuel to set the zero reference point on this new octane scale. On the other hand, if 2,2,4-trimethylpentane (also called isooctane) is used as the fuel in the standard test engine, no knocking is heard. So, pure isooctane was assigned the value of 100 on the octane scale. Then, various mixtures of n-heptane and isooctane were used as reference fuels to standardize the test engine. The amount of knock from, for example, 20 percent n-heptane/80 percent 2,2,4-trimethylpentane was assigned an octane value of 80. Because of the discovery of TEL and other additives, the octane scale now extends beyond 100—that is, gasoline mixtures have been formulated that burn more evenly than pure isooctane.

The octane number scale is an excellent way of evaluating gasolines. For example, if the compression ratio of the engine can be increased, then the power obtained from the combustion of the gasoline-air mixture in the engine can be increased. Increasing the compression requires work. The fuel supplies the necessary energy for this work, but the power produced exceeds the work required for the compression. Less heat is rejected to the cooling system, more heat energy of the fuel goes to useful work, and efficiency increases. A typical 1930 automobile engine had a 5:1 compression ratio; but a 1960 automobile had a 10:1 compression ratio. In the 1930 auto, the fuel had an octane rating of 70. The fuel was compressed to 100 psi before ignition, and during combustion, it created a maximum pressure of 400 psi on the power stroke. In the 1960 auto, with an octane rating of 100, the fuel-air mixture was compressed to 200 psi, and during combustion it created 900 psi on the power stroke. For the same amount of gasoline, the 1960 car delivered about one-third more power.

Box 6.4

Greater Efficiency: Armor and the Dubbs Process

To meet the increased demand for gasoline, more efficient methods of refining were required. In 1909, Jesse A. Dubbs, a Pennsylvania oil refiner, built a small skimming plant near Santa Barbara, California, to produce asphalt from heavy crude oil. A problem he faced was that the oil was emulsified (well mixed) with salt water during its formation in the nearby Santa Maria field. His three sons, including the aptly dubbed Carbon Petroleum (C. P.) Dubbs, worked with him on the difficult task of separating the crude oil from the water, which could not be done by ordinary distillation methods. The senior Dubbs invented and constructed a tube still that heated the oil under pressure generated by the vaporization of some of the oil. However, when he filed his patent application in November 1909 after operating the process for several months, he apparently did not realize a key point: that he might be cracking as well as demulsifying the oil. In a second patent application, filed in June 1911, he failed to claim any cracking upon heating to pressures between 50 and 600 psi. With the announcement of the Burton process in 1913, as well as the appearance of a visitor who wanted to purchase his patents, Dubbs began to view his process in a new light.

The visitor was Frank Belknap, who represented the large meatpacking firm of Armour and Company. Its president, J. Ogden Armour, was attracted to the new petroleum industry with a rather ingenious idea in mind. He saw a parallel between the improvements, based on Henry Ford's assembly line, that had been made in the meat processing industry and possible innovations in the petroleum industry. Profits were greatly enhanced in meat processing by changing from a batch process (one person butchering an animal) to a continuous one in which each worker performed one task on a continuous line. If the petroleum industry could make an analogous change, then greater profits might also result. This scheme required a change from cracking and distilling one charge at a time in a Burton unit to a new process with crude oil continuously flowing into a unit reactor, the same chemical reaction occurring at each point in the flowing system, and gasoline continuously flowing out.

Belknap was also an agent for Standard Asphalt, a company owned by Armour, and they purchased a group of patents from Dubbs for $25,000 in 1913. Some of these obviously related to asphalt production, but Belknap was also familiar with the Burton process and recognized immediately the similarities between the Dubbs process (patent filed in 1909 but not granted until 1915) and the Burton process (patent filed and granted in 1913). C. P. Dubbs, who was then working for Standard Asphalt, helped Belknap persuade the senior Dubbs to sell the patents. Belknap began a yearlong effort of amending the pending patents so that they broadly described the treatment of oil with heat and pressure.

This lead to the formation of Universal Oil Products, UOP, the patent holding company, which was soon transformed into a research and development laboratory to further improve the process. C. P. Dubbs was appointed the principal research engineer on the new laboratory staff because of his experience as a refinery operator and his energy and enthusiasm. At the Chicago laboratory, he conceived the idea of "clean circulation," which became his major improvement to his father's process. In the clean circulation process, a part of the cracked distillate (or reflux), which is heavier than gasoline, was circulated from the condenser of the fractionating

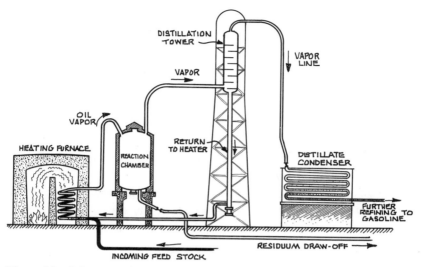

Figure 6.9 Schematic diagram of the Dubbs cracking still. (Adapted from *Petroleum Progress and Profits: A History of Process Innovation*, by John L. Enos, MIT Press, 1962, with permission of Shell Oil Company.)

tower back to the cracking tubes for additional cracking as shown in figure 6.9. This cleaner, lighter reflux diluted the incoming fresh feedstock and reduced the formation of coke.

One of the most important consequences of this innovation was that a new feedstock could be used, namely crude oil itself, because less coke formed. Recall that the Burton process could only accept gas oil or other distillates produced from the first distillation of the crude oil. This advantage, plus the continuous flow feature from Jesse Dubbs's first patent, were the two major improvements over the Burton process.

Two pilot plants were constructed in Kansas. The first followed Jesse's patent, while the second incorporated C. P.'s clean circulation concept. The feedstock was heated as it passed through a bank of horizontal pipes in the furnace and then rose through a split run-back pipe to the fractionating tower, a type of condenser. A portion of the condensed liquid from the tower was mixed with incoming feedstock and recycled for additional cracking, while the gasoline fraction was condensed and sent to storage tanks. Gas oil leaving the heating coils was held at a temperature of 820°F and 135 psi to undergo cracking. There were similarities to the Burton process, but there were also the important differences of continuous processing and of higher temperature and pressure.

In December 1919, Shell Oil's subsidiary, Roxana Petroleum, purchased a license to operate the Dubbs process with a guarantee that the Roxana unit would have a capacity of 250 barrels per day (compared to Burton at 89 barrels or Burton-Clark at 225 barrels per day). Construction of unit No. 1 at Roxana was completed in March 1921, but C. P. Dubbs could not reach the guaranteed output of 52.5 barrels of gasoline on the basis of 250 barrels of feedstock per day because they were unable to clean out the unit within the required 24-hour time period. Even worse, in 1921, the unit exploded and killed two men. This was a severe, costly blow to the young process design company, and C. P. felt personally responsible.

At this point he made a major break from his father's design of long, small diameter "soaking tubes" in which the starting material was heated and thus cracked. He decided that a single, large diameter reaction vessel would be less likely to become clogged with coke leading to ruptures, fires, and explosions. He also added two vents and a safety valve to the reaction chamber, a cylinder 4 feet in diameter and 30 feet long. The furnace was redesigned to contain nine rows of tubes, alternating between three and four tubes per row. The fractionating tower was raised higher to increase the pressure in the recycle leg.

The second unit at Roxana cost $60,000, had a 500-barrel capacity, and became the basis of a standard design for still units. Buying a standardized unit from what was now becoming a well-recognized process design company was a great aid to small refiners—they could attract investor capital by using standard units from UOP, and they could operate on any feedstock lighter than heavy gas oil. By the end of 1926 the Dubbs process was well established in the industry, and many companies working as licensees had contributed to its improvement. With other improvements, such as electric welding of the steel, the capacity of the Dubbs unit increased to 4,000 barrels per day in 1931 with a 21 percent yield of gasoline.

By 1926 great improvements, notably by Warren K. Lewis of MIT, had been made in fractional distillation so that the production of a specific gasoline from the cracking unit was possible. By cracking less severely and removing the cracked oil from the soaking chamber more quickly, refiners could reduce the coke formation and the down time for cleaning. The resulting residual oil was a better fuel and sold at a higher price.

New Chemistry: Catalysis and the Houdry Process

The breakthroughs of Burton and Dubbs depended upon what seem in hindsight as reasonably straightforward approaches to the problem of increasing the yield of gasoline. Burton did, however, uncover the chemistry of cracking, while the two Dubbses used engineering (some chemists might say "plumbing") to improve the process. In Europe, a quite different approach was tried involving a type of chemistry unusual for that time.

A French engineer and businessman, Eugene J. Houdry, became interested in the challenge of producing better fuels for automobiles and airplanes. He was an avid auto racing fan and a director of a company that manufactured auto parts. While visiting the United States in 1922, Houdry attended the Indianapolis 500 race and visited Ford's assembly plant in Detroit. Houdry recognized that further improvements in engine design could not be made until the quality of the fuel was enhanced. Also, from his World War I military experiences and observing the importance of the airplane, he knew that great quantities of high-quality gasoline would be essential to making France a stronger nation.

During and after World War I, many researchers had tried to make gasoline out of a variety of abundant starting materials. In particular, they tried the distillation of bituminous coal at low temperature, the hydrogenation (addition of hydrogen) of bituminous coal or lignite coal or oils derived from them, and the hydrogenation of carbon monoxide. Houdry's firm had invested in some of these processes, including those being developed by the group headed by E. A. Prudhomme. When the results of using metal catalysts (box 6.5), such as nickel and cobalt, in the reaction of carbon monoxide and hydrogen did not lead to a commercially viable process, Houdry bought out the Prudhomme group. He decided to work exclusively on catalysis research aimed at fuels. His background in engine fuels and in mechanical engineering prepared him for the evaluation of each new batch of gasoline and for the

Catalysis: Chemical Wizardry

Catalysis is normally used to speed up a chemical process by adding another chemical, called the *catalyst*, which allows the starting materials to undergo a chemical change more efficiently. The use of catalysts in industrial processes provides many of the products of our modern society. In biology, catalysis is essential to life itself; most of our biochemical reactions are controlled by catalysts known as enzymes.

The almost magical property of a catalyst stems from it ability to emerge unchanged from a chemical process which it influences so dramatically. Even though it is intimately involved in the chemical reaction, the catalyst is not used up and can be reused time after time. The enzymes that control digestion are examples of this chemistry.

One of the important technologies that uses catalysts is the catalytic muffler on all recent automobiles. The catalytic converter prevents unburned hydrocarbons, that is, gasoline, from getting into the atmosphere. If any gasoline escapes from the engine cylinders, as it goes through the exhaust pipe it is converted to water and carbon dioxide by a catalyst of rhodium metal. This element is quite expensive, but only a little bit is required to do the job.

The advent of the catalytic converter inspired another changed in the composition of gasoline. Up until the 1970s, TEL was used as an antiknock agent even though lead pollution resulted from its combustion. Lead, however, also "poisons" the catalyst in the catalytic converter. Therefore, TEL was essentially removed from gasoline, and different types of hydrocarbons were added to make the fuel burn smoothly.

Box 6.5

fabrication of the new equipment to handle the catalyst in the cracking vessels. Just as important, his family had the financial resources to fund such a major research effort, and he had the necessary energy and enthusiasm for the project.

Houdry formed a process company and employed a chemist, a mechanical engineer, and a civil engineer to begin their study of the chemistry of hydrocarbons and catalysis. A good deal of Houdry's time was spent as a promoter of the potential for the process to the French government. By 1929, Houdry had demonstrated a commercial process to produce gasoline from lignite coal. But the cost of the process was very high, and the French government withdrew its subsidy, forcing Houdry to close his plant. However, the group had also been working on catalysis using petroleum as the starting material. Prior to Burton's thermal cracking, which doubled the amount of gasoline produced from a barrel of crude, there was fear of worldwide shortage of crude oil. World War I renewed such fears. After the war, demand for gasoline, especially higher quality gasoline, again exceeded available supplies. Houdry realized that "catalytic cracking" might again increase the amount of gasoline that could be produced from a barrel of oil.

In 1925, little was known about catalysis in organic chemistry, especially the theory. By contrast industrial processes for making inorganic materials like ammonia, nitric acid, and sulfuric acid were well developed. Two known types of catalysts, metals used for hydrogenation and aluminum chloride for benzene reactions, retarded the cracking reaction for petroleum. Nickel, cobalt, and iron were known to initiate activity, but they produced large quantities of carbon that coated the catalyst and eliminated the activity. With determination reminiscent of Edison, who tried hundreds of substances as filaments for his electric light bulb, Houdry spent three years doing empirical tests of compounds as catalysts.

In 1927, early one April morning, he noticed that a charge of lowgrade heavy crude petroleum was producing a clear distillate from a column packed with oxides of silicon and aluminum. It tested well in his automobile on a nearby hill, meeting Houdry's high standards of performance. Of course, carbon was deposited on the catalyst during the reaction. So, having found a suitable catalyst, Houdry had to face the problem of regeneration of the catalyst.

Gasoline

He decided to take the simplest approach to removing carbon coating from the pebbles of silica-alumina. First, he removed the oil from the reaction vessel, added air, and heated the catalyst to burn the carbon to carbon dioxide. Paradoxically, this turned out to be advantageous: the cracking reaction is endothermic (requires heat), but the oxidization of carbon is exothermic (gives off heat). By using some of the heat liberated in the regeneration of the catalyst for the cracking process, the process became more economical because less fuel was required. Among the first oil companies to visit Houdry's laboratory after this success was Standard Oil of New Jersey. A number of design problems remained to be solved, and the competing hydrogenation process of the I. G. Farben Company in Germany seemed to have more promise. Thus, interest in the Houdry process was limited.

In the six years from 1923 to 1929, Houdry had used a large amount of money from his and his wife's families. Although the stock market crash and the resulting business recession might have reduced demand for automobiles and, thus, gasoline, demand did not decrease as anticipated. It was nevertheless difficult to obtain additional funds for research, so Houdry's dream had to wait until he came to America and to Sun Oil.

In the latter 1930s, as the United States and Europe emerged from the Depression, demand for gasoline and, therefore, petroleum increased (figure 6.2), and the need to replace thermal cracking as the standard process became obvious. Sun Oil, a small refinery that marketed a single premium grade of gasoline, wanted to be first with the new process to maintain its competitive edge. By 1936, Houdry's research at Sun Oil led to increasing the yield of gasoline from 23 percent of the crude oil to 43 percent using the catalytic cracking process. The octane rating increased from 72 for thermally cracked gasoline to 88 for gasoline from the catalytic process.

Based on these research results, both Sun and Standard Oil of New York (Socony, later to become Mobil) constructed catalytic cracking units. Socony constructed a Houdry unit (figure 6.10) in New Jersey that processed two thousand barrels of light gas oil per day in June, 1936. The following year Sun Oil's Houdry unit began processing six times that amount using heavy gas oil. By 1939, Sun and Socony had in operation fifteen catalytic cracking units having a

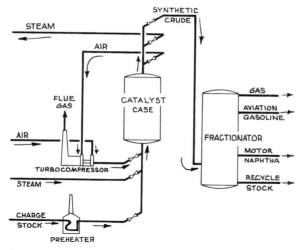

Figure 6.10 Schematic diagram of a Houdry cracking unit. (Adapted with permission of MIT Press from Enos, 1962.)

combined capacity of over two hundred thousand barrels per day. In three years these two companies increased the industry's total cracking by 10 percent, and the cost of cracking by the Houdry process was three-fourths the cost of thermal cracking in 1938.

Like the Burton process, the Houdry process was a significant technological advance. In the Houdry method the synthetic catalyst consisted of pellets formed by the intermingling of silicon and aluminum compounds. The petroleum mixture was then passed over the pellets, which were placed in "fixed beds." After cracking occurred on the surface of the catalyst, the gasoline product was continuously swept out by the incoming uncracked mixture. Thus, Dubbs's continuous process was linked to a more efficient chemical process to improve the yield of the primary product, gasoline. It was also financially lucrative because, with the patent protections, each company was able to invest large sums of money very quickly to capture a significant share of the market.

These fixed bed units operated by Houdry and others became crucial to the Allied war effort in World War II because they produced high-quality gasoline and, in the first two years of the war, 90 percent of the cracked aviation fuel. After the war, however, the Houdry fixed bed process would yield to a new innovation.

Squeezing More Gas out of the Barrel: Fluid Catalytic Cracking

The Houdry Company was asking a very high price to license its process, so seven large oil corporations based in three different countries entered into a cooperative research and development program to circumvent the Houdry patents. Unlike previous developments that came from very independent firms with a few engineers frequently working in secret and protecting their results with patents, this was an atypically friendly effort conducted by large groups of engineers and scientists from different companies, many of whom were specialists in very narrow areas. Whereas thermal cracking was marked by nasty, long-term patent litigations, patents and information were pooled from the beginning of the fluid bed research. The major problems were all technological and scientific in origin and did not include those of finance, administration, and promotion of the product as with thermal cracking (box 6.6).

Fluid catalytic cracking demonstrated the benefits resulting from the creation of research and development laboratories at all the major oil companies beginning in the 1920s and the steady accumulation of knowledge about process design. Developments in fluid catalytic cracking were accelerated by World War II and proceeded from pilot plant to large-scale commercial units in 1942 without constructing small-scale commercial test units.

Standard Oil of New Jersey had been working on catalysis in the 1920s, including catalytic cracking, but it concentrated on hydrogenation reactions because the raw materials—coal, heavy oils, tar, and hydrogen were in abundant supply at good prices. However, in the 1930s the company returned to catalytic cracking with the catalyst in a powdered form suspended in the oil vapors that undergo cracking. This mixture of catalyst and oil was called "Suspensoid." The powdered catalyst was propelled into a stream of vaporized oil by means of a rotating screw. The mixture was then fed into a cracking coil, where it was heated under pressure. In the next stage, the light cracked fraction was separated from the heavy, and the catalyst was filtered from the heavy fraction and discarded.

The process gave a higher yield of gasoline with a higher octane number than the thermal cracking process did. Of course, removing and discarding the catalyst were problems and increased the

Oil Refining and Conservation of Mass

From what you already know, logical first questions might deal with how much of a particular product can be produced, via some process, from a barrel (42 gal) of oil, and at what price can it be sold by the owners of the refinery to make a profit? Such questions bring to mind the picture of crude oil flowing into some reactor and the product flowing out the other side to satisfy some market demand.

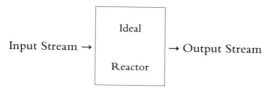

Of course, in the real world we would have to add at least one more stream, since it would be a rare process, indeed, that produced only the desired product in 100 percent yield.

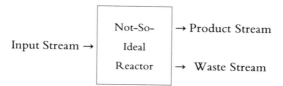

In a typical refinery, crude oil, as the input stream, is processed through a number of reactors to generate a series of product streams, most of which contain mixtures of chemicals. These mixed-product streams are further separated into product streams having fewer chemicals in higher concentrations and into waste streams.

The first, and absolutely essential, step in the analysis of an oil refinery process, or of any chemical process, is the careful accounting of this flow of material through the reactor. It is the key to monitoring the rate of production, the yield of the desired product, and the profits available from the sale of the chemical. This accounting process is called a *material balance*. It is actually a restatement of the law of conservation of mass—that matter cannot be created or destroyed, only transformed from one form to another in the reactor. From the masses (or volumes) involved, chemical engineers begin to get a feel for the overall size of the chemical process, and hence the size and cost of reactors, compressors, heat exchangers, distilling and fractionating towers, and other processing equipment necessary to construct the process plant.

Box 6.6

cost. By 1934, a 12,000-barrel-per-day unit had been designed but was not built. In 1940, Imperial Oil Company of Canada did convert a unit to Suspensoid cracking, and four more were constructed during World War II. The Suspensoid process, however, proved to be the direct antecedent of the fluid bed process.

In 1938 Professors Warren K. Lewis and E. R. Gilliland in the Department of Chemical Engineering at MIT suggested, while consulting for Standard Oil Development, that the reaction tubes be moved from their traditional horizontal to a vertical position. They were asked to investigate the behavior of finely divided particles in vertical tubes and thus began research on the modern fluid bed crackers. Under certain conditions, finely divided powders in contact with vapors can exhibit the properties of a fluid. The fluid bed will flow in any direction with a slight differential pressure; it is relatively compressible and with agitation will attain uniform temperature. Oil vapors mixed with the solid powder catalyst in the right proportion and with the proper pressure can cause the mass to flow just like a liquid along any desired path. After cracking, the catalyst can be easily separated from the oil vapors before they condense to form a liquid stream. Two MIT graduate students, John Chambers and Scott Walker, had used air and a catalyst with oil vapors. Standard quickly took over the project, and within six months it was producing gasoline in its Bayway Refinery laboratories. Full-scale production followed shortly.

On 25 May 1942, the first fluid bed unit was placed on stream at the Baton Rouge refinery of Standard Oil of New Jersey handling 12,000 barrels per day, and other units at Bayway and Baytown followed that same year. The U.S. military's demand for gasoline rose from 33,000 barrels per day in December 1940 to 507,000 in October 1942, and the fluid bed units helped to meet this demand for aviation gasoline in the 100+ octane range. The process, however, did not begin to dominate the industry until the postwar years. By 1956 seven refineries in the United States were processing in the range of 200,000 to 300,000 barrels per day at each location. The fluid catalytic cracker became the industry standard for cracking petroleum and continues to be so today. Thus, we see how research—in this case a collaboration between industry and academia—led to a technological breakthrough that improved the mode of production of a major industry.

A Modern Refinery: Unit Operations

A modern refinery appears bewilderingly complex when considered in its entirety. It becomes less forbidding when its functions are systematically broken down into more easily understood categories of fundamental processes. Sometimes these fundamental processs are called *unit operations* because only one type of physical or chemical step (i.e., operation) is performed in such a basic process.

Examples of unit operations involving a physical change include distilling, condensing, drying, extracting, evaporating, melting, and absorption. Examples of unit operations involving a chemical change include cracking, isomerization, oxidation, hydrogenation, combustion, and reaction with water. A series of these unit operations can be linked together to form a *unit process* that might produce a single chemical product such as kerosene or gasoline. In a like manner, a series of unit processes can be linked together to form a *plant process* such as a complex oil refinery that produces a large number of chemical commodities.

Thus, the modern refinery is analogous to the assembly line mode of production of machines and appliances. There is a continuous inflow of raw materials that is systematically changed and separated into products of higher value. The particular product or mixture of products depends upon the specific conditions chosen for the unit operations, how they are combined into unit processes, and the number of processes in a plant.

Box 6.7

The Refining of an Industry

The Bayway Refinery, operated by EXXON, USA in New Jersey, was founded in 1908. Today it operates the largest fluid bed catalytic cracker in the world. A modern refinery (box 6.7) can chemically convert as much as 40 to 50 percent of the crude oil into artificial, or synthetically made, gasoline as a result of a series of chemical engineering process design improvements that took place during the first half of the twentieth century. This is about a fourfold increase in the amount of gasoline available from a barrel of crude oil since Henry Ford sold his first Model T.

Other processes have been developed to further increase the amount of gasoline per barrel of oil. We have examined two of the most important ones, thermal cracking and catalytic cracking, and

Figure 6.11 A modern oil refinery—UNOCAL's Chicago refinery. (Courtesy of UNOCAL.)

have focused on one of the refinery's most important products, gasoline. But this refining process also set the stage for the new petrochemicals industry—a host of secondary and tertiary products flowing from the refining of oil and the production of gasoline. Products such as the gases (methane, ethane, hydrogen, etc.) go to petrochemical plants as the feed stocks for the synthetic reactions that produce other products such as polymers, plastics, and pharmaceuticals. Others, such as fuel oil, kerosene, and asphalt, are sold for direct use.

In a modern refinery (figure 6.11), the product mix is changing constantly to meet market demands. For example, in severe winter weather the refinery will increase its production of fuel oil, and during summer, when travel is heavy, the refinery steps up production of gasoline. Weather may also change significantly the blending operations for the gasoline. In winter, the automobile starts easier with a higher concentration of the low-boiling components, but in summer these lower boilers can lead to vapor lock. Very elaborate computer systems control the flow of crude oil to the pipe stills for distillation and to the other unit processes for the chemical synthesis of the desired product mix that will best meet market demand and maximize profit for the refinery.

Beginning with Burton's pressurized thermal cracking process, significant changes in the processing of crude oil occurred over the next fifty years, and the yield of gasoline manufactured from a barrel of crude oil increased from 10 percent to as high as 50 percent. Today, depending on the seasonal variations in the demand for gasoline, 40 to 50 percent of each barrel of crude oil is converted into gasoline. The remainder of the oil is converted into other products, such as heating oil and petrochemicals. In 1930, 18.5 billion gallons of gasoline were sold, and by the 1980s production had increased over fivefold to between 103 and 117 billion gallons. Whether another significant increase in the yield of gasoline will occur depends on a complex variety of factors including the world supply of crude oil, the demand for gasoline relative to other petroleum products, the capital costs required to improve refining technology, and, certainly not least of all, new developments in petroleum chemistry and engineering.

Exercises

1. (a) Explain the difference between petroleum, kerosene, and gasoline.
   (b) Why did the early automobile developers not choose simply to use kerosene in the engines of their machines?

2. Estimate how many gallons of gasoline are used each year in the United States. Compare your answer to the value given in the introduction to this chapter. Explain any significant discrepancy.

3. (a) What does *distillation* mean, and how is it carried out?
   (b) How does *thermal cracking* differ from *catalytic cracking*?

4. (a) Draw three of the five isomers of hexane, C_6H_{14}, found in gasoline.
   (b) What is the general formula—number of carbons and hydrogens—for all *oct*anes?
   (c) Draw the structure of isooctane.
   (d) Draw the structure of *cyclo*hexane and give the general formula for all cyclic alkanes.
   (e) Draw the structure of hexene (straight-chain form) and give a general formula for alkenes.

5. Explain the function of TEL and what problem it solved, as well as the problem it created.

6. Using the best yield (cited in the chapter) of gasoline from a barrel of crude, calculate the cost of gasoline based on a price of $25 per barrel of oil. Compare this to the price you pay for gasoline and give reasons for any difference.

7. Describe the changes in petroleum refining technology, starting with Rockefeller and going on to Burton, Dubbs, and finally Houdry. What advances occurred after Houdry's developments?

Bridge Design: Concrete Aesthetics

7

An Ancient Problem

The task of building a bridge poses an ancient problem of deceptive simplicity: how do you create a structure that spans a waterway, a roadway, or a valley and supports not only its own weight but the weight of people and vehicles that cross it? If the distance is short, the solution might be as simple as finding pieces of a suitably strong material and laying them over the obstacle (e.g., figure 7.1, foreground). But as the distance to be spanned grows longer, the problem becomes more difficult, beginning with the choice of materials.

Certainly materials for a bridge will be chosen for their strength in bearing loads, but availability and cost will also influence the selection. Wood, for example, offers an attractive combination of properties: it is relatively difficult to crush, that is, it is strong in *compression*; and it is relatively difficult to pull apart, that is, it is strong in *tension* (table 7.1). Wood is also readily available and inexpensive in many locales. These qualities might seem to make wood an excellent material for bridge construction, as indeed it is in many instances, but it has its drawbacks. A variety of insects eat wood or bore through it. Wood burns. Harsh weather takes its toll on wooden structures. Structures made of wood require a great deal of maintenance and often have disappointingly short lives. Besides, as strong as wood appears to be in tension and compression, it can not cope as well as some other materials with the huge forces that develop in very large structures.

Figure 7.1 Post Bridge (foreground) at Dartmoor, England. This bridge probably dates from Roman times. The stone arch bridge in the background is of much more recent origin. (Photograph by N. Copp.)

Stone offers relief from some of the deficiencies in wood as a building material. Stone structures last a long time with a minimum of care as evidenced by any number of ancient stone structures that are now tourist attractions. The outstanding structural asset of stone is its great strength in compression. Unfortunately, this strength is not matched by the ability of stone to withstand tension (table 7.1). Provided that the design of a bridge takes into account the weakness of stone in tension, it can make a very fine material for bridges, even fairly large ones. The arch form is particularly appropriate for execution in stone because it places all of the material in compression (e.g., figure 7.1, background). The beam very quickly ceases to be an appropriate form for stone bridges as the spans become longer because the unsupported middle sections of beams develop considerable tension along their underside. Despite its low cost, therefore, stone is not the material of choice for very long span bridges.

The longest spanning bridges in the world are made from steel. In contrast to stone, steel and its predecessor, iron, are renowned for their strength in tension (table 7.1). A form that takes great advantage of this strength is the suspension bridge. Aside from the columns that are in compression, the weight of the deck and its cargo in a suspension bridge is carried by steel cables in tension. One

Bridge Design

Table 7.1
Strengths of Materials (psi)

Material	Tension	Compression
wood	15,000	4,000
stone/concrete	600	6,000
high-tensile steel	225,000	60,000
mild steel	60,000	60,000

problem with steel as a structural material, in comparison to stone, is its high cost.

Arch bridges in stone and suspension bridges in steel—the design accommodates the material. But what about the beam? A horizontal beam must accommodate both types of stress: compression in the upper portion and tension in the lower regions. Without a material that combines the tensile strength of steel with the compressive strength and low cost of stone, the beam form seemed completely inappropriate for long spanning bridges. While bridge builders lacked such a material, a scientific understanding of beams was not crucial for building long spans. Only after the development of a new material did an understanding of this simple and ancient form become important. The rest of this chapter explores the development of a new building material during the Second Industrial Revolution, its use in constructing long spanning beams, and the possibilities it offers for a form of art called structural art.

A New Material, a Return to an Old Form

Concrete—simply a mixture of sand, gravel, water, and cement—has been used as a building material since Middle Eastern people constructed temples with it six thousand years ago. Part of the reason for its popularity lies in the fact that it is plastic before it becomes rock hard. Like stone, properly cured concrete is very difficult to crush. A one-inch cube of average structural concrete can support 3,000 pounds, the weight of a hippopotamus. Also like stone, however, concrete does very poorly under tension. Only about 300 pounds of force would be needed to pull apart a one-inch cube. These properties make concrete an excellent choice for building columns or arches, in which all of the material suffers compression, but a very poor choice for building beams that span any great distance. A long concrete beam's own weight would cause it to bend, sending cracks across its middle as the concrete along the underside of the beam pulled apart (figure 7.2a). Yet thousands of safe bridges have been built with concrete beams. There must be more than meets the eye in these structures. What is missing from sight is steel.

Steel is another old material, not as old as concrete, but older than Christianity. We think of it as new only because large quantities could not be produced rapidly until the latter part of the nineteenth century. Before Sir Henry Bessemer developed his famous steel-making process and began producing tons of the metal in a matter of minutes, the major steel companies required the better part of two weeks to create 50 pounds of it. Steel contrasts beautifully with concrete. It remains malleable throughout its lifetime, and it withstands tension very well. Some modern high-strength steels do not break until they are pulled on with a force greater than 400,000 pounds per square inch. You could suspend yourself with a thread considerably thinner than a human hair if you used steel as strong as this.

Concrete withstands compression. Steel does well under tension. Why not combine these materials and obtain both properties? The first person to take official credit for this apparently obvious idea was Jean Monier, a French gardener who obtained a patent in 1867 for adding wire to the concrete he used in making flower pots. But it required the insight of two engineers in the 1890s, G. A. Wayss of Germany and François Hennebique of France, to make Monier's technique blossom in structural engineering. Wayss and Hennebique realized that steel reinforcing in concrete beams would carry the tension that otherwise might cause a structural failure. Soon thereafter, Robert Maillart joined the ranks of the great structural engineers by using reinforced concrete to build some of the most elegant arch bridges ever constructed (figure 7.3).

Despite the great utility of reinforced concrete, it presented a few disadvantages. The major drawback is that the concrete in a reinforced structure must crack before the reinforcing steel encased within it can withstand the amount of tension it was designed to carry. Such cracks in reinforced concrete beams, if not properly

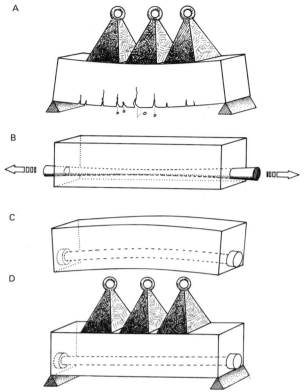

Figure 7.2 A concrete beam under a heavy load will crack along the lower surface as the amount of stress exceeds the capacity of the material to withstand tension, (*A*). Prestressing alleviates this problem by stretching a steel cable within the lower section of the beam to place the cable in tension, (*B*), then securing the cable against the concrete with buckles so that the beam is placed in compression along its lower edge, (*C*). Now, heavy loads will not cause cracking until the tension they create exceeds the compression produced in the lower part of the beam by the prestressing cable, (*D*).

Figure 7.3 The Salginatobel Bridge, a reinforced concrete bridge in Switzerland built by Robert Maillart in 1930. The bridge spans 90 meters. (Photograph courtesy of M.-C. Blumer-Maillart.)

controlled, allow water to seep in where it can corrode the metal reinforcing bars and weaken the structure.

A related solution to the problem of concrete's low tolerance for tension began to develop in 1903, the same year as the Wright brothers' most historic flight. A young French engineer, Eugene Freyssinet, started thinking about the problem after hearing a few lectures on reinforced concrete. Why not replace the tension that appears in the reinforcing bars with "previously imposed and permanent stresses of sufficient value ...," he asked. Hidden behind that opaque question lay "the single most significant new direction in structural engineering" (Billington 1976). Imagine a row of books lightly compressed between your hands to form a sort of bridge across open space. The bridge will stand up under its own weight, but any additional load will send the books crashing to the floor. The same books compressed powerfully between your hands will hold a substantial load before collapsing. The strong "precompression" applied to the books in anticipation of more weight counteracts their tendency to spread apart along the bottom when additional weight is added. This, essentially, is what Freyssinet understood.

Freyssinet spent much of the next twenty-five years thinking about how to apply the idea of precompression to concrete structures. The problem was made particularly difficult by the tendency of a compressed concrete beam to shorten, a tendency that threatened to erode the effects of any applied stresses. This behavior of concrete, called *creep*, doomed to failure early attempts at prestressing. After years spent studying the concrete in bridges he had designed and built, Freyssinet realized that permanent precompression could be imposed upon a concrete beam by adding steel cables to the beam, stretching them until they developed very high tension stresses, and then locking the tensed cables against the ends of the beam (figure 7.2, bottom). The success of this tactic lay in generating such large tension stresses that creep in the concrete would only erode part of the cable's beneficial effects. Fortunately for Freyssinet, the years in which he was developing his innovative ideas coincided with the years during which high tensile strength steel was developed.

Despite its apparent similarity to reinforced concrete, the idea of prestressing marked a significant departure from previous practices. Reinforced concrete featured the virtually independent actions of two materials. Prestressed concrete, however, emerged as a new material. With the concrete in compression and the steel in tension, the two components function together to produce a single building material with characteristics unlike any of its predecessors.

Prestressed concrete offers several distinct advantages over its predecessor for many applications, especially those involving long beams. The action of the prestressing cable places all of the beam's concrete in compression, a situation well suited to the properties of concrete. Reinforced concrete beams, however, are only in compression in that portion above the neutral axis; full advantage is not taken of the material in terms of its ability to withstand deflection without cracking. One consequence is that prestressed concrete beams can be thinner than reinforced concrete beams—it takes less material to achieve a desirable degree of stiffness in a bridge if prestressed concrete is used than if reinforced concrete is employed. Less material may also mean lower costs for prestressed concrete struc-

Figure 7.4 An early prestressed concrete bridge designed by Eugene Freyssinet and constructed in 1950 near Ussy, France. (Photograph courtesy of the Portland Cement Association.)

tures, although this savings may be offset by greater costs for high-strength steel and concrete. Another consequence is that prestressed concrete beams normally do not develop the cracks characteristic of reinforced concrete beams. This increases the useful lifetime and decreases the maintenance costs of prestressed concrete beams relative to reinforced ones. For many applications, prestressed concrete has proved to be the more economical and versatile material.

Despite these advantages that are now apparent to us, Freyssinet's business partner considered the idea of prestressing so unsound that he refused to use it. Undeterred, Freyssinet "decided to risk all that [he] had of fortune, reputation, and strength in making ... prestressing an industrial reality" (Billington 1976), not a decision to be made lightly by a man fifty years old. He patented the idea in 1928 and proceeded to apply it in the construction of bridges (figure 7.4) and other structures. A major public demonstration of prestressed concrete's effectiveness came in 1935 when Freyssinet used it to save a wharf that had been sinking into the harbor at Le Havre. His prestressed concrete structure did not crack, as reinforced concrete would have, and so avoided premature failure from the corrosive effects of saltwater on the steel.

With New Materials Come New Forms

Prestressed concrete offered more than durability. As a new material, it opened new possibilities for design. Early European innovations with the new material were closely followed by Gustave Magnel, a Belgian engineer, who was among the first to accept Freyssinet's ideas and explore their potential. In 1949, the city engineers of Philadelphia commissioned Magnel to design a bridge for a section of Walnut Lane that crossed over another road. They turned to Magnel after an earlier proposal for a stone-faced arch had to be rejected because the low bid of $1,047,790 exceeded expectations. Some of the city's engineers had seen prestressed concrete used in the construction of sludge tanks and wondered if an inexpensive bridge could be made of the same material. Magnel, as perhaps the most knowledgeable English-speaking person thoroughly familiar with Freyssinet's work, was the logical choice for the task.

Magnel's design for the Walnut Lane Bridge not only saved the city a considerable amount of money by eliciting a low bid of $597,600 but its simple form also proved to be visually appealing (figure 7.5). The art jury that evaluated plans for all of Philadelphia's major structures greatly admired the straight, slim lines of Magnel's proposed bridge. Magnel achieved this simplicity of design as well as economy by constructing the bridge from thirteen identical prestressed concrete girders that proved easy to make and inexpensive to install. The originality of Magnel's design may be somewhat obscured by the now-commonplace occurrence of this form, but in 1950 when the Walnut Lane Bridge was completed with its 155-foot span, it differed markedly from other American bridges spanning similar distances. Prior to 1950, bridges in the United States with moderately long spans were typically constructed as arches or trusses, longer spans as suspension bridges. Stone or concrete beams were certainly inappropriate because the material could not survive the tension stress that would develop along the lower portion of the beam. The new material of prestressed concrete eventually allowed the beam to become common in bridges of moderate span. Use of

Figure 7.5 The Walnut Lane Bridge, a prestressed concrete bridge designed by Gustave Magnel, built in Philadelphia in 1950. (Photograph courtesy of David P. Billington.)

the new material in the United States grew slowly, however, possibly because American engineers remained skeptical that a European technique could be adapted to American culture.

The Scientific Foundation

"Given a long enough lever and a place to stand, I could move the Earth," or so the Greek mathematician Archimedes is reputed to have boasted more than two thousand years ago. Archimedes recognized that, in a lever system, even a small force can move a massive object if the force is applied a long distance from the point about which the lever rotates. We apply this principle in less ambitious ways than envisioned by Archimedes when we use a bottle opener. To get a feel for this concept, try closing a door by pushing on a point near the hinges; the difficulty in closing a door this way, compared to pushing on the door knob, is surprising.

Isaac Newton expressed the same principle more formally in his great work, the *Principia*, published in the late seventeenth century. Like Archimedes, Newton understood that the angular acceleration of an object depends on the amount of force applied to the lever arm as well as the length of the lever arm, that is, the distance separating the point at which the force is applied from the axis of rotation. The combination of force and distance in this context is called *torque* by physicists and is given the Greek symbol Γ. Newton defined torque as the product of the object's moment of inertia, I, and its angular acceleration, α. (*Moment of inertia* is the measure of an object's resistance to a change in angular motion. *Angular acceleration* is the rate at which the object's rotation rate changes.)

$$\Gamma = I\alpha$$

This expression bears a striking resemblance to Newton's second law of motion relating force (F) to mass (m) and acceleration (a):

$$F = ma$$

This is no accident. Torque represents the rotational equivalent of the second law, which pertains to translational motion or movement in a straight line. These laws can not be proved but have been exhaustively confirmed by observation of macroscopic objects. (Atoms and molecules behave according to a different set of rules.)

Taken together, Newton's second law and his description of torque define the central problem of building a useful bridge: If a body is to remain stationary, then the net force and net torque on it must be zero. In other words, for all the forces that act in a vertical direction, the upward forces must exactly balance the downward forces. Similarly, for all horizontally acting forces, the ones directed to the right must equal the ones directed to the left. The same consideration applies to the moments: clockwise torques must equal counterclockwise torques. An object for which these things are true is said to be in *equilibrium*. The central problem in bridge design has always been to produce a structure in equilibrium. Because engineers who design bridges use the term *moment* in place of torque, we will follow that practice for the remainder of the chapter.

A Balancing of Forces and Moments: The Walnut Lane Bridge

In contrast to machinery, bridges are essentially stationary structures. Any condition other than a state of equilibrium produces undesirable movements in the bridge and may lead to its collapse. This argument for *external* forces applies equally well to forces and moments that exist entirely within the material of the bridge, that is *internal* forces and moments. One key to appreciating the structure of a bridge thus lies in understanding the external forces and internal forces and moments that act on the bridge.

The Walnut Lane Bridge employs three sections to carry traffic across a valley: a center span that stretches 160 feet between its supports and two side spans at 74 feet apiece. The long center span represents the most surprising element of this concrete structure and is the focus of the following discussion. The center span of the Walnut Lane Bridge consists of thirteen prestressed concrete beams aligned side-by-side and supported only at each end. Such an arrangement, where a beam is supported at each end and horizontal movement is permitted at one end, is called a *simple beam*. These beams must bear not only their own weight and the weight of the roadway (i.e., the *dead load*) but also the weight of the trucks, cars, people, snow, and other things that might travel across it or accumulate on it (i.e., the *live load*). These loads represent external forces on the structure.

The external forces acting on the Walnut Lane Bridge are best understood by considering one entire beam. The only external forces that we are concerned with here are those that act in the vertical direction such as the loads. Engineers must also consider horizontal forces, such as those created by winds and earthquakes, but our concern is only with the question of how a bridge manages to support its own weight and the weight of vehicles on it.

Considering only the dead load for the moment, two external vertical forces act on each beam in the Walnut Lane Bridge: a downward force due to gravity and counteracting upward forces at each of the supports. The downwardly directed vertical force on each beam is equivalent to the beam's weight of 272,000 pounds, or 272 kips in the parlance of civil engineers (1 kip, abbreviated k, represents 1,000 pounds or 1 kilopound). Because each beam is horizontal and its weight is evenly distributed along its length, then each support holds one half of the bridge's total weight or 136 k. In other words, the upward vertical force supplied by each support equals one half of the downward force due to gravity. The sum of these external vertical forces is thus zero. (The vertical forces contributed

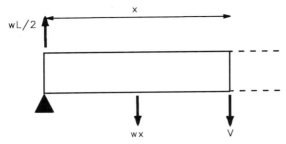

Figure 7.6 A schematic diagram of a portion (length X) of one beam in the Walnut Lane Bridge showing three vertical forces acting on this portion of the beam: $wL/2$ represents the vertical reaction at the support (triangle on the left) equivalent to one half of the beam's weight; wx represents the weight of the portion with w indicating the weight per unit length and x the length of the portion; and V represents the shear force.

---

by loads other than the beam's own weight will be treated later in this chapter.) Clearly, the supports must be up to the task of applying 136 k (one half of the beam's weight) for each end of the beam, but this alone is not sufficient for the bridge to work. The beam itself must not break. In order to set our minds at ease, we must determine the internal forces within the beam as it carries its loads.

The internal forces and moments acting on each beam in the Walnut Lane Bridge are more easily analyzed if only a short segment of the beam is considered in isolation from all other segments of the beam. The rationale for this approach is that if the entire bridge is to be in equilibrium, then all of its segments must also be in equilibrium. The benefit of this approach is that an understanding of the entire structure can be achieved by examining only a portion of the structure. Imagine a segment of a beam extending an unspecified distance, x, from the left support as shown in figure 7.6. Many forces act on this segment. If it is to remain in equilibrium, the total of the upward forces must equal the total of the downward forces, the sum of the forces to the right must equal the sum of those to the left, and the moments tending to rotate the section clockwise must balance those in the counterclockwise direction. We will consider the implications of each of these.

Three vertical forces act on the segment shown in figure 7.6: the downward force of the segment's weight, calculated by multiplying the segment's length and the unit weight of the beam (repre-

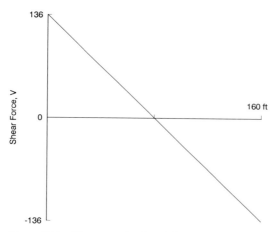

Figure 7.7 The magnitude of the shear force varies along the length (160 ft) of each beam in the Walnut Lane Bridge, reaching maximum values at either end.

sented by w; 1.7 k/ft in this case); the upward force of 136 k supplied by the support; and a vertical force contributed by the remainder of the beam. This last force is an internal force called *shear*. It is not hard to visualize that the weight of the segment adjacent to the portion of the beam under consideration will tend to shear the beam, or break it in a plane perpendicular to the beam's length.

The shear force, V, may be upward or downward. We will guess V is down, and if we guessed wrong the results will correct us by giving a negative answer. The upward force on the segment must equal the sum of the downward forces (note that L designates the total length of the beam):

$$\frac{wL}{2} = wx + V$$

or

$$V = w\left(\frac{L}{2} - x\right)$$

A graph of the shear force as a function of the length of the segment (figure 7.7) reveals that the situation is worst near each support where $V = 136$ k. Fortunately, for beams that are much longer than they are thick, as in the Walnut Lane Bridge, the shear force rarely poses a significant problem compared to the others we are about to

Bridge Design

encounter. Much more imposing problems arise from the moments tending to rotate the beam.

Bending Moments and the Walnut Lane Bridge

Common experience makes it obvious that when weights are added along the length of a beam supported at each end, it will probably break in the middle. As we will discover, this is because the stresses in the material are largest in the middle of a simple beam. Our tasks here are to understand how the beams in the Walnut Lane Bridge withstand these stresses, especially the tension, and then to understand how Magnel could have used concrete to build such a thin, yet safe, structure. Stresses in the beams of the Walnut Lane Bridge arise from internal forces in the concrete that are associated in turn with internal *bending moments*. By starting with bending moments, we will come to understand the stresses in the Walnut Lane Bridge, the benefits of prestressing, and the sense of Magnel's design.

Once again, reducing the problem to an isolated segment of one beam makes the analysis easier. Think of an imaginary axis, perpendicular to the page, passing through a portion of one beam as illustrated in figure 7.8. The vertical force contributed by the bearing at the left end tends to rotate the segment about the axis in a clockwise direction, while the weight of the beam segment tends to cause a rotation in the opposite direction. The strength of each tendency is given by the force times the distance from the force to the axis. The tendency to rotate in a clockwise direction equals the force at the bearing, $wL/2$, multiplied by x, the length of the illustrated portion of the beam. Because ths is the only moment in the clockwise direction, the total clockwise moment, M_{cw}, can be calculated as follows:

$$M_{cw} = \left(\frac{wL}{2}\right) \times x$$

The counterclockwise moment, M_{cc}, is given as $wx^2/2$: this is the weight of the illustrated portion, wx, times the distance from the support at which the weight acts, which we will assume to be in the middle of the portion, that is, at $x/2$.

$$M_{cc} = \frac{wx^2}{2}$$

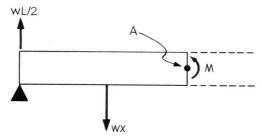

Figure 7.8 A schematic diagram of a portion of one beam in the Walnut Lane Bridge showing a torque or moment (M): $wL/2$ represents the vertical reaction at the support (triangle on the left) equivalent to one half of the beam's weight; wx represents the weight of the portion with w indicating the weight per unit length and x the length of the portion; A indicates an imaginary axis of rotation about which the illustrated portion of the beam can rotate; and M indicates the torque or moment about the axis of rotation that is produced by the weight of the unillustrated portion.

Surprisingly, these opposite moments do not cancel each other! For example, taking x as $L/2$ and substituting values from the Walnut Lane Bridge shows that the clockwise moment is considerably larger than the counterclockwise moment:

$$M_{cw} = \frac{wL}{2} \times \left(\frac{L}{2}\right) \qquad M_{cc} = \frac{w(L/2)^2}{2}$$

$$= 10{,}880 \text{ kft} \qquad = 5{,}440 \text{ kft}$$

The imbalance in these moments creates an unpleasant situation. It appears as if the bridge is not in equilibrium. The fact that the bridge has served Philadelphia well since 1950, however, suggests that something has been left out of our analysis. There must be another moment that acts in the counterclockwise direction in addition to the one calculated above. The additional moment, called a *bending moment* and labeled M, comes from the action of the remaining part of the beam on the segment under consideration. This moment is shown in figure 7.8. The clockwise and counterclockwise moments can now be balanced by adding the bending moment.

$$\frac{wLx}{2} = \frac{wx^2}{2} + M$$

The value of the bending moment, M, about point A is given by:

$$M = \frac{wx}{2}(L - x)$$

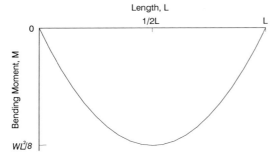

Figure 7.9 Diagram for the bending moment for a beam in the Walnut Lane Bridge that results from the beam's own weight. Following European rather than American custom, the bending moment diagram is drawn downward, the same direction in which the beam will tend to deflect. The curve is parabolic with the maximum bending moment occurring midway along the beam's length.

This equation shows that the bending moment will increase as the distance x increases up to a maximum at the middle of the span where x equals $L/2$ (figure 7.9):

$$M_{max} = \frac{wL^2}{8}$$

The maximum bending moment experienced by the beams in the center span of the Walnut Lane Bridge is thus:

$$M_{max} = \frac{1.7 \text{ k/ft} \times (160 \text{ ft})^2}{8}$$

$$= 5{,}440 \text{ kft}$$

This value represents the difference between the two moments calculated above.

A Bridge in Tension

The forces that create the bending moment, M, offer a pure tendency to rotate without imposing a net horizontal force on the bridge. This must be so if the bridge is to be in equilibrium because there are no horizontal forces in the concrete or on the beam to oppose net horizontal forces created by the bending moment. We can, however, imagine a pair of horizontal forces, called a *force couple*, one acting to the left in the upper section of the beam segment and an equal opposite force acting to the right in the lower

Innovation

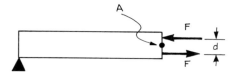

Figure 7.10 A schematic diagram of a portion of one beam in the Walnut Lane Bridge representing the moment at A as a force couple. The two forces in the couple are acting at points separated by a distance, d.

section (figure 7.10). Both of these forces tend to rotate the beam in a counterclockwise direction about A, so their effects are additive, not contradictory as the diagram might imply.

If we assume that the beam has a symmetric cross section so that A is midway between the upper and lower forces, and if the separation between the forces is d (see figure 7.10), then the total bending moment is:

$$M = \frac{dF}{2} + \frac{dF}{2} = dF$$

At the center of the beam where the bending moment reaches its maximum, the forces will also reach their maxima as calculated below for the center span of the Walnut Lane Bridge:

$$F_{max} = \frac{M_{max}}{d}$$
$$= \frac{5400 \text{ kft}}{5.84 \text{ ft}}$$
$$= 932 \text{ k}$$

This force, acting on the concrete in the beam, will generate stresses in the concrete. (A stress is equal to a force divided by the area over which the force is applied.) At the top of the beam where the arrow points to the left, the concrete will be under compression while the bottom of the beam, where the arrow points to the right, will experience tension. Since concrete does not withstand tension stresses very well, it is important to determine just how large this stress is. The force being discussed here is somewhat idealized. In reality it comes from stresses distributed over the entire cross section of the beam and so does not act only at two points along the beam's cross section. We will simplify the problem by assuming that the forces act only at two points within the more simplified, symmetric cross section shown in figure 7.11b rather than across the entire area shown in the actual cross section (figure 7.11a). For the simplified cross section of the Walnut Lane beams, the forces, F, are located at the center of the shaded areas and separated by 5.84 feet. Each shaded area represents 463 square inches. The stress, s, is thus:

$$s = \frac{F}{A}$$

$$= \frac{932 \text{ k}}{463 \text{ in}^2}$$

$$= 2.01 \text{ k/in}^2$$

or 2,010 pounds per square inch (psi).

That is, there will be approximately 2,000 psi of compression in the top of the beam and the same amount of tension in the bottom of the beam. The actual stresses are slightly different than these values because the forces are applied over the entire cross section of the actual beam rather than just over the top and bottom portions shown in figure 7.11b.

Two thousand pounds per square inch of tension is a great deal of stress for concrete, too much for the concrete in the Walnut Lane Bridge. The preceding calculations have shown that the bending moment needed to keep the bridge from rotating creates stresses in excess of what the concrete can bear. Yet the bridge stands. The resolution to this paradox resides in a steel cable.

Figure 7.11 Diagram of the cross section of one beam in the Walnut Lane Bridge as it actually exists (A) and as it is idealized (B) to simplify the calculations.

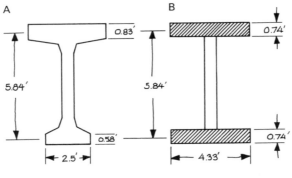

ACTUAL CROSS SECTION   SIMPLIFIED CROSS SECTION

Innovation

Prestressing in the Walnut Lane Bridge

In order to use concrete for the long spanning beams in the Walnut Lane Bridge, Magnel had to add steel cables to the lower section of each beam that, when stretched and secured against the ends of the beam, created a compressive force to counteract the dangerously high tension just discussed. In order to understand how a prestressing cable produces this benefit, we must turn again to a segment of one beam (figure 7.12). In keeping with the nature of prestressed concrete, and to clarify the analysis, we consider the concrete and the prestressing cable in the selected segment to act as a unit. External forces act on this segment as shown in figure 7.12.

The force labeled T in figure 7.12 is the force on the part of the steel contained in the beam segment that is exerted by the remainder of the steel tendon. As always, the total horizontal forces must balance. This means that an additional horizontal force acts to the left on the segment. This force is shown as C in figure 7.12 and represents the compressive force exerted by the remainder of the concrete on the segment under consideration. Its magnitude is equal to T, and it acts at the center of the beam. As before there is also the bending moment, M, within the concrete. The concrete adjacent to the segment being considered thus exerts two effects on the segment: it compresses the segment and it tends to twist it. As a result,

Figure 7.12 A schematic diagram of a portion of one beam in the Walnut Lane Bridge showing the position and effects of the prestressing cable. The cable (dark, thick line in the bottom half of the beam) is tensed and secured at the ends of the beam (only the left end is shown). The tensile force (T) in the cable leads to a compressive force (C) that acts on the concrete at the center of the beam. Because the cable is placed below the beam's central axis (dashed line), it also tends to bend the beam, that is, create a counterclockwise moment (M).

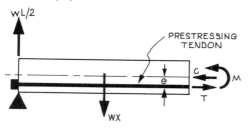

Bridge Design

the compressive force and bending moment are shown as external to the segment in figure 7.12, but they are actually internal to the beam.

These effects of the prestressing cable are not difficult to visualize. When an elastic steel cable is stretched before securing it against the ends of a beam, the beam will be compressed along its length. The tension created in the cable is exactly balanced by the compression that appears in the concrete as required for equilibrium. But that is not all. The cables in the Walnut Lane Bridge were laid below the center line of the beams. Tensing the cable under these circumstances causes the beam to bend in the same way that tightening a bowstring bends the bow. The question now concerns the magnitude of these effects in the Walnut Lane Bridge.

Magnel specified that the cables in the Walnut Lane Bridge be stretched until they experienced a tension of 2,000 k. Because the compressive force equals the tension in the cable, the amount of compression stress, C, produced by this force equals the tension in the cable divided by the total cross-sectional area of the beam:

$$C = \frac{2000 \text{ k}}{1354 \text{ in}^2}$$

$$= 1.48 \text{ k/in}^2 \quad \text{or} \quad 1,480 \text{ psi}.$$

This compression stress will partially offset the tension in the bottom of the beams in the Walnut Lane Bridge.

The bending moment generated by the cable, M_{ps}, is simply the tension force, F, multiplied by the distance at which that force acts from the center of the beam. This distance is referred to as the *eccentricity*, e. In the Walnut Lane Bridge, the eccentricity varies along the length of the beams from zero at the ends to 2.65 feet at the middle. (This variation was omitted from figure 7.12 for the sake of simplicity.) The maximum bending moment due to the cable is at midspan:

$$M_{ps} = Fe$$

$$= 2,000 \text{ k} \times 2.65 \text{ ft}$$

$$= 5,300 \text{ kft}$$

Because this moment will be in the counterclockwise direction (figure 7.12), it reduces the amount of bending moment, M, that the concrete must supply to balance the clockwise moment created by the support:

clockwise   counterclockwise

$$\frac{wLx}{2} = \frac{wx^2}{4} + M + M_{ps}$$

Without the prestressing cable, the concrete would have to supply a bending moment of 5,440 kft in order to keep the bridge in equilibrium, but this bending moment would create an unacceptable amount of tension. With the prestressing cable in place, however, the concrete need supply a bending moment of only 140 kft (5,440 kft − 5,300 kft) to obtain equilibrium. Shifting some of the counterclockwise moment to the cable helps the bridge to remain standing by reducing the tension stress in the concrete:

$$F = \frac{M}{d}$$

$$= \frac{140 \text{ kft}}{5.84 \text{ ft}}$$

$$= 24 \text{ k}$$

This force will yield stresses in the upper and lower sections of:

$$s = \frac{F}{A}$$

$$= \frac{24 \text{ k}}{463 \text{ in}^2}$$

$$= 0.052 \text{ k/in}^2 \quad \text{or} \quad 52 \text{ psi.}$$

The bending moment contributed by the cable thus reduces the tension in the bottom portion of the beam by a factor of nearly 40 from 2,000 psi to 52 psi!

When the two effects of the cable are considered together, it appears that all portions of the beam are under compression (figure 7.13). If this were true, it would represent an unnecessarily conservative design and a waste of material. The ideal situation would be if the prestressing left no net tension or compression in the bottom of the beam. The calculations completed above, however, failed to account for the weight of the roadway, the live loads, and the tendency for prestressed steel to lose a part of its tension.

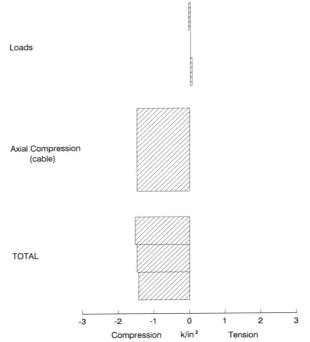

Figure 7.13 Stress diagram drawn for the center span of the Walnut Lane Bridge considering only the dead load and the prestressing cable. The stresses created by the dead load are mostly offset by the effects of the prestressing cable to leave only small amounts of compression in the top of the beams (upper portion of top graph) and tension in the bottom of the beams (lower portion of the top graph). When these stresses are added to the axial compression created by the prestressing cable (middle graph), the sum reveals a beam that is in compression at all points. The amount of compression in the bottom region of the beam suggests an inefficient use of material.

The roadway and sidewalks add .36 k/ft to the weight of each beam. The live loads—loads due to cars, trucks, and people—give an additional .45 k/ft per beam. These loads bring the unit weight of a beam to 2.51 k/ft (i.e., 1.7 + .36 + .45). The extra weight will of course increase the stresses in the beam. Another factor that should be considered involves the behavior of concrete. Concrete under compression gradually shrinks. This phenomenon known as creep is estimated to reduce the amount of tension in the prestressing cable to 87 percent of its original value or 1740 k. With these new values of w and T, the value of the bending moment at midspan becomes:

$$M = \frac{wL^2}{8} - eT$$

$$= 3{,}421 \text{ kft}$$

The force F becomes:

$$F = \frac{M}{d}$$

$$= 586 \text{ k}$$

and the stresses in the upper and lower sections increase to:

$$s = \frac{F}{A}$$

$$= 1.27 \text{ k/in}^2$$

To this we must add the uniform compression due to the reduced prestressing acting on the entire cross section:

$$s = \frac{T}{A}$$

$$= 1.29 \text{ k/in}^2$$

Using these figures to recalculate the stresses yields the diagrams in figure 7.14. The near absence of tension at the bottom of the beam indicates that the material in the Walnut Lane Bridge is being fully utilized. Even this small amount of tension probably reflects the approximate nature of the analysis conducted above rather than the actual behavior of the structures. By prestressing the concrete, Magnel succeeded in building a bridge with an unusually thin profile, a profile achieved in part by the parsimonious use of concrete.

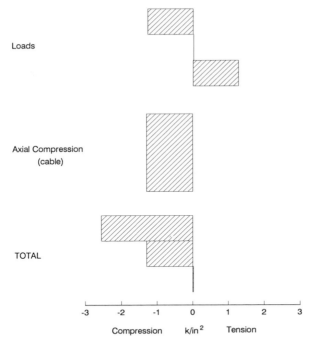

Figure 7.14 The stress diagrams for the center span of the Walnut Lane Bridge change when other loads and creep are considered along with the dead load and the prestressing cable. There is now more compression at the top of the beam (upper portion of the top graph) as well as more tension at the bottom of the beam (lower portion of the top graph) in response to the bending moments created by loads and the cable. When these stresses are added to the axial compression produced by the cable (middle graph), the sum reveals a beam that is in compression in its top and middle segments and experiences an acceptably small amount of tension at the bottom, suggesting an efficient use of material.

The Key Lessons of the Walnut Lane Bridge

The close look we have just taken at the reasoning behind the design of the Walnut Lane Bridge reveals a number of general points that can inform our appreciation of other bridges:

1. All forces and moments acting on and within a bridge must balance if the structure is to remain stationary and intact.
2. The bending moment created by the weight of the beam and the loads on it leads to very large internal stresses that must be controlled if the bridge is to survive without excessive cracking.
3. In bridges that are constructed as simple beams, the bending moment represents a tendency to rotate in the counterclockwise direction, indicated by a negative value on a graph of bending moment versus distance. This bending moment produces tension in the bottom of the beam and compression in the top. If you imagine a beam made of foam rubber and supported only at each end, you can see that the bending moment and the tension will reach their maximum values at the beam's center where it sags the most.
4. In other types of structures, where the bending moment tends to produce a clockwise rotation, the bending moment will take a positive value and the beam will crest in the middle with tension developing in the upper part of the beam and compression appearing in the lower portion.
5. In all cases, prestressing cables should be laid where tension is anticipated in concrete beams. The tensed cable adds a favorable compression and bending moment to the beam. Each of these effects helps offset the tension resulting from the various loads and allows an otherwise unsuitable concrete structure to support itself and its loads.

A New Way to Build Bridges

The Second World War left large areas of Europe in ruin. Bridges, especially those along the Rhine River in Germany, were among the hardest hit large structures because of their strategic importance during the war. Given the dependence of European commerce on railroads and trucks, the rapid redevelopment of Europe's bridges

Figure 7.15 An example of cantilever construction in which prestressing of the completed segment allows subsequent segments to be built outward from the free-hanging end. (Photograph courtesy of David P. Billington.)

posed a particularly important challenge to engineers. Prestressing had already provided the key that opened new possibilities for design in concrete. Following the war, engineers took advantage of prestressed concrete to develop a new method of construction.

A German engineer, Ulrich Finsterwalder, responded to the task by creating a daring method of construction called *cantilever construction*. Finsterwalder decided to start building bridges by starting at the supports and working toward the center without erecting scaffolding underneath the growing spans. He successfully applied this remarkable technique in constructing a 200 m span over the Rhine at Bendorf, Germany, a structure that remains one of the longest concrete beam spans in the world (figure 7.15).

Cantilever construction begins with erection of the foundations and piers in the usual way. Once these are completed, the deck is built outward from the supports segment by segment, each segment being supported by the previous one. The concrete segments may be manufactured away from the building site and lifted into place or poured into forms supported by the earlier sections. The partly completed deck extends out as a cantilever, from which the method gets its name.

The major problem associated with this style of construction is illustrated by a tale in the folklore of Abraham Lincoln. Lincoln gained wide acclaim as an unusually strong young man because he was able to grip a double-headed ax by the tip of its handle and hold it out at arm's length. The reason this stunt merits admiration is the same reason that makes cantilever construction a remarkable feat. Imagine gradually pushing a long rectangular slab of foam rubber lengthwise off the edge of a table. Where will tension appear? How will the amount of tension change as you extend the slab further off the table? As the span of the bridge grows, the weight of the structure increases as does the distance from the newest segment to the support. That is, the moment will be largest at the support and increase as construction proceeds. The value of the moment will be positive at the support and decrease toward the free end (figure 7.16a). An increased moment brings increased stresses, assuming that the cross-sectional area of the span near the support remains constant during construction. If the accumulating stresses, especially the tension developing at the *top* of the span over the support, grow unchecked, the structure will collapse.

Finsterwalder coped with the cantilever problem by passing prestressing cables through each segment as it was laid in place and then tensioning those cables to secure the new segment to the previous one and counteract tension in the concrete with compression. In this way, the loads are carried back to the pier without creating dangerously high levels of tension in the concrete. Construction continues until the cantilevers from adjacent piers meet at midspan. When the gap between the two converging cantilevers is closed, the structure becomes a built-in beam and loads added after closure contibute to the total stress as shown in figure 7.16b.

Prior to Finsterwalder's work, scaffolding built up from the ground was required to hold the bridge segments in place until the entire structure was completed and became self-supporting. Cantilever construction employs much less scaffolding, especially if the bridge is high. Because the cost of scaffolding typically represents about 30 percent of the cost of the bridge, Finsterwalder's method produced significant savings in the construction of new bridges, an important consideration anytime but especially in post–World War II Europe with its rising labor costs. A second advantage of cantilever construction is that it leaves the region underneath the

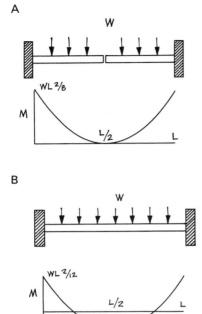

Figure 7.16 Bending moments in cantilevered beams. A: Before the ends are joined, cantilevered beams under a uniformly distributed load, w, exhibit a bending moment diagram that is parabolic in shape, as for simple beams, but has its maximum value at the supports and zero moment at the free ends of the beams. B: When the two cantilevered beams are joined across the middle, the bending moment diagram remains parabolic but shifts downward on the graph.

construction site open, a distinct advantage if the bridge is to cross a road or river carrying heavy traffic. A close examination of the Felsenau Bridge in Switzerland illustrates these advantages of cantilever construction and the possibilities for design offered by prestressed concrete.

The Felsenau Bridge: A Work of Structural Art

Christian Menn, a Swiss engineer, has been among the foremost designers of bridges for a number of years. Menn's visually striking bridges (e.g., figure 7.17) represent superb examples of an engineer's capacity for artistic expression within the constraints of cost and structural performance. They are works of structural art

Figure 7.17 The Ganter Bridge in Switzerland, a prestressed concrete bridge designed by Christian Menn and completed in 1980. The center span stretches 174 meters. The support on the right is 150 meters tall. (Photograph courtesy of William Case.)

characterized by three minimal conditions under which structural art thrives: efficiency, economy, and elegance (Billington 1983). Efficient structures incorporate the minimum amount of material needed to achieve the desired performance and safety characteristics. Economy means minimum cost, a goal aided by efficiency but also affected by the expense of construction. Elegance, or maximum aesthetic value, emerges from the engineer's artistic sensibility and affinity for simplicity over ornamentation. Structural art absolutely requires this third condition. We are surrounded by too many structures that are efficient, inexpensive, and ugly to think otherwise.

Menn's work beautifully satisfies the criteria for structural art. The shallow, thin-walled beams characteristic of his bridges reflect a striving for efficiency as well as elegance. He reduces cost by building with prestressed concrete, which is less expensive than many other building materials, and by using the cantilever construction method of building long spans with a minimum of expensive scaffolding. Finally, the aesthetic quality of his work is plainly visible in

Figure 7.18 The Felsenau Bridge in Switzerland, a prestressed concrete bridge designed by Christian Menn and completed in 1974. The cantilever construction method was used to build the longest spans (144 meters). (Photograph courtesy of David P. Billington.)

the finished structures. Works of structural art do not always emerge from strict adherence to these principles but from attempts to produce elegant structures within reasonable constraints of economy and efficiency. Menn, for example, occasionally sacrifices economy to a small degree in order to achieve greater elegance and artistic value.

When Christian Menn was invited to enter a design competition for a bridge to carry a six-lane expressway over the Aare River Valley in the small Bern suburb of Felsenau, he developed a seventeen-span structure with a total length of slightly over 1 kilometer (km) (figure 7.18). The longer spans, two of 144 meters and two of 96 meters, were to be constructed by the cantilever method while the others, ranging from 38 meters to 48 meters, were to be built

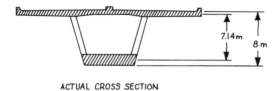

ACTUAL CROSS SECTION

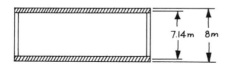

SIMPLIFIED CROSS SECTION

Figure 7.19 Actual and simplified cross sections of the Felsenau Bridge. The simplified cross section was used in the calculations described in the text.

using traditional methods. Menn's design owes much to the earlier work of Finsterwalder yet contains many elements that mark it as one of the most finely developed structures of its type. Six lanes make a very wide roadway and would normally require the building of two bridges side by side, each carrying three lanes. Menn chose to carry all six lanes with a single box deck, as shown in figure 7.19. This approach yielded a single bridge, rather than the clutter of two bridges standing side by side.

The elegance of this bridge is enhanced by the vertical taper of the box that allows a smooth meeting between the box and the piers while offering a relatively wide support where the box meets the upper part of the deck. The box is one monolithic structure along its entire 1116 meter length. As such it is quite likely the longest single piece of concrete that does not rest on the ground as a road. Menn's design was not only the most elegant entry in the contest but also among the least expensive. It won the competition.

Before the Spans Meet

In the cantilever method of construction, the stresses in the box at the piers will reach their greatest level just before the ends of the converging spans meet. It is not hard to visualize that the bending moment of a cantilever increases from zero at the free end to a maximum at the support. This contrasts with the simple beams of the Walnut Lane Bridge in which the bending moment was zero at the supports and reached its maximum at midspan. In terms of the Felsenau bridge just before completion, the maximum bending moment reached 106,000 Tm (ton-meters; figure 7.20). As always, the units for bending moment reflect a force times a distance, but in this case the force is given in tons (T, where 1 T = the weight of 1000 kg or 2205 pounds) and the distance in meters (m) to conform with the dimensions Menn actually used.

This huge bending moment leads to a force couple consisting of compression along the lower region of the box and tension at the top, a situation opposite the conditions prevailing in the Walnut Lane Bridge. The maximum force in the Felsenau Bridge before completion of the bridge is 14,800 T. This translates into nearly 33,000 k, a force more than thirty-three times greater than in the Walnut Lane Bridge!

Huge internal forces lead to huge stresses. The maximum stresses in the Felsenau Bridge can be approximated by considering a simplified cross section of the box (figure 7.19b). This simplification yields a value of 1,590 T/m^2. That is, the cantilever construction method applied to the Felsenau Bridge would be expected to produce 1,590 T/m^2 (2,300 psi!) of tension in the top section over the support and an equal amount of compression in the bottom section during the phase of construction just before connection at midspan. But the concrete used in the Felsenau Bridge could withstand only about 450 T/m^2 of tension. Obviously, something had to be done to ensure the structure's integrity. As in the Walnut Lane Bridge, prestressing cables rescued the situation.

Prestressing in the Felsenau Bridge

The prestressing in the box adjacent to the pier consists of a combination of cables applying an initial total force of 22,680 T to the concrete. As in the case of the Walnut Lane Bridge, the prestressing exerts two effects on the Felsenau Bridge: a pure compression acting at the center of the beam and a bending moment equal to the prestressing force times the vertical distance from the center of the box section to the location of the cables.

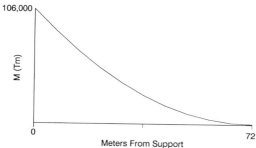

Figure 7.20 Just before the cantilevered beams in the Felsenau Bridge were joined (photograph), the bending moment diagram for one beam resulting from the dead load takes the form in the graph with a very large moment existing at the support and none at the free end. (Photograph courtesy of David P. Billington.)

Bridge Design

The pure compression created by the prestressing results in a compressive stress of 907 T/m^2 in the concrete. The cables are placed fairly high in the upper section where the tension is located. They lie about 0.1 meter below the top surface of the box and 3.9 meters above the center of the cross section. This positioning allows the cables to apply a bending moment of 88,500 Tm to the box. The force couple induced in the concrete by this bending moment equals 12,400 T. Assuming that this force is carried uniformly by the top and bottom sections of the box gives stresses of 1,330 T/m^2.

This time, however, the 1,330 T/m^2 of compression occurs in the *top* of the beam while the bottom of the beam experiences the same amount of tension. The cable, therefore, effectively offsets the stresses in the concrete created by the loads. The stress diagram in figure 7.21 shows that prestressing eliminates all of the tension at the pier that threatened the structure while the cantilevers remained unjoined. The entire beam now is placed in compression with the largest stress remaining well below the crushing strength of 4,500 T/m^2 for the concrete used in this structure.

The total load carried by the Felsenau Bridge rises from 41 to 55.4 T/m after the box is joined at midspan, the wearing surface is added, and an allowance of 11.4 T/m is made for the maximum expected live loads on the bridge. With this increase in load over the situation in the uncompleted structure, the bending moment at the supports rises to 131,000 Tm from the previous 106,000 Tm, and the bending moment at the middle changes from zero to 12,500 Tm (figure 7.22). The stresses in the box where it joins the pier now amount to 1,970 T/m^2 of tension in the top section and the same amount of compression in the bottom section. These stresses will be offset by the prestressing action of the steel cables, but as in the Walnut Lane Bridge an allowance must be made for creep such that the final prestress amounts to only 87 percent of the value given previously for the Felsenau Bridge. The corrected stresses and their sum for the completed bridge are presented in figure 7.23.

The compression of 1610 T/m^2 at the bottom of the beam lies well below the crushing strength of 4500 T/m^2 for the concrete used. Our numerical result shows some tension in the top section, but considering the approximations in the preceding calculations the figure depicts essentially zero stress. Besides, engineers normally

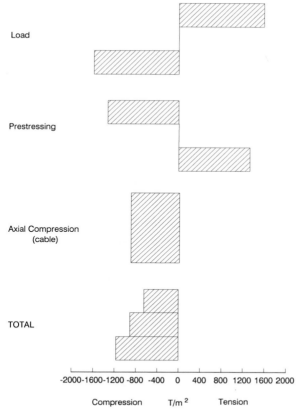

Figure 7.21 Stress diagrams for one cantilevered beam of the Felsenau Bridge before it is joined with its partner to complete the span. In contrast to the Walnut Lane Bridge, the bending moment due to the dead load creates tension in the beam's upper region and compression in its lower region at the support. The bending moment created by the prestressing cable partially offsets these stresses. When the stresses created by the bending moments are added to the axial compression created by the prestressing cable, the sum shows a beam entirely in compression. The presence of a significant amount of compression in the upper region of the beam suggests an inefficient use of material.

Bridge Design

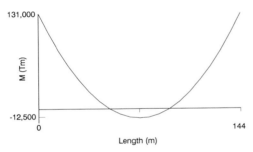

Figure 7.22 When the longest spans in the Felsenau Bridge are completed, the bending moment diagram reveals a smaller bending moment at the supports but a larger bending moment at center span than would have existed under full load just before the beams were joined.

assume that concrete can withstand about one tenth as much stress in tension as in compression, and the value shown in figure 7.23 is well below the 450 T/m^2 of tension allowed by this estimation. It is also true that a design may allow tension in a concrete structure under unusually large short-term loads. Engineers typically control this transitory stress by adding unprestressed reinforcing steel to the concrete. This solution is less costly than employing a large amount of high-tension steel.

The Result? Structural Art

Menn's use of prestressed concrete in the Felsenau Bridge allowed him to utilize a means of construction and a beam design that contributed to the structure's attributes as a work of structural art. The cantilever construction method, made possible by the prestressing of the concrete segments as they were added, eliminated the need for extensive and costly scaffolding during the construction of the Felsenau Bridge. The financial burden would have been especially heavy in this case because of the height reached by the completed structure. The continuity of the concrete box also helped reduce expenses by lowering maintenance costs; the absence of seams in the concrete meant that less water would seep in to rust the steel cables or freeze and contribute to cracking of the concrete. The choice of construction techniques, along with a relatively inexpensive building material and the decision to build one box instead of two, thus played significant roles in making the Felsenau Bridge economical.

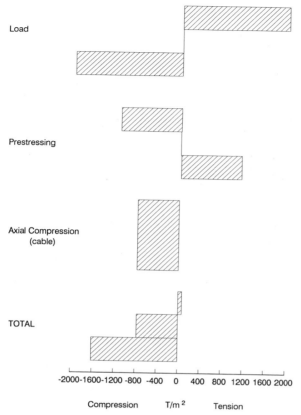

Figure 7.23 As seen for the bending moment diagram, the stress diagrams for the Felsenau Bridge change when the span is completed. Stresses resulting from the dead load bending moment, the prestressing cable bending moment, and the compressive force exerted by the cable on the beam's central axis sum to produce an acceptably small amount of tension in the upper region of the beam at the support, suggesting an efficient use of material.

The efficiency of the Felsenau Bridge is apparent in the design specifications that eliminated tension in the concrete without adding an excessive amount of compression (figure 7.23). One simplifying assumption that was made to obtain figure 7.23, however, was that the box has a uniform cross section along its length. The box in the actual bridge is about two and a half times deeper at the pier than at midspan. Menn designed the bridge this way to accommodate the large internal forces that appear in the deck where it joins the pier. This design reduces the internal force, which is given by $F = M/d$, by a factor of 2/5. Menn also gave the bottom portion of the beam a larger cross-sectional area near the pier than at midspan. This feature again decreases the stress, which is given by the force divided by the area of the bottom section, and helps hold the compression stresses to low levels. Another way to look at these decisions is to realize that Menn achieved efficiency by moving concrete from the middle of the long spans, where the forces were relatively small, to the piers, where the forces and stresses are greater.

These design features certainly contribute to the efficiency of the bridge, but they also lend it a striking visual appeal. Menn explicitly selected the single-box design for aesthetic reasons. Also, the long, gently arched spans give this huge structure a surprisingly light appearance, and the depth of the boxes at the piers convince us of its strength.

A Different Application

Magnel's Walnut Lane Bridge and Menn's Felsenau Bridge convincingly demonstrate not only the utility of prestressed concrete for long spanning beams but also the opportunities offered by the new material to create new forms. How have these lessons been applied in circumstances more closely related to our daily lives? One prominent example concerns the bridges in freeway interchanges.

Although the concept of the freeway arose in the American northeast, its most famous, or infamous, application has been in Southern California. The decline of the electric trolley companies in the Los Angeles basin after 1930 paralleled the rapid ascendancy of the automobile as the preferred mode of transportation. Greater privacy, a quieter ride, and more convenience all contributed to the appeal of the automobile over the doomed trolleys. As the trolley companies' profits declined, they began replacing tracks with roadways to carry more profitable buses, a process assisted by Firestone Tires, General Motors, and Standard Oil. The latter three companies were eventually convicted of conspiracy in undermining the city's trolley system, but the court imposed a fine of only one dollar, apparently recognizing that the decay of the trolley system began well before the conspirators intervened.

These events changed the face of the city. Concrete freeways replaced steel railroad tracks as the main arteries carrying people about the Los Angeles basin. The first freeway in Los Angeles opened in 1940 and connected the downtown area with the city of Pasadena ten miles to the northeast. By 1980, 725 miles of freeway crisscrossed Los Angeles county carrying more than 30 *billion* vehicle miles of traffic per year. The freeway system has become so extensive that it is visible in satellite photographs of the Los Angeles basin.

Bridges quickly became an indispensable component of the Los Angeles freeway system. Without bridges to route traffic smoothly from one freeway to another, the system would cease to be a network and become a pointless jumble of unconnected roadways. These huge structures must distribute automobile and truck traffic among as many as four freeways without unduly slowing the flow. This function requires bridges with spans long enough to permit an unimpeded flow of traffic along underlying roadways.

The first solution to the span problem came in 1949 with completion of the first freeway interchange in Los Angeles (figure 7.24). A simple beam made of reinforced concrete and seated on pillars no more than 52 feet apart carried traffic on the Hollywood freeway over cars traveling below on the Pasadena freeway. In a time when cars were fewer and slower than now, this solution worked well. But soon, driving conditions changed, and the short spans of the bridge made driving under the Hollywood freeway a slow and dangerous event as it remains today. Simple reinforced concrete beams were too heavy to use in long spans. Consequently, engineers in the California Department of Transportation (Caltrans) turned to steel when they built the Elysian viaduct near Dodger Stadium in 1962. The longest span in this bridge, 182 feet, permitted more and

Figure 7.24 Known as "the stack," these bridges compose the interchange between the Hollywood Freeway (uppermost level) and the Pasadena Freeway (lowest level extending from the foreground into the background), the first freeway interchange constructed in Los Angeles (1949). The beams are made of reinforced concrete and have a maximum span of 52 feet. (Photograph by N. Copp.)

wider traffic lanes and thus improved traffic flow on the freeway below the bridge. Steel eventually became too expensive for widespread use in the city's freeway system, however, as the cost of the basic material rose and as environmental regulations stiffened to require that Caltrans recover every speck of lead-based paint removed from steel bridges during their regular refinishing. Caltrans engineers eventually turned to prestressed concrete as their preferred building material for freeway interchanges.

The choices of Caltrans engineers to use prestressed concrete and a continuous beam design can be related to the need for economical, efficient structures, but they also have been affected by a desire for elegant bridges. Caltrans must operate within the constraints of the state's budget in building and maintaining thousands of structures around the state. Attention to budgetary limits is essential. For this reason, Caltrans engineers appreciate the low cost of prestressed concrete relative to steel ($13.10 per square foot vs. $16.50 per square foot in one application in Illinois). They also have come to appreciate its new possibilities for form as Califor-

Figure 7.25 The interchange between the San Bernardino and Barstow freeways, designed by engineers in the California Department of Transportation and completed in 1973. The longest spans are 273 feet. (Photograph by N. Copp.)

nians have begun to comment on the large structures visible from their homes. Designers of freeway bridges now emphasize elegant design, and prestressed concrete offers them considerable latitude for experimentation.

Not until 1963, thirteen years after completion of Magnel's successful Walnut Lane Bridge, was the first prestressed concrete bridge completed in Southern California's freeway system. Reasons for the delay remain obscure but probably stem from the reluctance of engineers in the United States to adopt a new technique not widely tested in American applications. This bridge, located about 12 miles west of downtown Los Angeles, is part of one of the busiest interchanges in the city's entire freeway system. Its largest span accommodates four wide lanes of high-speed traffic below, and its gentle arc allows two lanes of interchange traffic to move from one freeway to another with little reduction in speed. Completion of this structure marked an important step away from simple beams made of reinforced concrete toward the huge graceful arcs of more modern interchanges (figure 7.25).

The Southeast Connector

The Southeast Connector routes westbound traffic from the Santa Monica Freeway to the southbound lanes of the San Diego Freeway west of downtown Los Angeles (figure 7.26). It carries cars and trucks in two lanes along a trajectory of 2864 feet that rises 66 feet above the underlying freeway at the highest point. The gentle curvature includes twenty-three sections, most of which are made from reinforced-concrete girders employed as simple beams. The engineers relegated these girders to the shorter spans, spans of 131 feet or less. Two other sections of the connector pass diagonally over freeways requiring unbroken spans as long as 192 feet. None of the interchange bridges that preceded this one included spans of such length. The bridge's designers proposed not only to build a span of unprecedented length but also to make it shallower than previous interchange bridges: the depth of the 192-foot section of the Southeast Connector is only 1/27 its length, whereas the depth of its thinnest predecessor in the city's freeway system is fully 1/18 of the span, and the depth of the first interchange bridge in Los Angeles is

Figure 7.26 The Southeast Connector as seen from the southbound lanes of the San Diego Freeway. The 192-foot main span span is the central portion of a beam 466 feet long bounded by supports not visible in the photograph. (Photograph by N. Copp.)

1/8 of its span. The shallowness of the Southeast Connector produces not only greater efficiency and economy through the reduced need for materials but also an increased elegance. Our question is how the engineers achieved these goals.

The maximum span in the Southeast Connector is 192 feet, yet the longest girder is 466 feet. This girder rests atop four columns: one at each end and two others bordering a 192-foot span in the middle (figure 7.27). A beam constructed in this way is termed a *continuous beam*. The effect of the side spans in this case is to restrict rotation of the beam over the supports on either side of the center span. The side spans have the beneficial effect of reducing the magnitude of the bending moment at center span. This benefit comes at the cost of bending moments in the box over the supports, but the net effect is to permit thinner, longer structures. In contrast to simple beams and cantilevers, continuous beams develop significant tension stress in the top of the box where it passes over the supports as well as in the lower portion of the box at the middle of the center span.

The locations of the maximum stresses in the Southeast Connector are indicated in the bending moment diagram for the 466-foot beam (figure 7.28). While the calculations leading to this somewhat complicated shape lie beyond the scope of this chapter (but see Albano, Case, and Copp 1990), the behavior of the bridge that would result if it was allowed to react to these bending moments can be visualized by imagining that the bridge is made of a strip of foam rubber instead of prestressed concrete: the middle of the central span of the foam beam sags while the foam beam crests over the supports. Tension will be greatest along the bottom of the beam at the center of the longest span and along the top of the beam over the supports. The appearance of tension stress over the supports is one feature that distinguishes the behavior of the Southeast Con-

Figure 7.27 Schematic diagram of the 466-foot beam within the Southeast Connector.

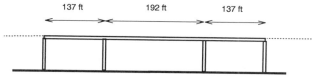

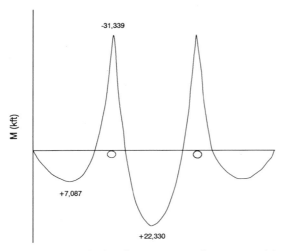

Figure 7.28 The bending moment diagram resulting from the dead load and live loads on the 466-foot beam in the Southeast Connector. The circles mark the positions of the supports that define the center span. The largest moments in a continuous beam occur at the supports, not at the center span as in simple beams (see figure 7.9). Following European convention, positive moments are shown below the line that denotes zero moment.

nector Bridge, and any continuous beam, from the simple beams of the Walnut Lane Bridge.

The bending moments shown in figure 7.28 loom ominously. They threaten to impose intolerable tension stresses on the center span and at the supports unless the design and material somehow accommodate them. Of course, prestressing is the answer, and you can probably imagine the path that the steel cable must take through the 466-foot beam. Careful positioning of the prestressing cable (figure 7.29) and generation of an appropriate amount of tension in it produce bending moment forces that partly counteract the tension that would otherwise appear in the concrete. In order to achieve this effect, the position of the cable must change along the length of the beam according to the size and direction of the bending moments imposed by the loads. The cable should run close to the central axis where the bending moment is small and far from the axis where the load-induced moment is large. Where the beam tends to bend downward under the load (i.e., where the bending moment is positive), the cable must lie below the central axis to generate a compensatory tendency to bend upward. The cable must be placed above the central axis where the beam tends to bend upward. At points

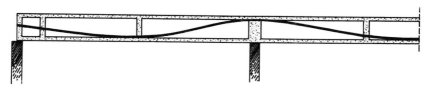

Figure 7.29 Diagram of one half of the 466-foot beam with one side cut away to reveal the average path taken by the prestressing cables (heavy line). In the actual bridge, there are eight cables on each side of the beam, each cable following a somewhat different path through the beam, but for the sake of simplicity these cables are treated as one.

where the beam shows no tendency to bend, that is, where the load-induced moments are zero, the cable should lie on the central axis. The cable must always impose bending moments that counteract the load-induced moments.

Surprisingly, the cable's bending moment diagram is not a mirror image of the load-induced bending moment diagram (compare figure 7.30a and b). The moment of 8,990 kft generated by the cable at the center of the 192-foot span falls short of the 22,230 kft needed to balance exactly the load-induced moment at that point. The difference between these two values corresponds to a tension in the bottom of the beam of 800 psi, still too much stress! These diagrams ignore, however, the axial compression imposed by the cable. Taking this into account lowers the residual tension at center span to about 190 psi (figure 7.31). The simplifications we employed to arrive at this value may have led to an overestimate of the amount of tension actually borne by the structure, but 190 psi of tension falls within the accepted limits for the type of concrete used in the Southeast Connector. Caltrans engineers must have anticipated excess tension under some circumstances, however, because they embedded steel reinforcing bars near the bottom of the beam at midspan.

Design Options

Several puzzling features of the Southeast Connector emerge from this brief look at its inner workings. Why did the engineers choose not to have the prestressing cable exactly offset the effects of the dead and live loads? Why was the beam not made thinner? Why does the beam's profile remain constant along its length despite a

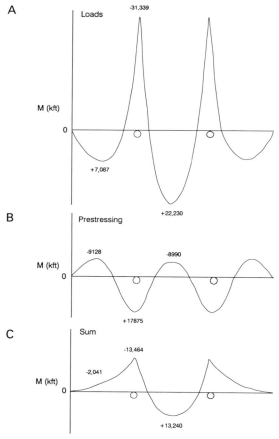

Figure 7.30 By carefully positioning the prestressing cables in the 466-foot beam of the Southeast Connector, Caltrans engineers were able to create a bending moment (*B*) that partially offset the bending moment created by the dead and live loads (*A*). The sum of these two bending moment diagrams (*C*) reveals some residual bending moment in the beam.

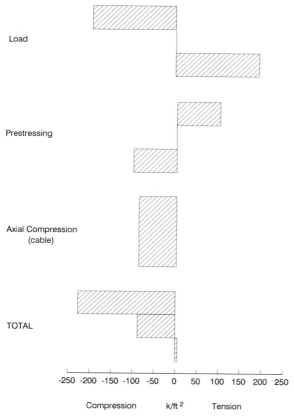

Figure 7.31 Stress diagrams for the middle of the 192-foot span in the Southeast Connector show that the prestressing cable creates stresses in the beam, through its bending moment as well as axial compression, that partially offset the stresses created in the beam by the loads. The result is a beam with compression in its upper region and such a small amount of tension in its lower region that there appears to have been an efficient use of materials in the Southeast Connector.

bending moment that varies? Doesn't this design waste material by making the beam needlessly deep where the bending moments are small?

We can not know the actual answers to these questions without interviewing the engineers involved in making the design decisions, but we can make reasonable guesses. Having the cable-induced bending moments exactly offset the load-induced moments was not needed in large part because of the axial compression contributed by the cable. Reducing the depth of the beam would have driven up the cost of the bridge by requiring more steel. The need to keep costs within a strict state-controlled budget probably accounts, in part, for the choice of design. The total bending moment diagram for the bridge suggests that designs other than a constant-profile beam could have been used. The method by which Caltrans constructs these bridges, however, makes a constant-profile beam an attractively inexpensive option. The girders in California's freeway bridges are cast in place. Concrete is poured on site into uniform wooden troughs erected above the underlying roadway. Less regular designs would require more complicated, expensive forms for casting the concrete. Like the cantilever construction method, the forms used by Caltrans engineers are erected in such a way as to avoid obstructing traffic below the bridge while it is being built. This eliminates the expense of rerouting traffic around the work site. Cast-in-place construction allows the engineers to make the long beams they need for freeway interchanges: the state limits the size of precast beams that can be transported on its roads to 125 feet. The economic advantages of a constant profile beam may have been the determining factor in its favor, but such beams also reflect one idea of visual appeal. Seen from a distance, the simple, uninterrupted arcs that make up a modern interchange in California create a degree of elegance not often expected from a freeway.

A Window on Technology

Unlike the other topics covered in this book, the relationship between science and the development of prestressed concrete bridges is a fairly distant one in that the new material did not arise immediately from a scientific discovery. Michael Faraday's scientific research on the relationship between electricity and magnetism paid clear dividends in terms of electric generators and transformers. The telegraph sprang quickly from the scientific work of Oersted and Henry. The Wright brothers' scientific approach to an engineering problem contributed to their success. It is more difficult to trace the link between science and prestressed concrete bridges. Concrete and steel are ancient materials that have been improved by modern science, but the idea of combining them to form prestressed concrete grew out of an engineer's search for a better building material, not from scientific research. Similarly, the roots of bridge design lie in long-known laws of physics and in practical experience with bridges that failed. Modern science and methods of analysis have given powerful tools to the bridge designer, but pinpointing a specific link between major principles of science and changing concepts of bridge design is difficult. Instead, an understanding of general scientific principles opens a window through which we might see bridges and other large structures in a new light. Just as knowing about the inner workings of the human body or the earth amplifies our appreciation of nature, so does an understanding of how a bridge works increase our ability to appreciate the technological environment.

Exercises

1. Imagine that the center span of the Southeast Connector (192 feet long) was built from prestressed concrete as a simple beam instead of as a continuous beam.
   (a) Draw the bending moment diagram that would result from the simple beam design.
   (b) Calculate the maximum bending moment that would occur in this 192 foot long simple beam given that the dead load of the Southeast Connector amounts to 9.7 k/ft. Compare your answer to the maximum bending moment that resulted from the continuous beam design.
   (c) Determine the amount of tension in psi that would result from the maximum bending moment calculated in (b) given that the distance from the center of the upper portion of the beam to the center of the lower portion is 6.5 feet and the area of the bottom section of the beam is 17.7 square feet. Compare your value to the allowable limit of 180 psi (10 percent of the allowable compression stress).
   (d) Draw the path that the prestressing cable should take through this simple beam to counteract the stresses created by the dead load. Assuming that the simple beam is constructed with the same depth as the

actual beam, 7 feet, how much tension would have to be generated in the cable to reduce the tension stress in the concrete below the allowable level? (The total cross-sectional area of the beam is 59 square feet. Assume that the maximum eccentricity is half the depth of the beam.)

(e) Caltrans engineers used steel in the Southeast Connector that could safely withstand 216 k/in^2 of tension. The total cross-sectional area of the steel used in the actual bridge was 32 square inches. How much would this cross-sectional area have to be increased to accept the amount of tension calculated in (d)?

2. Consider the distribution of stresses in the Southeast Connector as suggested by the bending moment diagram and develop two alternatives to the uniform beam design that Caltrans used for the 466-foot-long section of the bridge. Choose designs for which prestressed concrete would be an appropriate building material. Include in your designs a line showing the proper placement of the prestressing cable.

Risk

III

The interplay between scientific discovery and engineering innovation has produced new technological benefits, but it has also produced new risks. Risk is the probability of hazard. An understanding of basic rules of probability thus contributes to an understanding of risk. These two topics are presented together in the context of vaccines (chapter 8) and revisited in the more complex issues associated with the biological hazards of ionizing radiation (chapter 10). The concept of risk applies not only to an individual's concern for health but also to our society's concern for global effects that may attend widespread use of many modern technologies, as described in chapter 9. The uncertainty inherent in risk underlies the ambivalence with which many people view life in a modern technological culture. Opinions regarding the wisdom of a particular technology often vary according to reasonable differences in perceptions of the technology's benefits and risks. As a result, political, economic, and social factors become important elements of decisions regarding technological issues. Understanding these complex issues can be increased by appreciating the origins of modern technologies and the risks that now attend their use.

Vaccines: Good Intentions Are Not Enough

8

Controlling Nature

The citizens of Boston faced a grim prospect 270 years ago. In April of 1721, a sailor disembarked from an English ship anchored in Boston harbor and came ashore with a case of smallpox. Terrified of the dread disease, 1,000 of Boston's 12,000 residents evacuated the city. Virtually every susceptible person that remained, half of Boston's populace, fell victim to smallpox within the next 12 months. Almost 900 people died. Many of those fortunate enough to survive the scourge were permanently scarred or blinded by it.

Similar disasters struck cities in England and elsewhere during the same year as they had with disturbing frequency for more than a century. Smallpox surpassed the bubonic plague as the most feared disease of the seventeenth and eighteenth centuries. It has been claimed that smallpox took three million lives in a single epidemic in Mexico in 1520.

You will never have to face smallpox. The last case of the disease appeared in 1977 in Somalia. Smallpox is extinct, but the pathogen did not leave us of its own free will. It succumbed in a battle waged with vaccines. This landmark victory over a deadly infectious disease stands as a great achievement in our effort to control nature for medical benefit.

We strive to control nature when we fight disease with vaccines just as when we build hydroelectric dams, design airplanes, lay telegraph wires, or employ steam turbines. This is the engineering

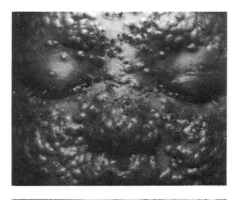
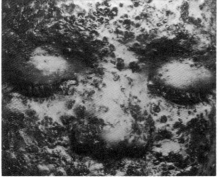

Figure 8.1 Smallpox produces pustules in the skin (top) that eventually erupt (bottom) leaving behind numerous scars. (Photograph reproduced, by permission, from Fenner, F. et al., *Smallpox and its eradication*. Geneva, World Health Organization, 1988, Plate 10.9, p. 440.)

impulse that has made our species so successful. The dream in preventing disease is creation of the perfect device: a vaccine that protects everyone completely without causing any adverse effects. This dream has driven vaccine development ever since Edward Jenner first used the new tool to control smallpox. After Jenner, progress toward that goal stalled for the better part of a century while our understanding of disease and immunity caught up with our desire to prevent disease. Empowered by the new scientific discoveries, the quest resumed, and vaccine production accelerated tremendously.

Among the early targets of the new, scientifically based vaccines was whooping cough, a potentially deadly childhood disease. The struggle to create and then improve whooping cough vaccine illustrates how scientific advances may guide technological developments. It also illustrates how we must make difficult decisions in using imperfect vaccines while we wait and hope for the perfect device in our fight against disease.

The Scourge of Smallpox

Victims of smallpox first experienced fever and vomiting followed within two weeks by an outbreak of small bumps on the skin. The bumps gradually grew in size and then erupted, spewing out a malodorous fluid filled with infectious agents (figure 8.1). If the unfortunate victim survived the attack, the bumps receded in a few weeks but left permanent, unsightly scars. A few survivors suffered blindness. Smallpox was highly infectious. Virtually all people in some locales could expect to contract it some time in their lives. Mortality from the disease reached 80 percent in some areas but averaged 25 percent. Smallpox epidemics swept through communities with distressing regularity, causing one tenth of all deaths in eighteenth century England. One in every seven children died of smallpox in seventeenth and eighteenth century Russia. Until 1721, no method of preventing this disease existed in standard medical practice in Europe or the American colonies.

Inoculation: A Daring Attempt at Prevention

Cotton Mather, pastor of Boston's North Church and famous witch hunter, solemnly recorded the onset of the 1721 Boston epidemic in his diary on May 21: "The grievous calamity of smallpox has now entered the town." Several years earlier, the very learned Reverend Mather had read a paper written by a Greek-Italian physician, Emanuele Timoni, on the subject of a technique called inoculation that was practiced in Turkey and was reputed to protect people from smallpox. Mather had vowed that he would try inoculation during the next epidemic in Boston. Fortune now presented him with the opportunity.

Inoculation involved intentionally inducing smallpox in a previously uninfected person by taking a small amount of pus from the oozing pustules of a mildly infected person and rubbing it into scratches on the skin of the courageous patient. Inoculators believed that this strongly counter-intuitive method would generate a mild case of the disease that somehow protected people from more severe cases. People at that time were well aware that surviving a case of smallpox meant never contracting the disease again, although they had no hint of the reason why. Variations on the technique had been practiced for centuries in China, Turkey, Greece, and Russia. In what must have been a most unpleasant experience, children were made to sleep in nightshirts smeared with the foul-smelling pus.

On June 6, 1721, Mather called on Boston's physicians to consider inoculating the local citizens but was rebuffed. The untested practice of inoculation was "likely to prove of dangerous consequence" the physicians claimed. Only Dr. Zabdiel Boylston, one of Boston's more highly respected physicians, responded favorably to Mather's plea. On June 26, 1721, Boylston attempted the first inoculation in the American colonies. Although apparently willing to use himself as the first test subject, he reasoned that he may already be protected from smallpox by virtue of having survived the 1702 epidemic. He inoculated his six-year-old son instead. He also inoculated one of his slaves and that person's two-year-old son. The first clinical trial of inoculation in America thus consisted of three people. Happily, they all survived the experience.

Boylston's action drew a sharp public outcry. Other physicians said he acted too hastily. Local newspapers published personal attacks against him. The protests were so great that Cotton Mather had Boylston inoculate his son in secret to avoid the damage that public disclosure would cause to his ministry. Despite the criticism, Boylston proceeded to inoculate 244 more people, 6 of whom died.

The epidemic was declared over on May 14, 1722; no new cases had appeared in the previous two weeks. The epidemic afflicted 5,759 people, roughly half of Boston's population and virtually everyone who was susceptible, and caused 844 deaths according to Boylston's records.

Was the Gamble Worth the Risk?

Was Boylston a hero or, in the extreme, a murderer? The decision depends on the balance between the risks associated with inoculation and the risks associated with smallpox. Most people believed inoculation to be a risky procedure at the time, but no quantitative esti-

Table 8.1
Results of Inoculation in Boston, 1721–1722

Persons Inoculated	280
Had the smallpox by inoculation	274
Had no effect	6
Suspected to have died of inoculation	6

mates of that risk had been made before Boylston's action. The idea of calculating risks was still new in 1721. Fortunately, Boylston was scientifically inclined enough to record all 244 of his own cases along with an additional 36 people inoculated by two other physicians during the same epidemic (table 8.1).

Bostonians faced a difficult choice between smallpox, a dreaded disease, and inoculation, a dangerous procedure. Boylston addressed this quandary directly; he hoped to apprise people "of the dangers in [naturally acquired smallpox], and the reasonable expectation they have of doing well in [inoculation]." Using his own figures, Boylston estimated the risk of dying from inoculation to be:

$$\frac{6}{274} \approx \frac{1}{46} \quad \text{or about 2\%}$$

In making this calculation, Boylston applied the definition of probability formally described more than half a century before by one of the first great theorists in probability, Jacques Bernoulli (see box 8.1): the *probability of an event* is the number of occurrences of that event divided by the total number of equally probable outcomes. The answer always comes out between zero and one: zero for events that never occur, and one for events that are certain to occur. The event of interest to Boylston was death following inoculation. He assumed, perhaps unwittingly, that the 274 patients had an equal probability of dying from inoculation, so the probability becomes 6 divided by 274. The key issue is whether all 274 patients had an equal probability of dying. If they did not, then the probability can not be estimated in this way. Boylston simply anticipated the common modern practice of assuming equal probability when it can not be proved otherwise.

Despite the high risk associated with inoculation, it compared favorably with the risk of dying from smallpox, which Boylston estimated to be:

> ## Origins of Probability
>
> Inoculation against smallpox provided one of the earliest opportunities to apply new concepts in probability to medical practice. Probability theory, emancipated in the 1660s from its roots in gambling, reached its first milestone in 1713 with the publication of the *Ars Conjectandi* by Jacques Bernoulli (1654–1705), a Swiss mathematics teacher. He laid the theoretical foundation for calculating probabilities in decision problems. Bernoulli argued that if the actual number of white balls in an urn filled with a mixture of white and black balls is unknown, he could estimate the proportion by reaching into the urn, taking samples of balls, and comparing the number of white balls that turned up to the total number of balls drawn from the urn. If he sampled enough times, the observed ratio would be very close to the actual proportion. His more general statement of this argument is known variously as the first limit theorem or the law of large numbers.
>
> Bernoulli went on to prove the "addition law": the probability that either of two mutually exclusive events will occur is simply the sum of their individual probabilities. If the chance of throwing a six on one toss of a six-sided die is 1/6 and the chance of throwing a one is the same, then the chance of throwing either a six or a one with a single toss is 1/6 + 1/6 or 1/3.
>
> Another useful rule in probability is the "multiplication law" described in 1718 by Abraham de Moivre, a contemporary of Isaac Newton. The multiplication law states that the probabilities associated with individual events are multiplied together to calculate the probability they will happen together. The chance of rolling a one and a six in a two-roll series is thus 1/6 × 1/6 = 1/36.

Box 8.1

$$\frac{844}{5{,}759} \approx \frac{1}{7} \quad \text{or about } 15\%$$

If the 274 people who reacted to the procedure had not been inoculated, 15 percent or 41 of them would have been expected to die from smallpox according to Boylston's accounting. Instead, only 6 died. It appears that Boylston saved 35 people and should be praised as a hero. He considered his trial to be a great success, and many other people, including Benjamin Franklin, agreed with him. Boylston was elected to the Royal Society of London in 1726 "for his intelligent use of inoculation for smallpox in 1721–22," only the fifteenth American to be honored in this way since the Society's founding in 1661.

Praise for Boylston and other inoculators was not universal, however. Inoculation carried other risks in addition to the risk of smallpox. An inoculated person could spread smallpox to other, susceptible people and exacerbate the epidemic that the procedure was intended to control. Boylston and a few other inoculators recorded instances in which people developed smallpox after handling an inoculation patient, but no one considered the important implications of these cases, possibly because it was not known at the time that "germs" cause disease. Another source of risk arose from the transfer of pus from one person to another. This method transferred not only smallpox but any other disease the donor was afflicted with at the time. No one considered this a serious risk in Boylston's time because physicians believed that the body could not be occupied by two diseases at once.

Boylston deserves our appreciation for realizing that arguments over inoculation should be decided by weighing the procedure's risks against the risks of the disease. All modern practice follows his example. But he failed to produce a completely convincing study of inoculation's value. Not only was his sample size small, he neglected to recruit a control group of people to whom the inoculated people could be compared, he did not determine that the deaths attributed to inoculation were anything more than chance events, and he failed to ask whether every inoculated person was really protected from smallpox. Without this information, no one in Boylston's time could be sure of the real benefits and risks of inoculation. Faulting Boylston for this, however, is a little like criticizing Christopher Columbus for not using a map to determine where he was in the New World. The scientific practice of medicine would not begin for more than 150 years, and the mathematical basis for calculating risk was barely 60 years old.

The Demise and Legacy of Inoculation

Early success with inoculation in the American colonies and simultaneously in England encouraged use of the procedure. Greater use, however, was inevitably accompanied by more inoculation-related deaths. Before long, opinion regarding inoculation was sharply divided. The College of Physicians in London pronounced it "highly salutary to the human race." But other physicians and clergymen pointed to inoculation-related deaths and denounced the practice. The Reverend Edmund Massey of London followed another, more novel line of reasoning in opposing inoculation. In his sermon ". . . Against the Dangerous and Sinful Practice of Inoculation," he self-servingly argued that "The fear of disease is a happy restraint to men. If men were more healthy, 'tis a great chance they would be less righteous. Let the Atheist and the scoffer inoculate" (Creighton 1965).

As the debate was waged, the practice of inoculation went in and out of favor but generally followed a course of increasing use. Concerns about the risks of inoculation fostered attempts to make the practice safer and more effective. Inoculation patients suffered repeated bleedings, purgings, prolonged reduction in food intake, and extended periods of isolation under their physician's attempts to improve the procedure. The good intentions of the physicians were not enough to save the doomed procedure. Without guidance from a scientific understanding of the disease and the body's immune system, physician's were unable to reduce inoculation's risks while leaving intact its benefit. Dissatisfaction with inoculation grew as its dangers persisted. Cities and nations in Europe began to outlaw inoculation in the 1760s even before an alternative method of preventing smallpox became available.

Regardless of its direct effect on smallpox, inoculation left an important legacy. It suggested that people could intervene in a dreaded infectious disease and influence the outcome. The promise of control over a ravaging aspect of nature was unprecedented. With this promise in hand, the high risks associated with inoculation prompted people to search for better methods, methods for fighting smallpox that produced greater benefit at lower risk. One of these people was Edward Jenner.

Jenner and the First Vaccine

Edward Jenner (figure 8.2) knew firsthand the rigors of inoculation. As a boy, in 1756, he endured the exhausting series of bleedings, purgings, and fastings that accompanied his inoculation. Not surprisingly, when he became a surgeon, he kept his eyes open for less

Figure 8.2 Edward Jenner (engraving by W. H. Mate after a painting by Thomas Lawrence). (Photograph courtesy of the Boston Athenaeum.)

brutal ways of preventing smallpox. Stories of how Jenner came upon the idea for the first vaccine differ, but he was undoubtedly not the first to discover a relationship between smallpox and cowpox, a related but milder disease that originated in cattle. Dairy farmers in eighteenth century England knew that people who had been sick with cowpox did not subsequently come down with smallpox. Cowpox passed from sick cows to dairy farmers and milkmaids when pus, oozing from sores on the cow's teats, contaminated cuts on the person's hands. In humans, cowpox did not often lead to serious consequences such as blindness or extensive scarring. Particularly daring or desperate dairy farmers exposed members of their families to cowpox hoping to ward off the deadly smallpox.

Jenner became famous and rich simply by paying attention to the dairy farmers' beliefs and testing them experimentally. He first became aware of the lore surrounding cowpox when he took his medical practice into the southern dairy country of England. Despite his boyhood experience, Jenner included inoculation in his surgical practice. Inoculation normally produced a mild case of smallpox, but a number of people inoculated by Jenner failed to

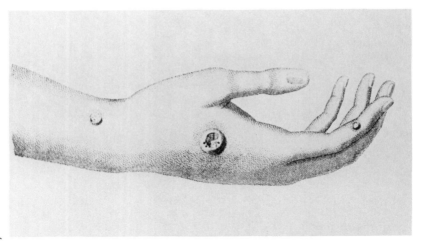

Figure 8.3 Drawing of Sarah Nelmes' hand showing an erupted cowpox pustule. Jenner drew the material for his first vaccination from Sarah Nelmes. (Reproduced, by permission, from: Jenner, 1798, as published in *Scientific Papers—Physiology, Medicine, Surgery, Geology*, vol. 38 of *The Harvard Classics*, C. W. Eliot (ed.), Grolier, Inc.)

show a substantial reaction. At least nine of these cases, recorded between 1791 and 1796, involved dairy farmers or milkmaids who reported having had cowpox. Several of these people had managed to avoid smallpox during one or more local epidemics. Bolstered by no stronger correlation than this, Jenner leapt ahead.

On May 14, 1796, Jenner treated a milkmaid named Sarah Nelmes for cowpox (figure 8.3). He set aside some of the pus from her sores and, later the same day, applied it with a needle to small scratches on the arm of eight-year-old James Phipps. Phipps developed the aches and fever characteristic of cowpox but soon recovered. On July 1, Jenner dramatically and apparently recklessly tested his hypothesis by intentionally infecting Phipps with smallpox. Fortunately for Phipps and Jenner, no disease developed. Several months later, Jenner again exposed Phipps to smallpox. Still no disease developed.

Jenner repeated the experiment, but this time he used the material from a child's cowpox pustules in inoculating one person who then developed cowpox and became a donor of pus for inoculating a second person and so on to effect a serial transfer of the cowpox. After taking one more step in the series and noting that all of the subjects had come down with cowpox, he and his nephew, Henry

Jenner, exposed two of them to smallpox through inoculation. Again, neither subject developed smallpox.

Perhaps with a keener eye for the scientific method than his uncle, Henry Jenner took the initiative in performing the obvious, but risky, control experiment. "To convince myself that the [smallpox material used in the test] was in a perfect state I at the same time inoculated a patient with some of it who never had gone through the cowpox, and it produced the smallpox in the usual manner" (Jenner 1798). The elder Jenner reported "much satisfaction" with these experiments because they showed that cowpox could be transferred from one person to another without losing any capacity to protect against smallpox. This was an important discovery for Jenner because cowpox appeared only infrequently in cows, making them an unreliable source of material for the new procedure. Jenner's discovery showed that people could serve, in effect, as culture chambers for cowpox. His use of the Latin term *variolae vaccinae* for cowpox (*variola*—smallpox, pockmarks; *vaccinus*—cow) led to the new procedure being called vaccination or, in some circles, variolation. (In modern usage, vaccination has come to encompass any procedure in which the body is exposed to a killed or debilitated pathogen, or an organism similar to an actual human pathogen, for the purpose of inducing a protective immune response. The term inoculation, which referred to exposure of a person to the actual, unattenuated pathogen, has largely disappeared from medical terminology.)

Having directly observed only three cases in which cowpox seemed to protect a person against smallpox and studied perhaps another fourteen cases that were entirely anecdotal, Edward Jenner stated hopefully in a pamphlet published in 1798 that ". . . I presume it may be unnecessary to produce further testimony in support of my assertion 'that the cow-pox protects the human constitution from the infection of the smallpox . . .'" Jenner appreciated that arguments should be supported by evidence, but he clearly did not see the need for a large sample size as would be advised in modern scientific practice.

Quantitative assessments also did not enter into Jenner's consideration of balance between the benefits and risks of vaccination. Jenner simply claimed that vaccination was less risky than inoculation. He then exaggerated the benefit of vaccination as providing "perfect security." In such statements, Jenner showed less sympathy for quantitative summaries of risk and benefit than his American predecessor, Dr. Boylston. Despite Boylston's example, calculations of probability remained rare in medicine during Jenner's day. Jenner attempted to publish his findings through the Royal Society of London but was denied—not for lack of evidence or quantitative arguments but because his procedure was "too revolutionary." He then published the pamphlet on his own.

Opposition to Vaccination

Like inoculation, Jenner's new procedure failed to gain immediate acceptance. Opposition to vaccination over the next century came from a number of quarters. Not surprisingly, a group of English inoculators objected that vaccination went against the established practice of inoculation, which, of course, it did. A new medical practice always faces an uphill struggle for acceptance when it displaces an existing, useful, and accepted practice. Various clergymen decried vaccination as contemptible interference in God's work. In a related and particularly novel criticism of vaccination, Thomas Malthus, Herbert Spencer, and others turned the benefits around and argued that vaccination should not be practiced because smallpox served a useful purpose in reducing the number of poor people; preventing the disease would only increase the burden on the poor. Florence Nightingale associated smallpox with the stench and filth of crowded living conditions among poor people and considered the disease retribution for living in such conditions. Her reasoning simply followed everyone's ignorance of the cause of smallpox.

Potentially the most devastating criticisms concerned the risks associated with vaccination. Antivaccination leagues sprang up in England and the United States nurtured by reports of injury or death after vaccination. These reports were not groundless. Jenner's method of transferring the material from cowpox pustules on the arm of one person to scratches on the arm of another person allowed other diseases including hepatitis, syphilis, leprosy, and even smallpox to be transmitted as well. The risks of vaccination appeared to be substantially greater than suggested by Jenner.

Despite these fears and objections, the practice of vaccination spread widely. By 1801, more than 100,000 people had been vaccinated in England alone. A large number of these people were British sailors—the British Admiralty had ordered every person in the fleet vaccinated. The Admiralty awarded a medal to Jenner in 1801. Emperor Napoleon I, not to be outdone by his archenemy, had his army vaccinated and awarded Jenner a medal in 1805. The lead taken by the military in recognizing the value of the new medical procedure was to be repeated later in the cases of sulfa drugs and penicillin during World War II. Between 1802 and 1807 the King of England rewarded Jenner with the then extravagant sum of £30,000 in recognition of his discovery. It seems hard to believe that these honors would have been awarded to Jenner had a great many people not come to share his belief that "the annihilation of smallpox will be the final result of this practice."

Vaccination: A Clinical Trial

One person quick to embrace vaccination was Benjamin Waterhouse, a British-educated physician who was a professor of "physic" at Harvard. An English colleague sent Waterhouse a copy of Jenner's pamphlet shortly after it was published in 1798. Waterhouse tried to advertise the new procedure by writing a newspaper article under the heading "Something Curious in the Medical Line." This less than inspiring title failed to arouse much enthusiasm for vaccination.

Unlike every other practitioner of vaccination at the time, however, Waterhouse decided to conduct a clinical trial. In 1801, he convinced the Board of Health in Boston to test Jenner's procedure on twenty-one boys, none of whom had had smallpox or been inoculated. In August, nineteen of them were inoculated with cowpox. Four months later, twelve boys from the "cowpox group" were inoculated with smallpox. Two other boys not previously exposed to cowpox were also inoculated with smallpox at this time. The trial thus created three experimental groups: (i) a group of seven boys who received only cowpox material; (ii) a group of twelve boys who received cowpox and smallpox material; and (iii) a pair of boys who received only smallpox material. Only the latter two developed cases of smallpox.

His experimental design allowed Waterhouse to learn that cowpox vaccine did not cause smallpox, that the vaccine indeed protected a person from smallpox, and that the material used to induce smallpox remained active during the experiment. His data pointed to the conclusion that the vaccine carried perfect benefit at no risk! In no case did the vaccine cause smallpox or any other serious adverse reaction, and in no case did it fail to protect against smallpox. Of course, we should not have complete confidence in this conclusion because the sample size is so small. Waterhouse did improve on the trials of inoculation, however, by including a control group.

Unimpeded by an understanding of probability theory, Waterhouse sent the results of the clinical trial to John Adams, then President of the United States, who responded with vague praise. Vice President Thomas Jefferson took a more active interest in the new procedure. With samples of cowpox material supplied by Waterhouse, Jefferson had a surgeon vaccinate his family, neighbors, and household staff, approximately seventy people in all. He must have been happy with the results because he wrote to Jenner that "medicine has never produced any single improvement of such utility ..." He continued, in an especially prescient phrase, that "... future generations will know by history only that the loathsome smallpox has existed, and by you had been extirpated" (WHO 1980).

Risks of Vaccination: A Quantitative Assessment

Thomas Jefferson and many others decided to use the new vaccine on evidence that was little better than anecdotal. Action had to be taken in the face of great uncertainty because the threat of smallpox was so great and no other evidence existed at the time. Having benefitted from the use of smallpox vaccine, we can now easily conclude that they made the correct decision. But inoculation also looked attractive on the evidence of the time, and people could only have had a qualitative appreciation that vaccination was safer. Just how much safer was it?

A clearer picture of the risks associated with vaccination emerged as the practice became widespread. The first complete accounting of the vaccine's risks came in 1968, after millions of people had been vaccinated, the viral nature of smallpox had been discovered, and scientific medical practice had matured to permit more careful discrimination of vaccine-related injuries and spurious effects. The survey revealed that improvements in the production, treatment, and testing of smallpox vaccine had reduced many of the risks associated with its early use but had failed to eliminate all adverse effects.

The 1968 study summarized data from more than 14 million vaccine recipients in the United States. The most serious complications involved transfer of the virus from the site of vaccination to other parts of the body. If a person scratched the vaccinated site and then rubbed an eye, for example, they may have developed a local infection that, on rare occasions, led to blindness. This is an example of *eczema vaccinatum*. *Vaccinia necrosum* occurred when the virus moved in the blood to invade and damage other organs such as the lungs, stomach, or esophagus. Vaccine-related damage to the brain is termed *postvaccinial encephalitis*. Each of these three complications could be fatal.

The risk to first-time recipients of the vaccine is determined by dividing the total number of applicable cases into the total number of complications. In the 1968 survey, 418 out of 5,594,000 first-time vaccine recipients exhibited vaccine-related complications. The risk of any complication from smallpox vaccine is thus:

$$\frac{418}{5,594,000} = \frac{1}{13,383} \quad \text{or about } 0.008\%$$

A risk of 1 in 13,383 appears to be a very slim one, but many people strongly objected when the Reagan administration proposed increasing the acceptable risk of cancer from 1/1,000,000 to 1/100,000 for some chemicals.

Two of the complications proved fatal in a few cases. Four of the 418 cases involved fatal encephalitis making the risk of dying from vaccine-induced encephalitis:

$$\frac{4}{5,594,000} = \frac{1}{1,398,500} \quad \text{or about } 0.00007\%$$

Two of the first-time vaccine recipients died from *vaccinia necrosum*. The risk of this outcome is:

$$\frac{2}{5,594,000} = \frac{1}{2,797,000} \quad \text{or about } 0.00004\%$$

The risk of first-time recipients dying from *either* of these causes is the sum of the probabilities associated with each of the two possible events, assuming that the two causes can never work together in producing a death—that is, that they are mutually exclusive:

0.00007% + 0.00004% = 0.00011%

These calculations show that Jenner was incorrect in believing vaccination to be perfectly safe. Compared to inoculation, however, vaccination represented a much safer way to prevent smallpox. The modern practice of vaccination proved 20,000 times less risky than inoculation (0.0001 percent versus 2 percent), although such a comparison should be made with caution given the difference in the size and reliability of the two data sets involved.

Benefits of Smallpox Vaccine

Whether the risks of vaccination are acceptable to you undoubtedly depends on how much benefit you might expect to gain. The benefit of smallpox vaccine has changed dramatically over the last 190 years. Smallpox was so virulent that, under crowded conditions, virtually everyone in a town could expect to be afflicted with the disease sometime in their lives. Records from a smallpox epidemic that swept through ten parishes in England during 1774 indicate that 93 percent of the residents (13,653 out of 14,713) had had smallpox at some time. Under some conditions, then, a person's chances of dying of smallpox approached the average mortality rate of 25 percent during the eighteenth century. The risks of vaccination look small indeed compared to the risk of smallpox.

Just as Jenner underestimated vaccination's risks, however, he also overstated its benefit. Vaccination does not provide perfect protection against smallpox. The World Health Organization surveyed the vaccine's impact on smallpox in India and found that a number of vaccinated people had contracted smallpox and that 6 percent of these people died from the disease. Although the vaccine had failed to protect these people, it remained advantageous because the mortality rate for smallpox among vaccinees was 25 percent, or one fourth, of the mortality rate among people who had not been vaccinated.

The Eradication of Smallpox

Well before careful studies quantitatively evaluated the risks and benefits of smallpox vaccine, the practice of vaccination became widespread. Vaccination gradually became compulsory across Europe after 1807. Soon, the average death rate from smallpox was approximately three times less in countries with compulsory vaccination than in countries without it. In 1958, the World Health Assembly of the United Nations called for the eradication of smallpox from the world. The promise of smallpox vaccine as envisioned by Edward Jenner and Thomas Jefferson was to be realized finally. The last case of smallpox appeared in Somalia in 1977 (figure 8.4), 179 years after Jenner introduced the first smallpox vaccine. After two years

Figure 8.4 Electron micrograph of smallpox virus particles isolated from the person with the last recorded case of smallpox. The particles appear approximately 100,000 times actual size. (Photograph reproduced, by permission, from: *The global eradication of smallpox. Final report of the Global Commission for the Certification of Smallpox Eradication*, Geneva, December, 1979. Geneva, World Health Organization, 1980, Fig. 2, p. 21.)

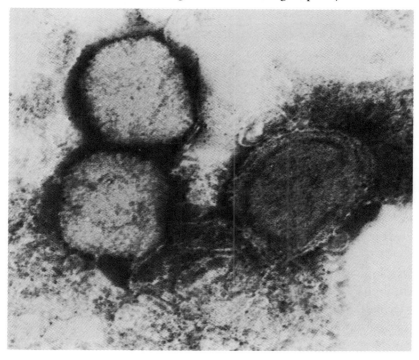

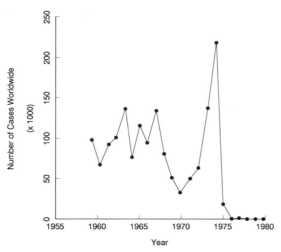

Figure 8.5 Results of the global smallpox eradication program begun in 1958 and declared a success in 1979. (Data from: *The global eradication of smallpox. Final report of the Global Commission for the Certification of Smallpox Eradication*, Geneva, December, 1979. Geneva, World Health Organization, 1980.)

of surveillance, the Global Commission for the Certification of Smallpox Eradication concluded that eradication had been achieved (figure 8.5). The eradication program was amazingly inexpensive, costing only $313 million. The smallpox virus now exists in only two laboratories, one in the United States and one in the Soviet Union. The commission recommended saving a few frozen stores of the virus and a large stock of the vaccine for use in research and in case the disease returns unexpectedly.

Smallpox Vaccination Discontinued

The last case of smallpox in the United States occurred in Texas in 1949. The vaccination program continued in the United States, however, because the disease still existed elsewhere in the world and could be imported into the United States. (The 1949 Texas case was, in fact, imported.) As the incidence of smallpox decreased, the willingness of people in the United States to assume the risks of vaccination also decreased. People grew less willing to support a vaccination program that produced a few deaths and cost approximately $150 million each year in the United States. In the period from 1950 to 1968, the vaccine probably led to as many deaths in the United States as were killed by smallpox in Europe (111). As a result, the government began to reconsider its smallpox vaccination policies in light of the palpable risk of vaccine-related injury.

This must have been a difficult decision. If vaccination continued well beyond the time when it could be expected to protect against smallpox, then some people would needlessly suffer vaccine-related injuries. Discontinuing the vaccination program prematurely, however, also carried risks in terms of smallpox cases. In 1971, the United States government suspended routine smallpox vaccination for school-age children although smallpox remained endemic to more than twelve countries at the time. The risks and costs associated with vaccination became unacceptable in the face of diminished benefits.

Jenner's Luck

Despite its imperfections, smallpox vaccine remains the major symbol of our ability to control, even eradicate, deadly infectious disease. It sprang from experiments following an observation, in good scientific fashion, as well as from a desire to improve upon previous technology, in good engineering fashion. Stimulated by this great example, ninteenth century physicians must have hungered after more vaccines to fight other dread diseases. But their good intentions were not enough. Jenner's fortune in discovering smallpox vaccine was not to be shared by others. The reason lies in the manner of Jenner's discovery.

Edward Jenner lived in complete ignorance of what caused smallpox. So did Zabdiel Boylston, Benjamin Waterhouse, and every other medical authority of the day. The ignorance was not restricted to smallpox but to all infectious diseases that were commonly and vaguely attributed to poisons produced in filthy, crowded living conditions. After all, epidemics struck with particular severity in impoverished communities where conditions were harsh. Jenner and others also knew nothing about the body's own defense mechanisms against disease. So when he reasoned that a case of cowpox would protect a person against smallpox because the two diseases resembled each other, he was simply guessing. It became clear how lucky Jenner's guess actually was after the cause of smallpox was discovered in the first third of the twentieth century.

Smallpox is a viral disease caused by *Poxvirus variolae*. Interestingly, Jenner used the term *virus* to describe smallpox, but he knew nothing about the organisms we recognize today as viruses. He used the word simply to indicate its Latin meaning of poison. Viruses consist of a protein coat surrounding a small amount of genetic information in the form of DNA or RNA. Genetically related viruses incorporate some of the same proteins in their coats. This is important because the coat proteins are the molecules most likely to trigger a protective immune response by the body. When certain white blood cells are exposed to foreign molecules (called antigens), such as the coat proteins of a smallpox virus, they respond by secreting proteins (called antibodies) that bind to the specific antigen in the blood and prepare it to be expelled from the body. Antibodies recognize their target by its specific three-dimensional shape. Any molecule assuming that shape will be attacked by the corresponding antibody.

Smallpox virus belongs to a group of viruses noted for their host-specificity as suggested by their names: cowpox, camelpox, raccoonpox, monkeypox, and buffalopox. It might seem that Jenner's vaccine worked simply because the cowpox virus is related to the smallpox virus and antibodies generated against one also attack the other. Jenner's luck probably produced a better solution. Jenner developed his vaccine from the pus that oozed out of cowpox sores. Naturally, he thought he was inducing cowpox when he vaccinated his patients, and he may well have been. But he may also have been infecting people with vaccinia virus. All modern smallpox vaccines were made from vaccinia virus, the origin of which remains a mystery. It may have been a fortunate contaminant of Jenner's original vaccines.

Vaccinia virus, also related to smallpox virus, stands out from the group because it is much less host-specific. It grows perfectly well in a wide variety of animals including cows, humans, and chick embryos. Jenner's likely good fortune, and ours as well, was threefold. First, the vaccinia virus somehow causes an immune reaction that is more effective against the smallpox virus than against itself, meaning that exposure to a small amount of vaccinia virus produces a very large protective response against smallpox. Second, the vaccinia virus usually multiplies only at the site of vaccination and induces an immune response without spreading to other areas of the body where it might cause a serious infection. Third, its lack of host-specificity means that it can be grown in other animals, such as chicken embryos, a necessary condition for producing the huge quantities of vaccine needed for worldwide eradication of smallpox. The match between the biology of vaccinia virus and the requirements for a vaccine is uncanny. Jenner, of course, was completely unaware of how good his vaccine really was, but luck has figured in more than one important discovery.

Strategies for Vaccines

Smallpox vaccination inaugurated the strategy, still in use today, of inducing an immune response against a pathogenic organism by exposing the body to a similar but less harmful organism. Applying that strategy to other diseases, however, required finding other organisms, like vaccinia virus, with just the right characteristics for a vaccine. Jenner had stumbled from incorrect reasoning to his discovery. Others were not likely to be so lucky.

The older practice of inoculation signaled another approach to vaccination. The actual pathogenic organism itself may be used to induce an immune response. Numerous vaccines, such as polio vaccine, now apply this strategy. The problem, of course, remained to somehow eliminate the disease-causing capacity of the pathogen while leaving intact its ability to trigger an immune response. The successful implementation of these two strategies evaded all practitioners until nearly the end of the nineteenth century.

Disease Loses Its Mystery

An astonishingly long time, eighty-four years, elapsed between Jenner's discovery of a smallpox vaccine and development of the next vaccine. Despite Jenner's example, vaccination proved to be a technology totally dependent on scientific understanding for its progress. In the absence of understanding about the nature of disease and immunity, all further attempts at vaccine production failed. This dependence of a medical technology on science contrasts sharply with the development of the steam engine, briefly described in chapter 5, but resembles somewhat the development of the electric generator (chapter 3) and the airplane (chapter 4).

The scientific approach to vaccine production began in 1876 when Robert Koch, a meticulous German bacteriologist and eventual Nobel laureate, proved for the first time that a disease, anthrax in sheep, was caused by a bacterium. The three-step sequence followed by Koch to obtain the proof that had eluded his predecessors is noteworthy because it has been followed ever since to link infectious diseases with their specific causes. Koch noted that the blood of sheep stricken with anthrax always contained many bacteria of a particular type, *Bacillus anthracis*. He succeeded in isolating this bacterium from diseased sheep and growing it in pure culture. Finally, healthy sheep developed anthrax after being injected with samples of the bacteria taken from pure cultures.

Other scientists had noted bacteria in diseased animals before Koch conducted his famous tests, but prevailing concepts of disease prevented them from believing that the bacteria actually caused the malady. For much of the nineteenth century, disease was considered a manifestation of some vague internal life force that had somehow become distorted. The major proponent of this view, Rudolph Virchow, argued that "Diseases have no independent or isolated existence; they are not autonomous organisms, not beings invading a body, nor parasites growing on it; they are only manifestations of life processes under altered conditions" (Taylor 1979). Virchow's view of disease largely replaced the earlier concept that diseases resulted when amorphous poisons or miasmas invaded the body, the belief ascribed to Florence Nightingale earlier in this chapter.

While people believed either that disease had no external cause or that amorphous substances mysteriously led to disease, progress on vaccines was impossible. The search for vaccines was bound to be futile until vaccines were seen as weapons directed against specific, identifiable, disease-causing agents. Koch helped to prove that such was the case, at least for anthrax. A large part of the credit for showing the broad applicability of the new concept, however, deservedly goes to Koch's French colleague and archrival, Louis Pasteur (figure 8.6). Pasteur's work on fermentation in the mid-nineteenth century established that microbes invisible to the naked eye caused the chemical transformations observed in fermentation and putrefac-

Figure 8.6 Louis Pasteur. (Photograph courtesy of the Library of Congress.)

tion. Having laid this important foundation, Pasteur quickly followed Koch's experiments on anthrax with some of his own that showed conclusively that it was the bacterium that caused the disease symptoms.

Pasteur and Koch thus established the "germ theory of disease." They and others rapidly extended the new concept to tuberculosis, diphtheria, and typhoid fever. By 1890, the list of diseases known to be caused by bacteria included pneumonia, plague, and gonorrhea. These findings helped push aside Virchow's less useful concept and established the foundation for the modern battle against this source of human suffering. (The one enclave where the germ theory seems not to have penetrated is in the remarkably tenacious belief that people "catch cold" by exposing themselves to drafts or cold weather.)

Application of the New Knowledge: Vaccines Return

Unlike Koch, who focused on developing techniques for culturing bacteria, Pasteur dedicated the rest of his career to understanding the relationship between bacteria and disease and how that knowledge might be applied to prevent disease. His next great discovery followed from that engineering impulse. In 1880, Pasteur had been growing cultures of the bacterium that caused fowl cholera, a fatal disease that strikes chickens, when he noticed that the older cultures had lost their capacity to induce disease. He injected some of the weakened bacteria into chickens hoping to recover bacteria with renewed virulence. With historic insight, Pasteur then injected these same chickens with strains of fowl cholera bacteria known to be virulent. None of the chickens developed the disease. Prior exposure to attenuated bacteria somehow protected the chickens from subsequent exposure to virulent bacteria. The first vaccine since smallpox vaccine had been discovered.

The source of Pasteur's insight remains a mystery, but two of his coworkers reported his keen interest in Jenner's writings. Years of considering Jenner's success led Pasteur to believe that diseases other than smallpox could be prevented by vaccination. Pasteur quickly followed his work on fowl cholera vaccine with a bold prediction that he could prevent anthrax in sheep with a vaccine made from attenuated anthrax bacteria. When the Agricultural Society of Melun requested a public demonstration of the new anthrax vaccine, Pasteur supervised the vaccination of twenty-four sheep, one goat, and six cows. Nearly a month later, he exposed the vaccinated animals and a similar control group of unvaccinated animals to virulent anthrax bacteria. The dramatic results were widely acclaimed: twenty-five of the twenty-nine control animals, but none of the vaccinated animals, died of anthrax. Shortly thereafter in 1880, Pasteur officially launched the new age of vaccine development by delivering a lecture at the International Medical Congress in London in which he paid homage to Jenner by adopting the terms vaccine and vaccination.

Just as Jenner encountered resistance to vaccination, so did Pasteur. One of Pasteur's bitterest critics was Robert Koch, co-developer of the germ theory that made vaccines sensible. Koch complained that Pasteur's laboratory technique was sloppy and his vaccines were contaminated by extraneous bacteria. He also believed the new procedure to be too risky to be worth the effort. Pasteur was stung by Koch's objections but ignored them and continued his work. Vaccination returned to human diseases in 1885 when Pasteur gave his new rabies vaccine to nine-year-old Joseph Meister. More than 2,400 people visited Pasteur within the following year to be vaccinated for rabies. The first Pasteur Institute was built in 1888 to cope with the influx of patients.

Pasteur made his vaccines from living, whole bacteria that he had attenuated—that is, caused to be less virulent by letting them age in culture dishes, by heating them, or by growing them in various animals. Ten years after Pasteur developed fowl cholera vaccine, however, Emil Adolph von Behring, recipient in 1901 of the first Nobel Prize in physiology or medicine, showed that protection from diphtheria and tetanus could be conferred by injecting toxins secreted by the bacteria instead of the living bacteria themselves. Thus, by 1890, vaccines based on live, attenuated pathogens and vaccines based on toxic fractions of bacteria had been added to Jenner's smallpox vaccine that used an organism similar to the pathogen to induce an immune response. These three methods remain the major forms of vaccination.

Vaccines Stimulate a New Field: Immunology

Shortly after discovering fowl cholera vaccine, Pasteur began to consider the mechanism by which vaccines worked. In the early 1880s, he briefly considered that the attenuated bacterium exhausted the supply within the human body of some nutrient crucial to the bacterium's growth. If this was true, then vaccines would have to be made from whole, living bacteria, Pasteur reasoned. The "exhaustion" theory of immunity died quickly, however, in the face of arguments marshaled by von Behring and a Russian researcher named Elia Metchnikoff among others.

Professor von Behring's work on diphtheria and, with Shibasaburo Kitsato, on tetanus showed that blood serum drawn from animals exposed to those bacteria contained an as yet unseen "antitoxin" (i.e., antibody) that protected the animals specifically against those diseases. This discovery in 1890 lent considerable credibility to those researchers, including Koch, who believed that immunity was conferred by substances, called *humors*, dissolved in the blood.

The "humoral immunity" camp was sharply opposed by another group, led by Metchnikoff, unified in the belief that cells are responsible for protecting the body against disease. Metchnikoff, while working in a marine laboratory near Messina, Sicily, had observed small pieces of wood being engulfed by cells within larval starfish. He later observed cells from other animals taking in bacteria. Although he was not the first person to observe what he termed "phagocytosis," he generalized from the observations and, in 1884, proposed a mechanism by which the body might rid itself of disease-causing bacteria.

Humoralists and cellularists waged a fierce debate, but the tide temporarily turned in favor of the humoralists when Paul Ehrlich, a German chemist, described an extensive series of experiments and ideas concerning antibodies. The central idea of his new "theory of immunity" was that "antitoxins" circulating in the blood bound to specific toxins by virtue of a "lock and key" fit between the two molecules. Specific binding between antibodies and antigens remains a cornerstone of immunology. Ehrlich's theory gave vaccines a firm theoretical base: their role was to promote the formation of specific antibodies. Although human immunity is now known to involve both cellular processes and blood-borne "humors," vaccine research is still guided by a need to understand the relationship between a vaccine and antibody formation.

Although the impetus for vaccine development came from Jenner's vaccine against smallpox, a viral disease, the major advances toward modern vaccines were all made on bacterial diseases. Suspicions that there existed disease-causing organisms smaller than bacteria surfaced in the early 1890s, but it would take another forty years to show that the causal agents were a previously unknown life form, viruses. Virus particles were seen for the first time with the aid of the electron microscope in 1939. The production of vaccines against viral diseases proceeded much as it had against bacterial diseases.

Whooping Cough: A New Challenge

Vaccine production blossomed in the wake of developments in bacteriology and immunology. One of the more interesting vaccines produced in the early part of the twentieth century was directed at whooping cough. The vaccine is interesting in this context because it so clearly illustrates the underlying drive for improvement characteristic of all vaccine production and, indeed, characteristic of engineering in general.

Whooping cough, also known as pertussis (from Latin for intense cough), has long been recognized as a major cause of childhood mortality. The benefits of a vaccine against this disease are potentially very large. Unfortunately, the degree of risk associated with pertussis vaccines has also been large relative to other modern vaccines. Whooping cough has not been eradicated, as has smallpox, so the need for a vaccine persists. But some parents have grown reluctant to expose their children to a vaccine they believe is dangerous. Governments of several countries have at one time or another made the same choice by suspending large-scale pertussis vaccination programs. Other parents and governments judge the risks of the disease too great to abandon the only effective vaccine against it.

What Can Be So Bad about a Cough?

A description of whooping cough from the epidemic of 1578 graphically illustrates what can be so bad about a cough: "The lung is so irritated that, in its attempt by every effort to cast forth the cause of the trouble, it can neither admit breath nor easily give it forth again. The sick person seems to swell up, and as if about to strangle, holds his breath clinging in the midst of his jaws—for they are free from this annoyance of coughing sometimes for the space of four or five hours, then the paroxysm of coughing returns." Children struggling with one of these coughing fits can only regain their breath when the spasm ceases. They then desperately suck air into their lungs creating a "whoop" sound.

The anonymous observer quoted above was describing the "paroxysmal stage" of pertussis. The disease actually begins earlier with typical "cold" symptoms: mild fever, watery eyes and nose, swelling around the eyes, and a mild cough. The cough gradually worsens over a period of about two weeks until it changes character and becomes the periodic, debilitating, whooping cough that may last for six weeks.

Pertussis can be lethal or permanently disabling. Most fatalities result from secondary infections such as pneumonia, but afflicted children may show convulsions and severe brain damage as a direct result of the disease. Just after the turn of the century, 17 of every 100,000 people in the United States could expect to die of pertussis. This mortality rate was less than half that of diphtheria but larger than that for either scarlet fever or measles. In 1934, for example, more than 200,000 cases of pertussis appeared in the United States with 7,000 proving fatal.

To make matters worse, pertussis spreads easily from person to person. Attack rates as high as 90 percent have been recorded. Young children are much more susceptible than older children or adults. Young children are not only more likely to contract the disease but also to die from it. In the late 1920s, the chance of dying from pertussis was as much as 130 times higher for patients less than five years old than for patients older than five years. Pertussis placed second only to diphtheria as a cause of childhood morality in the United States before 1940.

The highly contagious, lethal, and epidemic nature of pertussis made it a prime target for a vaccine in the early part of this century. A vaccine was forthcoming, but uncertainties in tracking the disease and ascribing injuries to the vaccine have caused some people to doubt its value.

Hope for a Pertussis Vaccine

The first hope that a vaccine could be developed for pertussis came in 1906 when Jules Bordet and his brother-in-law Octave Gengou succeeded in culturing the bacterium later named *Bordetella pertussis*. This feat was accomplished six years after they had isolated the bacterium from the throat of a child infected with whooping cough. Bordet was director of the Pasteur Institute in Brussels at the time and stood at the forefront of the newly burgeoning field of immunology. He later won the Nobel Prize for his work on immune reactions in the blood. Another pioneer in this field, the intimidating Sir Almroth Wright, had shown ten years earlier that vaccines could be developed from killed whole bacteria. Wright's vaccine was for typhoid fever, and its success triggered his nearly fanatical belief that human disease should be fought by assisting the body's own defense system.

These rapidly unfolding events in immunology and bacteriology sparked a flurry of activity to develop a pertussis vaccine. Controversy surrounded these efforts from the beginning. Researchers disagreed over the nature of the disease. Some argued that pertussis, like diphtheria and tetanus, was caused by toxins released by the bacterium. Others claimed that the disease sprang from the whole bacterium, not from a few of its parts. The strategies for vaccine production split along predictable lines. One group emulated Wright and made killed, whole bacterium vaccines while the other group tried to produce vaccines from fractions of the bacterium, as von Behring had done with diphtheria and tetanus.

The manufacture of "fractionated vaccines" involved subjecting the *Bordetella* bacterium to a variety of physical and chemical insults in an attempt to shatter it and then concocting a vaccine from the resulting pieces or "fractions." One of the most popular vaccines of this type was undenatured bacterial antigen (UBA), which consisted of bacteria ground up in a mill. Other fractionated vaccines were prepared by extracting the bacterium in various solvents to try to isolate the toxic element from the factor that imparted immunity. These attempts uniformly failed or produced such weak vaccines that they were not pursued. Ironically, this strategy has been resurrected in the most promising recent efforts to produce a low-risk pertussis vaccine.

The first killed, whole bacterium pertussis vaccine was produced by Louis Sauer at Northwestern University in 1913. Many others quickly followed. The typical procedure began by growing large numbers of *Bordetella* bacteria in culture, a task made possible by Bordet and Gengou's research. The bacteria were then killed by heating them or soaking them in formalin or phenol. Spinning the killed bacteria in a centrifuge separated them from the formalin or phenol in the latter cases. The isolated, killed bacteria were suspended in a weak salt solution, which then became the vaccine.

> ## The Mouse Test
>
> The community that produced pertussis vaccine differed radically from the one in which Jenner worked. Since Jenner's discovery, experimental medicine had begun in earnest, Louis Pasteur had established the science of bacteriology nearly single-handedly, the "germ theory" of infectious diseases had been established, and the concept of a protective immune response in humans was growing. The old practice of using humans as guinea pigs in initial vaccine tests gave way to using mice. Before pertussis vaccine was ever used on a human, it was tested on mice in controlled laboratory experiments. Mice do not develop whooping cough, but they do show brain disorders when injected with large numbers of *Bordetella* bacteria. A typical experiment involved injecting pertussis vaccine into a group of mice, exposing them and an unvaccinated group of mice to *Bordetella* bacteria, then counting the number of mice in each group that showed disorders such as partial paralysis. The level of the vaccine's protective effect was indicated by the lessening of disorders in the vaccinated group. The validity of the test hinged on believing that the vaccine's capacity for protecting mice from brain damage signified its capacity for protecting humans from pertussis. Tests conducted during the 1960s justified this assumption and established the "mouse intracerebral challenge protection" test as the basis for developing international standards of pertussis vaccine potency. The mouse test has not removed all uncertainty regarding the vaccine's effects on people, but it has provided one way of considering whether disastrous outcomes are likely.

Box 8.2

Unfortunately, Sauer's vaccine failed to work much of the time and produced unwanted adverse effects.

Pearl Kendrick, a researcher at the State of Michigan Health Department, finally produced a convincingly effective pertussis vaccine from whole, killed bacteria. Between 1934 and 1937, she and her colleague, Grace Eldering, tested their vaccine by giving it to 1,815 children under the age of six and comparing the proportion of pertussis cases in this group with the proportion of cases in a controlled group of 2,397 children who were not vaccinated. Fifteen percent of the control children but only 3 percent of the vaccinated children caught pertussis. Among those children known to have come in contact with pertussis, 69 percent of the control group came down with the disease while only 13 percent of the vaccinees developed it. Kendrick's advance grew out of her realization that populations of *Bordetella* bacteria differed markedly in their virulence and suitability for a vaccine. She and Eldering standardized their bacterial cultures and achieved greater reliability in their vaccine as a result. Kendrick combined her vaccine with diphtheria and tetanus toxoids to produce the DTP vaccine that has been in widespread use ever since.

The new combination vaccine gained immediate acceptance as the preferred weapon in the battle against pertussis, but a number of other vaccines remained in use. The effectiveness of the other vaccines was distressingly variable because the methods of preparation differed widely. In an important new development, the United States government stepped in and established standards for the manufacture and testing of pertussis vaccine. The standards took effect in 1947, the date commonly recognized as the beginning of a large-scale vaccination program directed at pertussis in the United States. Britain followed suit by establishing DTP vaccination as national policy in 1957. Forty-one of the fifty states in the United States require pertussis vaccination for entrance into elementary school. With the advent of federal standards and animal testing (see box 8.2), what vaccine could have seemed safer than pertussis vaccine?

Problems Arise with Pertussis Vaccine

Reports of injury followed closely on the widespread use of pertussis vaccine. Minor reactions such as swelling, fever, and localized pain appeared more frequently in association with pertussis vaccine than any other vaccine. On rare occasions, reports of more severe reactions surfaced. These included excessive sleeping, convulsions, a generalized inflammation of the brain called encephalitis, and bouts of high-pitched crying. Most of these reactions faded away without a trace, but a few people died or became paralyzed after being vaccinated.

The rare serious side effects attributed to pertussis vaccine drew the attention of the medical community, but the vaccine continued in use because of its effectiveness against the disease. As the number of vaccine recipients increased, however, so did the number of people reputed to have been severely injured by the vaccine, and so also did the concern over the wisdom of a large-scale vaccination program. People began to question whether the benefits of the vaccine were worth the apparent risks.

Following the 1974 publication of an article describing neurological problems associated with DTP vaccination, the Association of Parents of Vaccine-Damaged Children was formed in England to lobby Parliament for compensation of injuries suffered by their children apparently as a result of the pertussis vaccine. Some people argued that vaccination against pertussis, although voluntary in England, should be banned altogether. The English government responded by suspending all pertussis vaccination programs, and the number of people who elected to use pertussis vaccine fell sharply. Similar actions were taken by the governments of Sweden and Japan.

A large-scale study of the vaccine's adverse effects was undertaken in Britain between 1976 and 1979, and the results added fuel to the fire. The National Childhood Encephalopathy Study, as it was called, reported that children admitted to a hospital for encephalitis, encephalopathy, coma, or convulsions were twice as likely to have received pertussis vaccine within one week of becoming sick as children hospitalized for other reasons. The study produced risk estimates of 1:110,000 for "acute neurological problems" and 1:310,000 for permanent brain damage following injection of pertussis vaccine. The study was frequently cited in subsequent years as evidence of risks associated with pertussis vaccine.

A television program broadcast in the United States in 1982 claimed that one in every 700 children vaccinated for pertussis would show "serious reactions." The threat of injury galvanized a group of parents to form Dissatisfied Parents Together (DPT) and demand federal compensation for vaccine-related injuries or outright banning of DTP vaccine. The House Commerce Committee of the Congress called for a study of the vaccine's safety, but the government continued its support for pertussis vaccination. Who was right?

Benefits of Pertussis Vaccine

As mentioned earlier, up to 250,000 people per year, mostly children, suffered from pertussis in the 1930s, and approximately 7,000

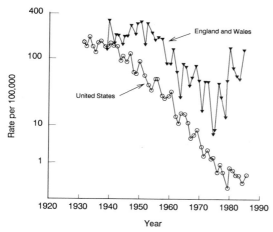

Figure 8.7 Effects of national vaccination programs on the incidence of whooping cough (pertussis) in the United States and England and Wales. The national pertussis vaccination program began in 1947 in the United States and ten years later in Britain. In 1977, when people in Britain began to reject pertussis vaccination in large numbers, the incidence of the disease promptly increased. (Graph redrawn from Cherry, 1989.)

people died from it each year in the United States. After 1947, when the DTP combined vaccine was standardized and put into routine use in the United States, the case rate soon fell by a factor of 100 to approximately 2,000 per year with fewer than a dozen deaths annually (figure 8.7). Approximately one in every 100,000 people now contracts pertussis in a year in the United States. To put this in perspective, gonorrhea is 400 times more common than pertussis and the plague is fifty times less common than pertussis these days. As alarming as the incidence of gonorrhea is, the point of the comparison is that pertussis is no longer very common in the United States.

Some people argued that the vaccine played no part in this dramatic reversal of pertussis, basing their arguments in part on observations that the death rate from pertussis began to decline well before the onset of pertussis vaccination programs. Other people have countered that many factors influence death rates, such as nutrition and public health, citing the steady decline in death rate from all respiratory infections during the century preceding pertussis vaccination. The benefits of pertussis vaccine in terms of a decrease in death rate may have been obscured against this backdrop of an increasingly healthy populace, although mortality from pertussis in the United States decreased more rapidly after introduction of the vaccine. The incidence of pertussis cases provides a clearer picture. The controversy that exploded in Britain regarding the safety of pertussis vaccine caused parental acceptance of the vaccine to drop from 75 percent in 1974 to 30 percent by 1978. The number of pertussis cases reported in England and Wales promptly jumped from less than 20 per 100,000 children in 1976 to over nearly 150 per 100,000 in 1977–78. As the acceptance rate for pertussis vaccine declined, the number of children who developed pertussis increased. A similar trend followed the temporary suspension of the pertussis vaccination program in Japan.

The growing epidemic of pertussis in England during the late 1970s hit hardest those children who had not been immunized; the incidence of pertussis was ten times higher among unimmunized than immunized children in England during 1977. Skeptics desperately argued that the vaccine's benefits were being exaggerated because an increase in upper respiratory infections resembling whooping cough could have led to false, high estimates for the prevalence of the disease. But the number of cases in which the *Bordetella* bacterium was isolated increased right along with the epidemic of 1977. The benefit of pertussis vaccine is underscored by noting that between 370 and 600 of every 100,000 people came down with pertussis in countries without vaccination programs while fewer than 29 in every 100,000 people caught the disease in countries with vaccination programs. These coincidences leave little doubt that pertussis vaccine confers substantial benefit in reducing the incidence of the disease.

The Difficulty of Learning Risks

Even Zabdiel Boylston in the eighteenth century knew that it is not enough to establish the benefits of a new procedure. The risks associated with a new vaccine must also be estimated. We want to know whether the alleged reactions to pertussis vaccine are actually caused by the vaccine or simply represent coincidental associations of essentially independent events. What is the incidence of such reactions? How serious are they?

Answers to these questions have been difficult to come by for pertussis vaccine, and agreement is rare. The risk of permanent paralysis from pertussis vaccine has been estimated to be as high as 1 in 6,500 and as low as 1 in 500,000 vaccinees! An increasing number of investigators claim there is no risk of serious, permanent neurological disorder associated with DTP. Is it any wonder that people disagree over the advisability of using this vaccine? A complicating factor is the fact that the DTP vaccine is actually a combination of three vaccines. How can reactions to vaccination be confidently assigned to any one of the three components under these circumstances? Furthermore, none of the reactions ascribed to pertussis vaccine are unique. All of them occur at some low rate in the general population independently of vaccination, but their base rates of occurrence are not well known. Some of these reactions, such as encephalopathy, may be difficult to diagnose with certainty, and others go unreported by parents.

Confounding factors such as these have made it difficult to assign risks to pertussis vaccine using only retrospective studies as we did for smallpox. *Retrospective studies,* as the name implies, look only backward after the fact in an attempt to discern the causes of observed events. But people forget or make spurious associations, and some useful data may be missing. In cases of brain damage, it would be interesting to know if any predisposing conditions existed prior to vaccination, but this information is often lacking in retrospective studies. It is not enough simply to monitor millions of vaccinations and count the number of reactions temporally associated with the procedure. Other evidence is required if a convincing link is to be made between a vaccine and a serious reaction, especially in controversial cases such as pertussis vaccine.

The preferred alternative is a *prospective study* in which a carefully selected group of people is divided into an experimental group that receives the vaccine and a control group that receives a placebo instead of the vaccine, they do not know which. People in both groups are followed closely before and after the treatment, and differences in the number and timing of "reactions" are noted. Each of these groups should be large. The medical histories of all participants should be thoroughly researched. The people who deliver the vaccine and collect the data should also be unaware of the particular treatment received by each subject; that is, the study should be conducted double blind. Physicians must be able to recognize unambiguously all reactions attributed to the vaccine. The background rates of all observed reactions must be known or at least estimated with confidence.

This is a tall order. Prospective studies of vaccines provide better estimates of risks and benefits, but they are difficult and expensive to do. They also present the ethical dilemma of withholding a potentially beneficial procedure from a group of people at risk in order to learn whether the treatment really is beneficial. This particular issue lies at the heart of a controversy regarding experimental tests of treatments for acquired immune deficiency syndrome (AIDS).

Despite the difficulties, a group of physicians at the University of California at Los Angeles used a prospective study to determine that, as suspected, the pertussis component of DTP vaccine is linked to alarming but temporary adverse effects such as seizures or shock. They followed two groups of children, one of which received DTP vaccine while the other received only the D (diphtheria) and T (tetanus toxoid) components. Adverse effects appeared more frequently in the DTP group than in the control group implicating pertussis vaccine as the cause of these side effects. The question still remains, however, whether pertussis vaccine causes serious brain damage leading to permanent paralysis or death in rare cases. The UCLA study reported no such tragic outcomes. In the absence of data from prospective studies, some physicians have interpreted a temporal correlation between vaccination and the onset of a severe neurological disorder or death as evidence of the vaccine's risks. To date, however, no direct evidence of a cause-effect link between pertussis vaccine and paralysis or death has been demonstrated. Many of the temporally associated cases involved children with neurological conditions that predated vaccination with DTP. The uncertainty remains. This proved to be true in the National Childhood Encephalopathy Study: in the face of reasonable alternative explanations for the observed instances of serious neurological damage following pertussis vaccination, the causal link between the vaccine and the effects could not be sustained.

Decisions Despite Uncertainty: Individual and Social Perspectives

The recent turn of medical opinion toward a more favorable view of pertussis vaccine has not yet affected large numbers of people who continue to perceive the vaccine's risks as greater than its benefits. What if we take the skeptical view that pertussis vaccine actually does cause tragic adverse effects? What might that mean in terms of the balance between the vaccine's risks and its benefits? We cannot offer detailed arguments in response to these questions here, but we can provide a bit more structure to the problem by taking widely published risk estimates for pertussis vaccine and distilling them down to an illustrative table (table 8.2). These figures reflect the commonly made assumptions that 90 percent of all children in the United States receive the full pertussis vaccination series and that the vaccination series confers full protection against the disease in 90 percent of the vaccinated people. They also pessimistically assume that pertussis vaccine can cause brain damage and death.

Imagine the difficult decision you would face in considering pertussis vaccine. If you want to minimize the possibility of any adverse effect no matter how minor or severe, you should be vaccinated. Accepting the vaccine, however, means that you may be accepting a slightly higher risk of permanent brain damage than if you had taken your chances with the disease. Avoiding the vaccine, on the other hand, increases the risk of dying substantially. Assuming that most people would choose in such a way as to minimize the possibility of the worst possible outcome, death, then individuals should elect the vaccine.

Table 8.2
Summary of Risk Estimates for Pertussis and Pertussis Vaccine (per million)

Adverse effect	Pertussis	Vaccine
mild reaction	66,700	10,000
temporary encephalitis	11	18
permanent brain damage	6	9
death	83	9

That decision is removed from those people who live in one of the forty-one states that require children to receive DTP vaccine before enrolling in public school. Given the adverse reactions that may occasionally follow DTP vaccine, a mandatory vaccination program places a heavy burden on a state's citizens by inflicting tragedy and hardship on some families. Are these forty-one states acting in the best interests of their citizens, or have the other nine states made the better decision? Should parents choose not to have their children vaccinated regardless of state requirements? These are difficult decisions.

In the absence of perfect vaccines, vaccination programs can be expected to produce injuries. An average of one out of every 100 children who go through the entire series of five pertussis vaccinations will experience convulsions or high-pitched crying or shock. This translates into 10,000 children for every million who are fully immunized. These adverse effects are worrisome but leave no permanent damage. Approximately 3 million children are born in the United States each year. If 90 percent of them receive the full pertussis vaccine series, we can expect 48 children to die or suffer permanent brain damage from the vaccine, assuming published risk estimates are correct. An additional 8,300 children will develop pertussis, with an unknown fraction of these dying or becoming brain damaged. Of the 300,000 children not vaccinated each year, 27 can be expected to either die from the disease or become permanently brain damaged. From the social perspective, ironically, the success of the vaccine in reducing the incidence of the disease creates a situation in which the number of vaccine-related injuries exceeds the number of disease-related injuries. Even under these circumstances, however, a pertussis vaccination program remains in the society's best interest because the alternative is less desirable. If millions of people avoid the vaccine, the number of pertussis cases and disease-related injuries will increase by a factor of ten over the current situation.

State or federal governments must consider the welfare of a large population when deciding on the advisability of a mandatory vaccination program. A few unfortunate vaccine-related injuries or deaths are regrettable but may be tolerated when the vaccine prevents an even greater number of disease-related injuries and deaths.

The balance between the risks and benefits of a vaccine are often presented in terms of how much the country would have to spend on medical care with and without a major vaccination program. A recent analysis of pertussis vaccination in the United States revealed that the benefits of a vaccination program exceed costs by a factor of 11 when certain common assumptions are made about factors such as the efficacy of the vaccine, the number of people who use it, and the costs of medical care. The ratio remains in favor of vaccination even when the calculation depends on much more pessimistic assumptions. More people become seriously ill or die in the absence of a program than if the number of vaccinated people is high. The country could be expected to spend between $25 million and $54 million on health care for whooping cough patients in the absence of a vaccination program, but only $10 to 18 million if 90 percent of the children receive the vaccine.

This dollars-and-cents argument may be appropriate for public policy analyses, but it still disturbs us when we are discussing human lives. Nevertheless, estimates of dollar costs and benefits provide a handle by which legislators may grasp a difficult issue, although individuals will almost certainly use other criteria in deciding whether or not to use a vaccine. Based on the dollar costs and benefits quoted above, the consensus among public health officials remains that society as a whole is better off when most people get vaccinated for pertussis.

Lowering the Risk: A Better Pertussis Vaccine

In providing pertussis vaccine for everyone through mandatory vaccination programs, we accept the burden of vaccine-related injuries. These injuries, although rare, produce substantial consequences in a large population and inflict financial and emotional burdens on families of vaccine-injured children. In 1991 the federal government took a step toward helping these people by amending the Public Health Service Act to extend programs that provide financial compensation for vaccine-related injuries. Compensating victims of vaccine-related injuries is a highly unsatisfactory way to cope with risk. It is like leaving the barn door open then paying the farmer for the horse that escaped. Everyone would prefer to prevent the horse from escaping in the first place, or in the case of pertussis, prevent adverse effects of the vaccine.

Ever since Jenner, research on vaccines has been driven by the desire for maximum benefit at minimum risk. The struggle has been especially fierce in the case of pertussis vaccine because of the controversy surrounding it and because the *Bordetella* bacterium has proven until recently very difficult to study. The key to reducing the risks associated with other vaccines lay in depressing the toxic effects of the organism used in the vaccine while leaving intact its capacity to stimulate a large, protective immune reaction. This was the approach taken, with notably little success, by investigators who explored "fractionated pertussis vaccines" in the 1920s and 1930s. The reason for their failure emerged only after new techniques were developed in molecular biology, a field that did not even begin in earnest until the structure of DNA was discovered in 1952.

Early workers were limited by their techniques to separating classes of compounds, such as fats and proteins, in their attempts to separate the toxic fraction from the immunogenic fraction of the pertussis bacterium. They would have succeeded only if those two properties resided in different fractions. Molecular biologists now look with tremendously increased resolution into different regions of a single protein. When they turned their powerful new tools on the *Bordetella* bacterium, they were startled to find the properties of toxicity and immunogenicity residing on the same molecule.

The key molecule is a protein known as pertussis toxin because it has long been associated with the toxic effects of *Bordetella* bacteria. The bacterium possesses other toxins as well, but pertussis toxin is one of the major ones. It poisons the body by interfering with the ability of many cells to respond to hormonal signals and altering the internal signaling machinery of cells. Cells affected by the toxin can no longer coordinate their actions with their neighbors or other tissues in the body. Among its other actions, pertussis toxin promotes the release of insulin from the pancreas and thus lowers blood sugar level. It also enhances the body's response to histamine, a chemical in the body involved in allergic reactions. Heightened allergic reactions in the presence of pertussis toxin may underlie the cases of encephalitis that sometimes follow pertussis vaccination.

A single enzymatically-active region of the pertussis toxin molecule has been blamed for these toxic reactions. Molecular biologists have isolated the gene that codes for pertussis toxin, determined its structure, systematically modified its code, and reinserted the modified genes into *Bordetella* bacteria to create strains that produce altered pertussis toxin. By examining the effects of various mutations, they determined that the toxic region of the protein lay some distance from the immunogenic region. They then directed mutations only to the toxic region leaving the immunogenic region intact. This marks a new era of vaccine development, one in which the distinction between scientist and engineer is blurred even more than previously. Scientists can now intentionally design vaccine molecules in a way analogous to how engineers design bridges, for example. A vaccine produced from the new, engineered *Bordetella* bacteria proved to be both safe and effective when tested on mice.

The goal of separating toxicity from immunogenicity has finally been achieved for pertussis vaccine, more than fifty years after Pearl Kendrick introduced DTP vaccination. Has the new pertussis vaccine also achieved the elusive goal of being risk free? It is too early to tell. Human trials have not yet been completed (see box 8.3). Besides, rare vaccine-related injuries will not be expected to appear in limited trials on small groups of people. If rare adverse effects do follow the new vaccine, reasonable estimates of risk will not be developed until a very large number of people have been vaccinated with it. Our only recourse is to live with the uncertainty and continue to hope for the best.

New scientific understanding and techniques bring new hope for a better pertussis vaccine, but vaccines, like any other technology, are shaped by a variety of constraints. Even when the capability for making a vaccine exists, economic and political concerns may stifle production and distribution. Pharmaceutical companies are reluctant to take on the costs of vaccine research and development if they are not likely to recoup their investment through vaccine sales. For this reason, diseases that do not afflict very many people are unattractive candidates for vaccine development. A disease may be common, however, and still fail to attract vaccine development if the afflicted populace is unable to afford the vaccine as is true in many developing countries. Again, good intentions, even when combined with scientific capability, are not enough.

Science vs. Engineering in Vaccine Development

The distinction between scientist and engineer is blurred in vaccine research and development. The creation of new vaccines has been so closely tied to scientific discoveries in bacteriology, virology, and immunology that much of the "design" of vaccines has been carried out by the scientists doing the basic research.

Substantial scientific and engineering problems persist after the scientists create a new vaccine in the laboratory. Once it is clear that a vaccine can be made, it must be tested to determine its utility—the balance between its beneficial and adverse effects. These tests normally proceed in three stages: tests on animals only, combined tests on animals and a few humans, and tests on humans only. An entire series of trials for a new vaccine typically requires five to seven years to complete.

If the testing indicates a useful vaccine, then production must be scaled-up from the small batches used in the scientific work to the massive amounts required for large vaccination programs. In the case of bacterial vaccines, this means growing sufficient numbers of bacteria to create enough vaccine for the millions of children born each year. In the case of viral vaccines, this means incubating the virus in a large enough number of chicken eggs to generate millions of vaccine doses each year. It takes a year or two to work out all the glitches of large-scale production. The nine years that are required to get a vaccine from the laboratory bench to the market testify to the magnitude of the engineering problems that exist after the basic research has been done. Even after all of this work, the benefits and risks of a vaccine can not be known until it has seen widespread use.

Box 8.3

Exercises

1. The risk estimates made for pertussis vaccine assume that 90 percent of newborns are vaccinated, but in some inner city areas the compliance rate is probably no more than 35 percent. What effect does this difference have on the various risks associated with using or not using pertussis vaccine?

2. Given your answer to question 1, what is the responsibility of the federal government regarding large-scale vaccination programs? Should the government take an active role in promoting vaccination or simply rely on the state governments or individuals to "do the right thing"?

3. Records kept by the Centers for Disease Control (CDC) show that there have been approximately 1,800 cases of pertussis and 10 pertussis-related deaths per year for the last ten years in the United States. Using these figures, the risk of dying becomes about 1/180, not 1/800 as used in calculating the risks reported in table 8.2 of this chapter. How does this information affect the balance between the risks associated with vaccination and the risks associated with not being vaccinated? How might this influence an individual's decision regarding the vaccine, or a bureaucrat's decision regarding public health policy?

4. Using the multiplication law of probability, calculate the probability of obtaining a straight flush by drawing the first five cards from a 52-card deck.

5. The oral polio vaccine (OPV) now causes most, if not all, cases of polio in the United States. Use the data presented below (from the CDC) to estimate the risks associated with using or not using OPV. Calculate the probabilities of the various outcomes. Would you recommend polio vaccine for your child? as a product to be manufactured by your pharmaceutical company? as an element in a mandatory national vaccination program? What assumptions must you make to arrive at estimates of the various risks?

Polio (1973–1983)

1. 183 cases of polio reported in this ten-year period
   - 88 not vaccine-related
   - 95 cases were vaccine-related
     - 67 of these occurred in people who had not received OPV but had contacted someone who had received it.
2. 250 million doses were delivered
   - each person receives 3 doses on average
3. Outcomes of polio:
   - "natural" polio
     - 1 to 10 of every 1,000 cases end in paralysis
     - 6 to 8 of every 10,000 cases are fatal
   - vaccine-related polio
     - 2 of every 10,000,000 cases end in paralysis
     - 1 of every 100,000,000 cases is fatal

4. Assumptions and data that may be useful:
    - OPV is 95 percent effective
    - 50 percent of U.S. citizens have been vaccinated with OPV
    - there are approximately 3 million births per year in the United States
    - The average population in the United States during this period was approximately 225 million

6. Everyone would like to see an improved pertussis vaccine on the market, but there are difficulties in reaching this goal. What important scientific and policy considerations make it difficult to assess and introduce a new pertussis vaccine? (Jenner faced some of these difficulties when his smallpox vaccine replaced inoculation.)

The Greenhouse Effect: Revolution Involves Risk

9

A Climate of Change

We expect, almost demand, that technology will change with time. Compare the everyday world of the children of the post–World War II baby boom to that of their children to see how rapidly things evolve in a generation. As the twentieth century draws to a close, the speed of change is accelerating even more, and people accept this even though they may experience difficulties in keeping up with the rate of change. New products and methods arrive without any apparent letup, and we somehow integrate them into our living patterns.

However, what we have not been concerned with in the past is that our technological thirst will have a large-scale impact on the planet. To be sure, we certainly appreciate the concrete, demonstrable impacts of clearing forests for agriculture, the problems of regional air and water pollution, and the changes in temperature from paving over large parts of cities. By contrast, we have little experience with more abstract long-term effects on the dynamics of global processes.

Although the prime qualification for the TV weather forecaster seems to be a talent for stand-up comedy, this belies the host of high-tech methods that can be brought to bear upon short-term estimates of what the weather will be like. Nevertheless, people seem to relish a certain skepticism about these predictions and their precision. But the overall pattern and general range of temperatures and precipitation is expected to be generally constant. For those of us who live in temperate zones, we expect the summers to be hot, the winters to be cold or worse, and the spring and autumn to be more or less pleasant. We attribute stories of our parents or grandparents of how much hotter or colder it was in the past to either faulty memories or romanticizing the good (or bad) old days.

Thus, when we are now called upon to consider the possibility that we as a species are in the process of altering the patterns of climate, it seems like some science fiction plot. This chapter seeks to deal with this unsettling proposition by examining the physical and chemical principles behind such claims and by examining some of the evidence used to support such a drastic scenario.

Prelude to Climate Change?

In the mid-eighteenth century, British inventors and entrepreneurs began what is commonly called the First Industrial Revolution in which coal, iron, and steam power were to play major roles. A major watershed was reached in 1769 when James Watt introduced a more efficient steam engine for industrialists to use in place of water power. Along with their European and American competitors, they also unwittingly launched a global climate experiment with potential consequences that we are just beginning to appreciate over two centuries later. A subsequent phase of industrialization, dubbed the Second Industrial Revolution, accelerated the pace of economic growth in the late 1800s. The fact that such human activity might affect the presumably vast and immensely powerful processes of the earth itself was not considered or, more likely, not even imagined.

The fuel for the industrial engines was primarily coal along with similar substances like lignite or peat. These were burned to run the steam engines that powered machinery, such as looms in the great textile mills of the British midlands. Prior to this new technology, the water wheel had been the chief source of nonanimal power from the Middle Ages onward. These fossil-fuel-powered engines replaced water wheels as the prime movers for manufacturing because mills and factories were no longer constrained to locations by streams and rivers. By the beginning of the nineteenth century, the

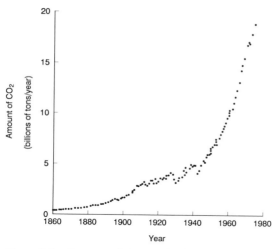

Figure 9.1 Exponential growth of carbon dioxide production from human activities since the late ninteenth century. The levelling-off between 1910 and 1940 reflects the effects of the First World War and the Depression. (Adapted and reprinted from *The Challenge of Global Warming* by permission of Island Press. Copyright (c) 1989 Island Press.)

burning of carbon-containing fuels was beginning its pattern of exponential growth (figure 9.1) except for the 1915 to 1945 period, which encompassed two world wars and a global economic depression. When coal and other fuels like wood burn, the carbon in them forms carbon dioxide gas (box 9.1), and it is reasonable to ask where the carbon dioxide ends up after spewing out of the smokestack.

The great French mathematician and physicist, Jean Baptist Fourier, first proposed the idea of the earth's atmosphere acting like the glass on a greenhouse in 1827. Later on, other scientists began to understand that gases in the air could selectively absorb heat. John Tyndall, the English physicist known for his work on light scattering, felt that water vapor was the main absorbing agent. Others, including the French scientists Lecher and Pertner, thought that carbon dioxide was also involved. This simple gaseous molecule, also a product of decomposition of organic matter, has the property of being able to trap heat or infrared radiation.

The Swedish scientist Svante Arrhenius was a world renowned chemist who made major contributions to our understanding of acids, rates of chemical reactions, and solutions of ionic materials like salt and was the third recipient of the Nobel Prize in chemistry in 1903. History, however, may eventually know him best as a

Carbon Dioxide: Description of a Suspect Molecule

The greenhouse phenomenon has taken on some aspects of a criminal trial with an alleged "crime" (global warming) about to happen or already in progress, a victim (the Earth), and a chief suspect (carbon dioxide). Scientists, government officials, industry representatives, economists and others play the role of lawyers, judges, and juries. Under the circumstances, it seems wise to record a detailed description of this allegedly malevolent molecule.

Carbon dioxide was the first constituent of air to be isolated and characterized by the Scottish chemist Joseph Black in 1754. This is rather remarkable since it forms only 0.03 percent of the gases in the atmosphere, compared to oxygen, which is about 20 percent. As a chemical it is rather benign but intimately involved in many natural processes—such as respiration. Just as burning carbon-containing molecules leads to the combination of carbon with oxygen, so does metabolism, but in a more complex way. We, of course, take in oxygen continuously and then exhale carbon dioxide continuously. A typical overall example of "burning" up calories is summarized by the chemical equation:

$$C_6H_{12}O_6 + 6O_2 \rightarrow 6CO_2 + 6H_2O$$
sugar

Decomposition of minerals called carbonates, such as calcium carbonate (chalk, marble), also produces carbon dioxide.

$$CaCO_3 \xrightarrow{heat} CaO + CO_2$$

The analysis for carbon dioxide is relatively straightforward and can be done simply by weighing a substance which absorbs carbon dioxide before and after being exposed to the gas. For more sensitive determinations involving small amounts of carbon dioxide, a precision electronic instrument called a mass spectrometer is employed. With this latter technique, one part carbon dioxide in a million parts of air can be accurately measured.

When carbon dioxide is cooled, it does not form a liquid but instead goes directly to the solid state, forming "dry ice" at about $-60°C$. This is used for a variety of cooling processes, and since it "sublimes" (goes directly to gaseous carbon dioxide) there is no mess left behind. When the gas dissolves in water, it forms carbonic acid, which gives the fizz to soft drinks, champagne, and other carbonated beverages. Escaping carbon dioxide also forms holes in some cheeses during aging and causes bread to rise.

Thus, carbon dioxide has been used by humans for thousands of years in various technologies, as well as being an intimate part of our own biochemistry. Now, we have developed technology on such a large scale that carbon dioxide is being regarded in a less friendly light. Some fundamental physical properties of the molecule now become very crucial to our understanding of how it may have global effects.

climatologist. He became interested in the causes of ice ages and in 1896 published a paper on the effects of carbon dioxide concentrations on the temperature of the earth's surface. His calculations indicated that reasonable variations in carbon dioxide (for example, doubling the concentration) could lead to enough cooling and warming to account for the ice ages and warm interglacial periods.

However, the potential link between the burning of fossil fuels and possibly higher atmospheric concentration of carbon dioxide was not clearly defined until the late 1930s by the English scientist G. S. Callendar. According to his estimates, a doubling of the carbon dioxide concentration could lead to an increase in the temperature at the planet's surface of about 2°C (or 4°F). In a later paper, he noted that only in the twentieth century had human activity been on a scale large enough to "disturb nature's slow moving carbon-balance," and he very cautiously stated that "some slight amelioration in climate" might be expected.

The seriousness and impact of this issue was not fully grasped until the late 1950s, when it began to be systematically studied. Roger Revelle and Hans Seuss, geoscientists at the Scripps Oceanographic Institute near San Diego, reiterated the greenhouse hypothesis and warned of a "great geophysical experiment" (S. Schneider 1989a). Revelle was also the director of the first internationally coordinated effort to study the earth, the International Geophysical Year Project of 1957–58. Their concerns led Charles Keeling to instigate a carbon dioxide monitoring system on the slopes of Mauna Loa on the island of Hawaii. Since 1958, a systematic record of the atmospheric carbon dioxide has been kept in Hawaii and at other locations around the globe.

In terms of the general public's awareness of this issue, the matter did not really arise until the mid-1980s. It probably is fair to say that the greenhouse effect was science fiction in the 1970s, emerged as a scientifically respectable issue by the mid-1980s, and became a matter of public policy on many levels including the international scene in the late 1980s and early 1990s. Reports of increasing levels of carbon dioxide in the atmosphere and a small increase in average global temperature, coupled with several unusually hot summers, have piqued the public's interest.

This increased attention to a phenomenon that may alter the ecology of the planet is not surprising and leads to many important questions. Are the predictions true or are we experiencing some random fluctuations in temperature? How would global warming affect us as individuals, as well as on a national basis? How will we recognize the signal of its onset—is the carbon dioxide increase the "smoking gun"? Are there other potential causes of the greenhouse effect? Is the effect preventable or inevitable and, if so, irreversible? Are there technical fixes for this situation? What can or should be done—if anything—and who should lead the way?

The purpose of this chapter is to investigate these questions and to sort fact from fiction. In order to do so, we must link scientific principles with the technology that may be leading us into the global risk business.

The Atmosphere as a Cover

If there were no gaseous layer surrounding the planet, the heat would be lost and the earth would be very cold. The moon, for example, has no atmosphere, and its temperature can drop to $-180°C$ ($-290°F$) in the absence of sunlight. However, chemical compounds like water and carbon dioxide can absorb some of this heat (as infrared radiation, see box 9.2) and thus maintain a moderately warm shell around the planet somewhat akin to the glass or plastic exterior of a plant hothouse or greenhouse (which actually prevents breezes from removing heat). These compounds not only store energy but they can also transfer it downward by collisions with other molecules nearer the earth's surface and warm them up as well. A molecule can store energy in various modes: it can rotate, vibrate, and move through space. Temperature can be described as the average energy of the molecules, so as this energy gets distributed around the planet, an average surface temperature is established.

The so-called greenhouse gases are found in the lower part of the atmosphere called the troposphere, which extends from the surface of the planet to about 12.5 kilometers up (about 40,000 feet). In the troposphere weather processes occur, and thus climate is determined in this region. Above the troposphere we find the stratosphere where some of the substances implicated in the greenhouse effect are also present. However, they may fulfill different roles at this altitude, for example, the ozone layer within this region absorbs dangerous higher energy ultraviolet light.

If there are any doubts regarding the *theory* behind this proposal, we can look to our nearest neighbors in the solar system to see how planetary temperatures can be affected by gases in the atmosphere. There exists for the triad of Earth, Venus, and Mars a phenomenon dubbed the Goldilocks problem. Venus has a thick atmosphere (ninety times denser than Earth's) heavily laden with carbon dioxide (over 75 percent), so that its surface temperature is "too hot," 460°C. Mars has a very thin atmosphere (1/100 that of Earth) with very little water and carbon dioxide and is, therefore, "too cold" with a surface temperature near $-60°C$. The Earth is "just right" for living organisms as we know them, including ourselves. Without the carbon dioxide and other gases in the atmosphere, our planet's surface temperature would be 35°C colder on average, considerably below the freezing point of water. Life would be problematic for the currently known flora and fauna, which flourish at a balmy average temperature of about 15°C worldwide.

For a clear understanding of the global warming issue, two terms must be defined: average surface temperature and greenhouse effect. The former is straightforward and allows for the fact that different regions on the planet experience different temperatures depending on location and season. However, a few degrees change in the *global average* is a massive effect because it involves such a large area. The second term, on the other hand, is used as a shorthand expression. We are really concerned about the appearance of an *excess* greenhouse effect resulting in a gradual warming of the planet. Furthermore, this would result from anthropogenically produced gases resulting from our adoption of certain technologies.

Energy Flow and Molecular Heat Traps

The basic principles behind the so-called greenhouse effect emerge from two scientifically important phenomena: the main source of the earth's energy is the sun (with internal heat of the planet being the other) and molecules can absorb heat as well as other forms of radiation. As depicted in figure 9.2, 30 percent of incoming solar radiation is reflected by the earth's albedo. The other 70 percent penetrates the atmosphere because the sun's rays are mostly visible light and the gases in the air do not absorb it. Once it is absorbed by the earth's surface (including living organisms), much of the energy

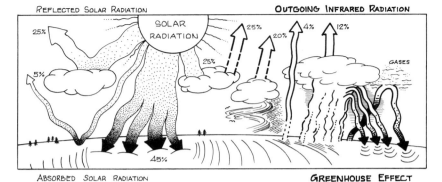

Figure 9.2 A schematic diagram of the greenhouse effect starting with the incoming rays of the sun (left) and ending with trapped infrared radiation (right). (Adapted with permission of Sierra Club Books from *Global Warming*, by Stephen H. Schneider. Copyright (c) 1989 by Stephen H. Schneider.)

in the visible light is used for processes like photosynthesis. Since no process is completely efficient, some energy will be reemitted as heat.

The various forms of radiation ranging from high-energy, and thus dangerous, X rays to low-energy radio waves form a continuous spectrum of radiation. (See the detailed discussion of high-energy radiation in chapter 10.) The visible range is a very small part of this spectrum sandwiched between the ultraviolet and the infrared regions. One property of any radiation is its wavelength, and it is important to note that the *smaller* the wavelength, the *greater* the energy. Each type of radiation has a characteristic energy associated with it. For example, blue light is more energetic than red; we can imagine a process that absorbs blue light, uses some of the energy, and emits the leftover energy as red light or as light that is even lower in energy, infrared (IR). This excess energy then can radiate out into space or be captured by some other molecule.

A molecule will absorb a particular type of light (radiation) because the energy in the light will result in some change in the molecule. For example, a solution of a red dye has that particular color because the dye has absorbed the other colors of light, such as green and blue, and let the red pass through. This process occurs because electrons in the dye take up the light of the energy associated with blue and green. Infrared energy, on the other hand, gets absorbed because of other molecular processes (box 9.2).

Absorbing Heat: Infrared Spectra

Absorption of infrared radiation is associated with the phenomenon of chemical bonding. Imagine the structure of carbon dioxide as three balls (one carbon and two oxygen atoms) stuck together with two springs (analogs of the bonds). The relative positions of the balls can be altered by stretching or bending the springs to change the distances or angles between the balls. This is what chemical bonds do when they absorb energy to which their particular types of bonds respond. For example, a carbon dioxide molecule might undergo a symmetrical stretching of its two bonds:

O—C—O + infrared radiation → O ← C → O

Although these bonds are composed of electrons, infrared energy is only strong enough to wiggle them around a bit. The atoms oscillate about an average distance, called the *bond length*. When radiation of the right energy is absorbed, the amplitude (distance) of oscillation increases. The angles between various atoms can also be distorted, and these movements also increase in frequency as energy is absorbed.

By plotting the percentage of the light transmitted (the complement of that absorbed) at each wavelength, you obtain an infrared spectrum, or IR spectrum (figure 9.3). The vertical axis of the graph depicts the transmittance—something with a value of 100 percent would be totally transparent to infrared radiation and would have no absorption at that particular energy. Depending on the types of atoms and how they are arranged in a molecule, radiation of different energies is absorbed in a very distinctive pattern for that molecule. The pattern of absorption bands (the downward peaks) is like a set of fingerprints for each molecule, so the IR spectrum is a very useful method used by chemists to characterize a chemical compound and then to test for its presence or absence.

The IR spectrum of carbon dioxide in figure 9.3 was obtained from measurements on the gas. As you can see, there are several quite distinct absorption bands. For comparison, the spectrum of methane, CH_4, also shown in figure 9.3, is very different because its molecular structure involves four hydrogens connected to a single carbon atom. As a result, the main band of methane is located at the higher range of infrared energy (shorter wavelength) compared to carbon dioxide.

Box 9.2

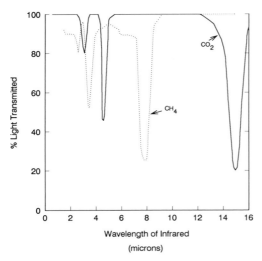

Figure 9.3 Infrared Spectra of carbon dioxide, CO_2 (dark line) and methane, CH_4 (light line). (Adapted with permission from *W. W. Coblentz, Investigations of Infrared Spectra*, 1905 Carnegie Institution, reprinted in 1962 by the Coblentz Society and Perkin-Elmer Corporation.)

Excess Carbon Dioxide?

The general source of the increased levels of heat-absorbing gases in the earth's atmosphere involves the vast developments in technology that humans have created in the last two centuries. We are now able to dump carbon dioxide into the atmosphere at a rate that cannot be offset by natural absorption processes of the geosphere and biosphere. For each human, over four tons of carbon dioxide are emitted into the troposphere each year. Remember that we are burning materials that took millions of years to form as well as trees that grew for decades or longer. A subsidiary cause is the explosion in population—fossil fuels rapidly burned by five billion people, rather than one billion, obviously have a greater impact.

The primary sources of extra carbon dioxide are postulated to be fossil fuel use and deforestation. The reasoning is based upon some fairly simple chemistry. When a carbon-containing organic substance burns, it combines with oxygen, the ubiquitous oxygen in air. In most cases, there is excess oxygen present, so that the carbon is fully oxidized to carbon dioxide. (In the event of limited oxygen, there is a good chance of forming carbon monoxide, a deadly poison, but that is another problem.) The hydrogen in these organic compounds also combines with oxygen to form water. For example, the combustion of gasoline to provide energy for transportation can be summarized by the following (where octane, C_8H_{18}, is used to represent a typical gasoline molecule):

$$C_8H_{18} + 12.5\ O_2 \rightarrow 8\ CO_2 + 9\ H_2O + \text{energy}$$
$$114 \text{ tons} +\quad 400\quad \rightarrow\quad 352\quad +\quad 162$$

514 tons reactants → 514 tons products

The important point to remember is that all of the atoms in octane and oxygen must end up, in rearranged form, as carbon dioxide and water. This is based upon the atomic theory and is called the law of conservation of mass, which governs the balancing of chemical equations, and is more simply stated as "mass of products equals mass of starting materials." In contrast, energy produced in nuclear power plants is obtained from the conversion of elements into different elements, and in the process some matter is converted to energy according to Einstein's famous equation, $E = mc^2$.

The preceding chemical equation also describes the *ratios* of the octane, oxygen, water, and carbon dioxide involved. Any unit of weight (or mass) can be used, such as grams, pounds, or tons. So, if 57 grams of octane burned completely, it would produce 176 grams of carbon dioxide. Note that the two numerical values have been halved, and a similar calculation would give the amounts of oxygen used and water produced.

Energy is released and absorbed by the breakdown and formation of chemical bonds. Here the products's bonds are more stable than those of the reactants, so energy is given off as thermal energy. This then "does work," as in propelling an automobile, but some of it is released as waste heat. Not all the energy is converted to work, because the second law of thermodynamics "prohibits" real world processes that are perfectly efficient. Different fuels yield differing amounts of carbon dioxide for the same amount of energy produced. For example, burning methane generates less carbon dioxide than burning an equivalent amount of coal.

It is also well documented that each year more carbon dioxide is being put into the atmosphere by massive burning of fossil fuels, as well as from large-scale deforestation especially in the tropics. Currently, human activities emit about 11 billion tons of carbon dioxide into the atmosphere each year. Deforestation is estimated to contribute about 15 percent of this total because trees are often burned outright to clear the land for farming. By way of comparison, the atmosphere holds an estimated 2.7 trillion tons of carbon dioxide.

Another source of carbon dioxide is slow decay of organic materials caused by bacteria acting on them. This process is a sort of "combustion in slow motion" and is important in deforestation where large quantities of vegetation, if not burned, are left to rot. Insects, fungi, and microorganisms feed on the wood, primarily cellulose, and release carbon dioxide from their metabolic processes. These chemical reactions are also subject to conservation of mass.

One of the richest potential contributors of carbon dioxide is the globe's system of oceans, which are estimated to contain 140 trillion tons of carbon dioxide, or about fifty times that found in the atmosphere at present. The oceans's role as a great reservoir for carbon dioxide is not clear, and they might be net producers under certain conditions. For example, if warming does indeed occur, the solubility of carbon dioxide in water would decrease, similar to a warm soft drink going flat. The oceans would then become net producers of carbon dioxide, and thus act in a positive feedback mode to warm the climate even more.

Carbon Dioxide Sinks

The biosphere, geosphere, atmosphere, and oceans are all interconnected, so that a process occurring in one produces consequences in another. For example, acid rain results from burning sulfur (in coal), leading to airborne sulfur oxides that dissolve in rain to form sulfuric acid that can enter lakes, rendering them acidic, and thus kill off fish and other organisms. This illustrates an imbalance caused by a process that rapidly adds more of a substance than the ecosystem can handle.

Carbon dioxide, on the other hand, which has been generated by animals, decay, and volcanic eruptions for millions of years, is absorbed by plants (photosynthesis), rocks, and water to maintain equilibrium concentrations in the atmosphere over periods that may last millions of years before a new balance is established. This complex set of processes is known as the *carbon cycle*, and it is still not known what alters its operation and exactly how it relates to ice ages and warmer epochs. Then along comes humankind, which harnessed that great carbon dioxide-producing technology, fire, and ultimately developed it to generate carbon dioxide at rates possibly faster than the natural sinks can cope with.

One of the biggest unknown factors with the greenhouse phenomenon is the role of the oceans and how they might act to absorb the excess carbon dioxide. A major problem in predicting the action of the oceans is the lack of detailed knowledge about the dynamics of ocean currents and how the atmosphere and surface waters interact. By comparison, the atmosphere is much better understood, but the way carbon dioxide is transferred between the atmosphere and the top layer of the ocean is not well understood. Also, the rate of transfer is not rapid.

Based on present information about the carbon cycle, estimates can be made for removal of carbon dioxide. Calculations suggest that even if all anthropogenic emissions of carbon dioxide were halted tomorrow, it would require 300 years for the sinks to absorb the existing *excess* carbon dioxide.

The Case for Global Warming

Since this topic might seem to lend itself to an element of alarmist hysteria (to paraphrase that excitable meteorologist, C. Little, "The sky is warming! The sky is warming!"), we need to examine the evidence, both direct and circumstantial, for a purported global warming.

The clearest evidence, albeit indirect, is the undisputed record of rising concentrations of carbon dioxide in the northern hemisphere (figure 9.4). The results of the monitoring in Hawaii have provided a clear pattern of increasing carbon dioxide with an 11 percent increase in the thirty years up to 1988. The data also show the seasonal fluctuations caused by plants absorbing more carbon dioxide during the summers and more fuel being burned in winters in the northern hemisphere. Monitoring at several other remote

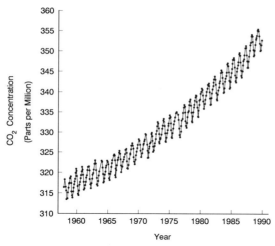

Figure 9.4 Concentration of atmospheric carbon dioxide in ppm of dry air versus time as observed by infrared analysis at Mauna Loa Observatory, Hawaii. Dots indicate monthly average concentration. (from Keeling et al., "A Three-Dimensional Model of Atmospheric CO_2 Transport Based on Observed Winds" in *Aspects of Climate Variability in the Pacific and the Western Americas*, Geophysical Monograph, American Geophysical Union, vol. 55, 1989 (Nov).)

sites, including the South Pole, has shown very similar trends for these years. Samples are taken in largely uninhabited areas to avoid contamination by local sources of carbon dioxide, such as power plants and motor vehicles.

Other carbon dioxide data from ice cores have shown a 25 percent increase since pre–Industrial Revolution times with a rise from 280 parts per million to 350 parts per million in 1989. When rain and snow freeze in glaciers, pockets of air containing the representative concentrations of carbon dioxide and other gases are trapped, thus forming a record of how the atmosphere was composed then. As the cores are taken deeper down, a profile over many thousands of years can be developed. Investigators have drilled 2,000 meters down in order to obtain a record that goes back over a thousand centuries.

Estimates of temperature suggest a strong relation between carbon dioxide and temperature. Over the last 160 thousand years, global temperatures follow the pattern of variation in the atmospheric carbon dioxide concentrations reasonably well (figure 9.5). Note that the temperature was about 2.5°C warmer about 130

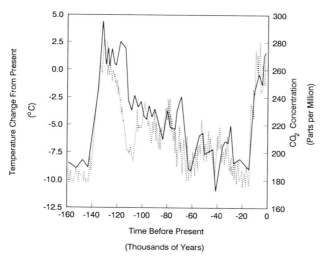

Figure 9.5 Correlation between CO_2 levels (heavy line and right axis scale) and temperature (left axis) over 160,000 years. (Adapted with permission from *Nature* vol. 329, pp. 408–414. Copyright (c) 1987 Macmillan Magazines Ltd.)

---

thousand years ago when the carbon dioxide was at its peak. These temperatures were estimated from the ratio of hydrogen to deuterium found in the ice. Since the ice is frozen rain that originated in the oceans, it should reflect the composition of the sea water that contains a tiny fraction of "heavy water" composed of deuterium and oxygen. The relative rates of evaporation for normal water and heavy water change slightly with temperature. Warmer periods are indicated by "enrichment" of the samples in deuterium. The end of the last ice age began with a dramatic increase in temperature starting about 20 thousand years ago.

Studies of more recent temperature profiles also support the greenhouse proposal. These observations form the basis for a claim that the earth's temperature has risen about 0.5°C since the late 1800s (figure 9.6). Even more dramatically, there is the fact that the 1980s was the hottest decade in a century including the three warmest years (1981, 1987, and 1988) in the last one hundred. Some observers have interpreted these data to be the signal for the start of a long-term warming caused by greenhouse gases. Most scientists are more cautious and are waiting for more evidence before proclaiming that global warming is actually under way.

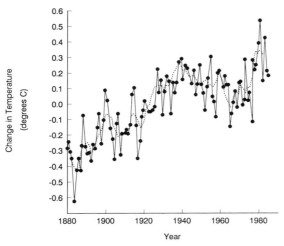

Figure 9.6 Global temperature changes since 1880 relative to the average temperature between 1951 and 1980. (Adapted with permission from J. Hansen and S. Lebedeff, *Journal of Geophysical Research*, vol. 92, no. D11, pp. 13,345–13,372. Copyright (c) 1987 American Geophysical Union.)

A third type of physical evidence includes claims that the ocean levels have risen about one centimeter a decade over the last century, and perhaps even faster in the last fifty years. As with the temperature data, however, the exact cause of the rise is unclear.

Finally, there is an even more speculative approach to obtaining results about global warming: computer modeling. One of the most difficult challenges facing humanity is the prediction of the weather. Prognostications of more than several days are still fraught with uncertainty, even with data from weather satellites, first introduced in the 1960s. Climatologists, on the other hand, seek to predict the overall global weather patterns on long time scales—years or even decades. The advent of powerful computers has given climatologists hope that they can indeed begin to tackle these questions. In a sense, these scientists are modern-day inheritors of LaPlaces's "determinism" by which the future can be calculated from the present assuming that all the variables such as mass, position, speed, and all the rules or equations are known. However, they are also forced to reckon with probability by including statistical factors in their calculations.

In terms of global warming, sophisticated global-circulation models involving increasing levels of greenhouse gases have shown significant changes in temperature and rainfall for various regions of the planet, particularly on the scale of continents. The scientists have checked their methods by successfully predicting seasonal changes in temperature, simulating the climates of past ice ages, and calculating temperatures on other planets like Venus. Climate modelers acknowledge that, despite some successes, there are still great uncertainties in their approach, especially when trying to predict weather patterns for areas as large as continents, much less for specific areas like the Los Angeles basin or the state of Iowa. The next generation of models will involve coupling atmospheric models with better models of the oceans, once those data become available and more powerful computers are developed. However, whether these calculations will be accurate enough to predict the future climate is not known at this time.

The Forecast is Cloudy

Legitimate uncertainties regarding the existence of the greenhouse effect have affected the development of policies to combat it. The temperature rise claimed by several researchers is very modest and may be a phenomenon perhaps caused by random solar fluctuations. In fact, the data even show a decrease between 1940 and 1975. By way of contrast, studies from ocean surface measurements over the last century do not show an increase in temperature.

The correlation between carbon dioxide levels and temperature over millions of years is enticing, but it is a classic cause-and-effect situation: which caused which? For example, a warming trend may have led to the oceans releasing carbon dioxide. And there are other explanations for ice ages besides reduced carbon dioxide levels; for example, the sun's energy output may have decreased for a time.

It is also not clear how the oceans will behave in absorbing or emitting carbon dioxide or what net effect more clouds will have on the temperature. Another problem involves estimating the effect of particles sent into the atmosphere by industrial processes, dust storms, fires, and so forth. Some believe this has produced enough particles to prevent significant amounts of sunlight from reaching the earth. In a similar way, sulfate aerosols, tiny droplets of water formed around sulfate particles, may be enhancing the size of high clouds that reflect sunlight and thus increase the earth's albedo.

The complex computer models have been criticized for using certain assumptions while not including others. New information about the ocean-atmosphere interaction may significantly change the results of the models's calculations. For example, a very recent set of calculations suggests that the Antarctic ice cap, instead of melting, will actually absorb excess water in the atmosphere and help prevent sea levels from rising.

Last, there is the question of recognizing the signal, if indeed there will be a clear sign, or there may be just a slow series of incremental changes. If the greenhouse effect involved only carbon dioxide, we could focus on its sources and sinks and perhaps gather enough new information to confirm whether a problem really exists or not. But, the situation is not that simple, and neither is the complex array of technologies used to support our modern ways of living.

More Greenhouse Gases

There are also well-documented data showing that trace greenhouse gases, such as chlorofluorocarbons (CFCs) also implicated in the ozone layer depletion, have increased dramatically. Methane, nitrous oxide, and ozone are other gases whose concentrations are rising. The common greenhouse gases and their estimated effects on global temperature are given in table 9.1.

Methane, CH_4, is a prime candidate for becoming a particularly serious problem. Our appetite for meat and dairy products has led to large-scale industrialization of agriculture, so that beef and dairy cattle and other ruminants, like sheep, have become major producers of methane. Other sources are rice paddies, termites feeding on plant material, releases in the drilling for petroleum, decomposition of waste in landfills, decay in wetlands, and leaks in natural gas lines. Studies of ice cores indicate that the levels of atmospheric methane have doubled (increased 100 percent) in the last one hundred years and possibly quadrupled since the last ice age.

Although it is two hundred times less abundant in the atmosphere than carbon dioxide, the insidious problem of methane is that

Table 9.1
Common Greenhouse Gases and Their Estimated Effects

Gas	1986[a]	2030[b]	Warming[c]	°C/ppm[d]	Ratio
CO_2	346	112	1.0	.009	1
CH_4	1.65	1.2	.35	.3	30
N_2O	.305	.05	.20	4	400
O_3	.035	.03	.3	10	1000
CCFC-12	.0004	.0005	.08	160	18,000
CFC-11	.00023	.0005	.07	140	15,000

a. 1986 concentration in parts per million (ppm)
b. estimated increase (ppm) by year 2030
c. estimated temperature rise (°C) due to increase in each gas
d. warming divided by change in 2030
Data from Abrahamson 1989, pp. 141, 222, 232, 244

it absorbs infrared radiation about twenty times more efficiently and at higher energy than carbon dioxide. It also strongly absorbs radiation with a wavelength near 8 microns (figure 9.3), where carbon dioxide and water have a "window", that is, they absorb very weakly, as shown in figure 9.7.

So, increases of a couple ppm of methane have a more pronounced effect than in the case of CO_2, as shown in table 9.1. If we examine the structures of methane and carbon dioxide,

CH_4 O=C=O

we note that methane contains one single bond between the centrally placed carbon and each hydrogen. These four C—H bonds vibrate at higher frequency (higher energy) than the two double bonds between each oxygen and the carbon in carbon dioxide. Thus, methane can potentially store more heat than carbon dioxide. (Note that each bond, symbolized by a dash, represents a pair of electrons shared between the two atoms.)

Nitrous oxide, N_2O, is more commonly called laughing gas, a general anesthetic used in dentistry. The main sources of excess atmospheric nitrous oxide are the burning of coal, especially in power plants, and agriculture in which nitrogen from fertilizers combines with oxygen and is released. Deforestation also produces some nitrous oxide. It is present at less than a thousand times the

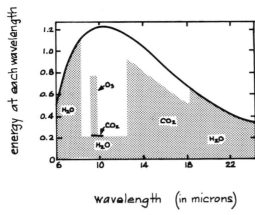

Figure 9.7 Profile of infrared absorption by Earth's atmosphere. The heavy curved line represents the top of an "envelope" of radiation that could be absorbed, while the hatched area shows the areas of radiation actually absorbed. The open spaces show radiation vulnerable to absorption by increased greenhouse gases. (Adapted and reprinted from *The Challenge of Global Warming* by permission of Island Press. Copyright (c) 1989 Island Press.)

carbon dioxide concentrations and also absorbs heat of different energies.

Another greenhouse gas that results from combustion of fossil fuels is ozone, albeit indirectly. Gasoline and other transportation fuels are produced by refining petroleum or in some countries from coal. In the classic smog situation, such as the Los Angeles area, the ozone molecule results from reactions of unburned fuel and/or oxides of nitrogen in the presence of abundant sunlight. Ozone is a very reactive substance, hence the damage to plant and animal tissues it causes. This ozone is confined to the stratosphere, where it is dissipated by chemical reactions. However, we keep producing a continual and increasing amount of it from transportation and also power plants.

The last main category of heat-absorbing gases are the CFCs, compounds of carbon, chlorine, and fluorine and commonly called *Freons*. Freon-12, CF_2Cl_2, can be used in refrigeration and the manufacture of styrofoam, while Freon-11, $CFCl_3$, is used as a propellent for aerosol sprays. Although their scientific names are complicated, their structures are simple variations of methane—chlorine and fluorine atoms merely take the place of the hydrogen atoms. These compounds are present in very minute concentration, less than one part per billion, but they are extremely efficient at absorbing heat and also are very long-lived. One reason that they are better absorbers is that they are more complex molecules, so one CFC is roughly equivalent to 10,000 carbon dioxide molecules.

The initial reason for the popularity of CFCs was their almost inert chemical reactivity, thus making them relatively safe. Their ability to absorb radiation is classified as a physical property, while their reactivity in the stratosphere, which endangers the ozone layer, is a chemical property (box 9.3).

Although the most emphasis in discussions of the greenhouse effect is usually placed on carbon dioxide, the importance of the trace greenhouse gases cannot be neglected. There is general agreement among scientists in this field that these other gases combined can account for an increase in temperature equal to that of carbon dioxide. Besides the fact that these gases seem to be increasing at a faster rate than carbon dioxide, they also tend to absorb infrared radiation where carbon dioxide has an absorption window. In figure 9.7, the solid curve at the top shows the profile of infrared energy emitted by the Earth's surface. The hatched area under the curve indicates that existing water and carbon dioxide already absorb most of the IR. However, there is a large gap between 8.5 and 12.5 microns, where heat can leak through except for a narrow absorption by ozone. It is in this region where CFCs, nitrous oxide, and methane can play a key role. Relatively small *absolute* changes will lead to significantly more absorption in this window.

In contrast, much larger changes in the carbon dioxide levels will have less of an effect, because carbon dioxide already is absorbing most of the radiation in the region where it is a strong absorber (12.5 to 18 microns). In other words, that region of the spectrum is nearly saturated by absorbing molecules. This is not to say that carbon dioxide emissions are not a serious problem, but we must be careful not to ignore other substances that can absorb heat.

Sinks for Trace Gases

Methane has a lifetime of about ten years in the atmosphere, so what we produce today will be absorbing heat up to a decade from now. It is removed by photochemical reaction with reactive trace mole-

Depletion of the Ozone Layer

In the mid-1970s, Rowland and Molina proposed that the synthetic compounds called Freons, used as propellants and refrigerants, were a potential threat to the world's ozone layer. This layer of gas, composed of O_3, ozone molecules (a cousin of O_2, molecular oxygen), is located in the stratosphere about 20 km above the Earth's surface. The Freons, or chlorofluorocarbons, are composed of carbon, chlorine, and fluorine, as implied by the name. Their chemical formulas are CF_2Cl_2 (Freon-12) and $CFCl_3$ (Freon-11).

These molecules and others of similar composition were valued because of their lack of chemical reactivity under ordinary conditions. Their volatility also made them useful as propellants and refrigerants. Later on, after they had been manufactured and used on a large scale, it was found out that they interacted with ultraviolet light to produce very reactive chlorine atoms.

Atmospheric scientists discovered in the mid-1980s that the ozone above the South Pole was severely depleted by as much as 50 percent in the early spring. This observation was repeated several years in a row by the late 1980s. Along with the seasonal drop in ozone levels, significant concentrations of the ClO molecule were found. The proposal was made and subsequently verified that the ClO resulted from the chlorine atoms that had come from the unreacted CFCs drifting upward. Moreover, it was also shown that ClO was a remarkable catalyst for the decomposition of ozone. Each ClO molecule caused the destruction of perhaps thousands of O_3 molecules before it was destroyed.

The fact that scientific proof was able to be developed led in 1987 to international agreements aimed at limiting the production and release of CFCs and halons, another class of synthetic chemicals. These agreements, named the Montreal Protocols, contained a schedule for phasing out or minimizing the use of ozone-destroying compounds over a period of about twenty years.

The key difference between the ozone depletion scenario and that for global warming is that in the former case definite and irrefutable evidence was obtained to support a theory proposed about a dozen years earlier. The victim and the smoking gun were both found in the actual depletion of ozone layer and the simultaneous appearance of the reactive ClO molecule. With these results, government officials could proceed with negotiations to fashion effective international policies.

Box 9.3

cules. In contrast, ozone survives for only several months. This reflects the fact that it is very reactive and also undergoes photochemical reactions.

Nitrous oxide is destroyed by uv light, but this is not an efficient process—it takes 150 years for it to be eliminated. The CFCs also are long-lived—Freon-11 lasts 65 years and Freon-12 survives for twice that time, 165 years. They and other CFCs are destroyed in the stratosphere by photochemical processes that then result in the destruction of the ozone layer. Thus, the aerosol used today will persist and could have warming effects well into the middle of the twenty-first century.

Water Everywhere

Water has not been portrayed as a particularly dangerous suspect because atmospheric water is in rapid equilibrium with the oceans and other sinks. So excess water produced from combustion will probably not affect the balance significantly. However, in a hotter climate the air will be able to hold more water vapor, and consequently more heat could be retained. Also, the role of water as the stuff of clouds is another uncertain prospect. Some types of clouds have a net cooling effect, whereas others aid warming, and the projected mix of cloud types is very nebulous.

In terms of its absorption spectrum in the infrared region, water already present in the atmosphere is absorbing essentially all the radiation in the region where water absorbs strongly. If additional water vapor was produced, there would be some heat absorbed in the window in which water absorbs weakly. So, more water would have a minor but not insignificant effect.

Consequences of Global Warming

What are the consequences of a few degrees increase in the average global surface temperature? Those of us living in northern temperate latitudes of the United States often complain about cold winters and move to Florida or the Southwest, so a little balmier weather would be welcome. If that was all there was to it, the greenhouse effect might be just some interesting scientific problem. Unfortunately, when the problem involves even small changes but on a large scale, the consequences are potentially vast and vary from one region to another.

The two main risks involve the rise of the ocean levels and the shift in weather patterns. These would then produce indirect effects that would have social, economic, and political consequences. There is also the danger of "positive feedback" that could exacerbate the global warming by increasing the rate at which greenhouse gases enter the atmosphere.

Let us first deal with the potential rise in sea levels. The most obvious cause would be the melting of ice packs on land, such as glaciers, the Greenland ice shelf, or the West Antarctic ice sheet. Melting snow packs and glaciers are most likely to be the initial problem, and this would add more water to the oceans and geosphere. The other, more subtle phenomenon at work here is a very simple principle: water expands when heated and its density decreases. Although this effect over several degrees Celsius is small (0.1 percent increase in volume from 15°C to 20°C), it becomes a major change when applied throughout the oceans worldwide.

The combination of expansion and glacial melting would result in the ocean surface's rising one-third meter (1 foot) or even as much as 1 meter by the middle of the twenty-first century. This would cause serious problems for low-lying nations like Bangladesh and Holland, and low areas and cities throughout the world would be at risk. New Orleans, Miami, Venice, and many others might need to cope with displaced homes, businesses, industries, and transportation systems. Disruptions to fisheries and agriculture systems, such as the Nile Delta and Chesapeake Bay, would occur.

Of course, even more severe damage would occur if the ice sheets in Greenland and Antarctica would start to melt. Then we would be faced with rises of 5 to 8 meters. Half of the people of Bangladesh live on land that is 5 meters or less above sea level. The displaced population, along with the impact on agriculture, would have enormous consequences for that country and its neighbors with the food and refugee problems that would be created. In the United States, most of south Florida would be under water, and several million people would be displaced.

The other consequence of warming could be even more devastating. Changes in rainfall, temperature, and humidity over large regions of the world could have immense and unprecedented results. For example, if the American Midwest became significantly drier, the vast wheat and corn growing areas may be crippled, perhaps even wiped out. These grain belts may shift northward even into parts of Canada. These agricultural shifts could have enormous economic ramifications on the U.S. balance of payments and standard of living.

Hotter weather in the midlatitudes of the United States might also mean greater discomfort with more hot, humid summers in some areas like the Southeast. This would mean increased power demands to run air conditioners and even to maintain present levels of refrigeration. At the same time, other regions might experience more rainy weather, which could assist their agriculture production but could also cause negative impacts, such as flooding. Increased rainfall, such as has occurred in the Great Lakes region in the past two decades, could cause troublesome rises in large lakes and rivers.

Some forecasters claim that warmer climates will allow more vegetation to grow and thus consume excess carbon dioxide to establish a new equilibrium for the globe. This is an example of negative feedback acting to regulate a complex system, a well-known process in physiology and engineering. However, some fear that a process of positive feedback might just as well occur in which the warmer climate leads to more rapid decay of organic matter, thus increasing the atmospheric carbon dioxide and leading to further warming, which produces even more carbon dioxide. Melting of Arctic permafrost may release more methane, which will enhance the possibility of warming. The climate would then warm up even more, leading to more decay, and so on.

Whatever and wherever the impacts, it is clear that world leaders need to confront this issue seriously. An international conference of environmental ministers met in late 1989 in the Netherlands. They reached no specific goals or deadline except to concur that there was a need to adopt some guidelines in the near future.

The Political Question: Is Action Needed?

The greenhouse effect poses the type of problem that governments and international organizations probably like to deal with least of all. Most governments are reasonably quick in acting to cope with disasters, imminent crises, external threats of war, and clear-cut national needs, such as transportation and health care. The environmental arena poses a host of scenarios with worst-case impacts that may be disastrous, with present negative effects that are minimal, and with future consequences that are very uncertain. As a result, governments are slow to take action and are constantly looking for more proof by urging a seemingly never-ending series of research programs, review panels, and policy reports. This is especially true when they are confronted with a "slow-motion crisis," which will not bring doom tomorrow. When the problem requires international cooperation, then the complexities multiply.

Witness the long struggle to reach some agreement for concrete action on the acid rain question between Canada and the United States. The Canadians were frustrated time and again by Washington's insistence on more data gathering. Of course, governments are to some extent under the influence of powerful special interests to pursue one course or the other. In contrast, the rapid agreement called the Montreal Protocol regarding cutbacks on the use and production of CFCs was reached in 1987, not many years after definitive data were obtained showing "holes" in the ozone layer over Antarctica. Some critics suggest that this pact was also too slow in coming, since the theory behind the effect was first proposed in the mid-1970s.

Even when there is general consensus concerning a particular problem, it is still unclear who should lead the way in reacting to the threat. On global environmental matters should the United Nations be the key player? Or should the nations most responsible for the perceived danger be the ones to make policy, as with the Montreal Protocol? Or should nations unilaterally decide to act, based in part upon moral principles? Of course, there is always the worry that by taking action, the wrong approach will be used or that the situation was not as grave as earlier perceived. In the meantime, citizens of one country or region are put at a disadvantage compared to those of other nations.

One pitfall to avoid is the one in which some countries think that they will be winners because their weather and agriculture is predicted to improve. The unpredictability of the regional consequences of global warming are bad for every country because of the instability it creates. Another danger involves pitting the industrialized counties against the developing ones. In the process of creating richer economies, energy use is almost certainly to increase. Plans must be devised to allow the developing nations to move along while the richer nations become more efficient.

Nevertheless, we may be in the situation of a commander who has been advised that an enemy attack might be imminent in one area. Does the commander respond immediately and divert forces from another area that then might become vulnerable? Or is it better to wait until the point of attack is certain? Time is a critical factor here, and action may have to begin before the information is complete. Or is the doomsday aspect of the problem overstated, and we really do have the luxury of waiting for more data?

Strategies

Several general approaches have been outlined by various writers on this topic. We will deal with three of them: (1) do nothing; (2) adapt to climate changes; or (3) prevent the changes.

The first approach essentially goes back to the "give us proof" attitude by funding continued scientific research on the various aspects of global warming. It is favored by those who are skeptical that anything at all is going to happen in terms of global warming. The operational effect of this approach is to wait until unambiguous evidence is available and then try to deal with any resulting crises.

Supporters of this view are also legitimately concerned with the immense amount of money that could be wasted if the dire warnings prove false. Scrubbing the carbon dioxide out of the exhaust gases of power and industrial plants in the United States would cost nearly $1 trillion or about $3000 per American. If Brazil were to invest in high-efficiency electrical equipment (motors, lights, appliances), it would cost about $4 billion, but it would save $19 billion in costs for power plants that would not have to be built. To develop additional energy technologies, as well as improve existing ones, would require $1 billion per year for the near future.

The adaptation strategy is based upon the ideas that we, as a species, are very clever and have adjusted to a great variety of habitats and climates already. Therefore, we should be able to cope with events that will slowly (we hope) develop over a quarter of a century or more, especially given the power of our modern technologies. It does not deny the possibility that global warming will occur or that it may already have begun. This approach embodies the notion that technological fixes can solve the problem. We might also employ economic approaches, such as "carbon taxes," to provide disincentives for using fossil fuels, thus slowing down the warming trend.

The prevention strategy is the most activist and rejects the idea of waiting before taking significant actions because that could lead to more severe consequences that will be more difficult to deal with. It would include a wide array of policies and agreements designed to attack the sources and practices that yield the greenhouse gases, especially carbon dioxide and the CFCs. These are not really attempts to prevent global warming as much as they are actions to minimize the consequences of a worst-case outcome. There are different versions of strategies ranging from assessing carbon taxes to substituting alternative energy sources for fossil fuels.

The more moderate version of prevention involves implementing practices that in themselves are beneficial and, as a bonus, also help to ameliorate the warming possibility. For example, the United States already faces serious problems directly related to fossil fuel dependence. These range from balance of payments deficits, to health-damaging air pollution in many cities, to environmental catastrophes on land and sea. Thus, decreasing our dependence on these energy sources would seem to provide benefits in addition to warding off the specter of global warming. However, adopting such a strategy may call for major and rapid changes in lifestyle, such as car pooling, use of mass transportation, building smaller homes, switching energy sources and fuels, and so forth. Industries and businesses will be forced to change and perhaps be put in decline while new ones rise up. Disruptions are not popular politically, and so many public officials will resist. Modest increases in the gasoline taxes simply to pay for improved highways and mass transport are viewed as anathema by many with a "no new taxes" viewpoint.

Another prevention activity involves a massive reforestation program including the "urban forest." The average growing tree absorbs on the order of 10 kilograms of carbon dioxide each year in the process of photosynthesis. Trees also have been documented as filters for air pollution and as aids to cooling, thus cutting air conditioning energy use. One power company has agreed to plant fifty million trees in Central America to offset the carbon dioxide production from its new power plant in the Northeast.

One policy that is already in place is the Montreal Protocol aimed at reducing the use and production of CFCs. Although the primary aim of this agreement was protection of the ozone layer, an equally important effect will be a decrease in greenhouse gas levels. Thus, if global warming does actually occur, this plan will include a built-in bonus.

On the other hand, some policies to protect the environment already in place may not be appropriate for the greenhouse problem. For example, the proposal to decrease acid rain by scrubbing smokestack exhaust involves a drop in efficiency, so that more coal must be burned, thus generating more carbon dioxide. A better solution may be to use a cleaner fuel like methane to begin with or to build more efficient power plants to cut coal use. A further complication concerns the claim that the acid rain is offsetting the warming trend by creating more cloud cover. Some have proposed that we reconsider nuclear power as an option for energy production, but there are environmental problems, such as waste disposal, associated even with trouble-free reactors.

An Overview

Steam power was the engine that began driving us down the road leading to this predicament, but it was later abetted by the internal combustion engine, deforestation, and an expanding population demanding industrial goods, abundant food, and comfortable housing. Scientific theory and research helped us to understand the relationships among different forms of energy, thus enabling us to design new and more powerful engines and machines. Government, business, and financial institutions provided an atmosphere for large-scale industrialization. Then, in the twentieth century, science showed us that there may be serious problems caused by these applications. The "great experiment" continues with us as part of it rather than as external, objective observers.

Scientific research is often regarded as the panacea for our problems and the basis for seemingly miraculous technologies. However, knowledge and wisdom are not the same things, and we must be ready to act prudently while we wait for a better understanding of nature. It is not necessary to know all the details before we begin to formulate policies. We can embark on cost-benefit analyses of the strategies just described and adjust the results as new information becomes available. We can implement some policies that target other problems as well as global warming.

Although technology seems to have gotten us into this predicament, it is likely that new science-based technologies and engineering developments will help us to attack the problem. Risks associated with technology include how we choose to use it, or abuse it, as well as the innate risks of the technology itself. The automobile is a classic example of a technology that is all but essential and yet carries with it well-known risks.

In the final analysis, some global consensus will likely be needed in order for leaders to devise effective policies. However, like detectives working on a case with no apparent victim, we still don't know if a crime has been committed. And we may lack the luxury of enough time to keep sifting through the evidence before discovering that something drastic is indeed occurring. Reliable scientific data and theories will be essential in guiding policy-making and determining whether to act, when, and how. The scientific consensus now seems to be that a version of the greenhouse effect will occur, but questions abound regarding when, where, and how much. We will very likely know by 2010 whether this scenario is more like fact or fiction, but, if it does come to pass, there will be no doubt about the true perpetrators of the deed—us.

Exercises

1. Briefly describe the scientific basis for the greenhouse effect. Explain what the term *greenhouse gas* means and give examples. Why is the greenhouse effect both good and bad for our planet?
2. Using the following data, plot a graph and use it to estimate what the average global temperature might be in the year 2040.

Year	CO_2—ppm
preindustrial	280
1860	270–290
1958	310–315
1970	325
1975	330
1982	339
1988	350

Assume that a doubling of atmospheric carbon dioxide compared to the pre-industrial level will cause the temperature to rise 3.0°C. What other assumptions have you used in extrapolating your graph? Why is it likely that your estimate is low?

3. (a) Based upon the chemical equation for the combustion of octane given in this chapter, estimate the number of tons of carbon dioxide produced each year by automobiles in the United States.
   (b) How many trees would have to be planted to offset the carbon dioxide produced by cars in the United States? Assume each new tree uses 10 kilograms of carbon dioxide per year.
   (c) Another strategy for reducing carbon dioxide output from transportation is to increase the fuel economy for automobiles. How much carbon dioxide would be "saved" each year by requiring the nation's cars to average 50 miles per gallon?

4. Of the two approaches for lowering carbon dioxide described in exercise 3, which one would you choose? Clearly state both the quantitative reasons and the value judgments for your decision.

5. Besides transportation, what other major sources of carbon dioxide are there? Name three and rank them in order of importance.

6. Suppose an international agreement was reached on strict controls on fossil fuel use and deforestation, but with no specific provisions for other greenhouse gases. What would be the benefits of this strategy? What is deficient about it?

7. Suppose that a 200,000 kW power plant is scheduled to burn natural gas instead of coal to lessen air pollution. Determine whether this approach will increase or decrease carbon dioxide emissions by estimating the emissions for each fuel for the same amount of electrical energy produced. Some relevant information: each kilogram of coal produces 2.2 kilogram of carbon dioxide and 7.4 kWh, while one kilogram of methane generates 2.8 kilogram of carbon dioxide and 14 kWh.

8. Discuss the differences between the "adaptation" and "prevention" strategies. Cite three reasons for using each approach. As a public policy advisor, which one would you recommend? Explain your position.

9. Suppose that the United States set a goal of carbon dioxide emission reductions of 40 percent from the current 5 tons per capita annually and

that China wished to increase its GNP to 20 percent that of the United States. China's current emissions per capita are 0.4 ton per year at most, and it relies mostly on coal as an energy source. What would the net effect on world carbon dioxide emissions be?

10. One proposal for reducing air pollution that is being considered very seriously is a plan to replace current fuels with ones that already contain some oxygen, so that they would burn more completely. The most popular scheme would use methanol, CH_3OH, as a liquid fuel in place of gasoline, C_8H_{18}. One immediate difference between these two fuels is that the miles per gallon of methanol is about half that of gasoline, so you would need to fill your tank more often. But methanol is proportionately cheaper, so the cost per mile driven is about the same as with gasoline. Compare the amounts of carbon dioxide produced by both gasoline and methanol in traveling 100 miles. Assume your car gets 25 miles per gallon for gasoline. Methanol weighs 6.6 pounds per gallon, and it combusts according to:

$$CH_3OH + 1.5\ O_2 \rightarrow CO_2 + 2\ H_2O$$
$\quad\ \ $|$\qquad\qquad\qquad\quad\ \ $|
$\ $6.6 lbs$\qquad\qquad\qquad$9.08 lbs

Gasoline reacts with oxygen according to:

$$C_8H_{18} + 12.5\ O_2 \rightarrow 8\ CO_2 + 9\ H_2O$$
$\quad$|$\qquad\qquad\qquad\qquad$|
$\ $5.6 lbs$\qquad\qquad\qquad$17.3 lbs

As a policy analyst, would you recommend the proposed change? Explain the basis for your decision.

Atomic Power: Difficulty in Estimating Cancer Risks

10

"It Worked"

At 5:29:45 on the morning of July 16, 1945, an unnaturally brilliant flash seared the New Mexican desert at a remote site known as Jornada del Muerto, the Journey of Death. Eleven pounds of plutonium, a baseball-sized lump, had released in less than one millionth of a second enough energy to equal the explosion of 37 million pounds of TNT. An awestruck observer twenty miles away described the event: "The flash of light was so bright at first as to seem to have no definite shape, but after perhaps half a second it looked bright yellow and hemispherical with the flat side down, like a half-risen sun but about twice as large (figure 10.1). Almost immediately a turgid rising of this luminous mass began, great swirls of flame seeming to ascend within a rather rectangular outline which expanded rapidly in height ... Suddenly out of the center of it there seemed to rise a narrower column to a considerably greater height. Then as a climax ... the top of the slenderer column seemed to mushroom out into a thick parasol of a rather bright but spectral blue ..." (Rhodes 1986). Isidor Rabi, Nobel laureate physicist and reluctant consultant to the Manhattan Project, tersely captured the historic transition: "This power of nature which we had first understood it to be—well, there it was" (Rhodes 1986). The atomic age was born.

Frank Oppenheimer, the brother of Robert Oppenheimer who directed the Manhattan Project, also witnessed the first atomic bomb

Figure 10.1 The first atomic bomb explosion, code-named Trinity, near Los Alamos, New Mexico, on July 16, 1945. The photograph was taken just 53 thousandths of a second after detonation. (Photograph courtesy of Los Alamos National Laboratory.)

explosion and felt a profound "sense of this ominous cloud hanging over us. It was so brilliant purple with all the radioactive glowing. And it just seemed to hang there forever ... It was very terrifying ... I wish I could remember what my brother said, but I can't ... I think we just said 'It worked'" (Rhodes 1986). Other observers shared the sense of terror and menace, but these feelings mixed with awe for the "grandeur" of the explosion, elation over an arduous task successfully completed, and pride from having gained a "new control ... over nature." The pride was well justified. Some of the greatest nuclear physicists had converged on Los Alamos, New Mexico only twenty-nine months earlier to work on a problem many people thought could not be solved. Staggering difficulties in nuclear physics and engineering stood between them and an atomic bomb. The July 16 detonation, code-named Trinity by Robert Oppenheimer in obscure reference to a John Donne poem that begins "Batter my heart, three person'd God," testified not only to Nature's power but also to the remarkable power of human reason.

What had begun as a drive to understand one of the great mysteries of nature had come to that deceptively simple phrase: it worked.

According to Robert Oppenheimer, they "knew that the world would never be the same. A few people laughed, a few people cried. Most people were silent. I remembered the line from the Hindu scripture, the *Bhagavad-Gita*: Vishnu is trying to persuade the Prince that he should do his duty and to impress him he takes on his multi-armed form and says, "Now I am become Death, the destroyer of worlds."

Oppenheimer's thoughts proved prophetic with undeniably horrible effect on August 6, only three weeks after Trinity, when the atomic bomb named Little Boy exploded over Hiroshima. Three days later, the new destructive force was unleashed on Nagasaki, an alternate target chosen at the last minute because of unfavorable cloud cover over Kokura, the primary target. One hundred and forty thousand of Hiroshima's residents, more than half of the city's population, died by the end of 1945. An additional 70,000 people died at Nagasaki in the same period. The blasts vaporized people near ground zero. Others died more slowly from burns and radiation sickness. More than half of all the buildings in Hiroshima were completely destroyed.

Few could have foreseen the destructive power of these bombs, but the physicists involved in their production certainly appreciated the unprecedented magnitude of the force better than the politicians who figured in their use. Einstein's famous equation, $E = mc^2$, means that two ounces of uranium release more energy through fission than is produced by burning three and one half *tons* of coal! The profound political and social consequences of tapping this great pool of energy seemed to go almost completely unappreciated by President Franklin Roosevelt and Prime Minister Winston Churchill, at least until August 6, 1945. Churchill and a number of others skeptically viewed the new devices merely as "bigger bombs." Control over nature's immense energy source, long hidden in the atomic nucleus, meant more than bigger bombs, however. It meant a new age, one that combines terrible responsibility with great promise. A number of nuclear physicists, most notably the great Danish scientist Neils Bohr, fervently hoped that atomic bombs would pose such an unacceptable threat that war would become unthinkable. Conflict between fear and hope marked this new age from the beginning as one of great uncertainty.

President Dwight Eisenhower saw another promise in nuclear fission. His 1954 initiative, dubbed "Atoms for Peace," sought to promote the atom as a source of energy for generating electricity. Peaceful uses of nuclear fission would not only provide an apparently limitless source of energy for the world's populations but would also open channels for negotiating limits to the spread of nuclear weapons, or so he hoped.

The first successful nuclear reactor was built in the service of the atomic bomb project. Enrico Fermi, a physicist who had fled fascism in Italy and came to the United States, and his coworkers assembled 93,000 pounds of uranium and 771,000 pounds of graphite into a roughly egg-shaped pile that produced the first sustained, controlled nuclear chain reaction on December 2, 1942. Fermi's reactor demonstrated the feasibility of nuclear chain reactions, but it produced pitifully little power, only a few watts. Nuclear reactors would have to produce much more power than that to be commercially useful. Another nuclear reactor constructed at Hanford, Washington in 1943 produced electricity, but its main initial purpose was to supply plutonium for the Manhattan Project. The first nuclear reactor to generate electricity as its primary function appeared in 1954 in the propulsion system of an American submarine.

The Submerged Atom Goes Ashore

A submarine is not the most comfortable or convenient place to spend a long time, and there is also a problem of refueling. Conventional submarines must come to the surface to refuel, not the submarine's strongest position. Suppose, however, a submarine's power plant had an almost inexhaustible supply of fuel. It could stay submerged and remain virtually undetectable for months. Admiral Hyman Rickover believed in this concept and adopted it as a personal crusade. His enthusiasm and determination led to the launching of the *Nautilus*, the first nuclear powered submarine, in 1954.

Meanwhile, the Atomic Energy Commission (AEC), which had been established in 1947 to oversee the development of atomic energy, was trying to convince electric utilities to use nuclear power plants. Some critics claim that this effort was actually one to make military R&D seem more respectable, or at least make atomic power seem less ominous. Regardless, utility companies balked at the government's invitation because the economic prospects for nuclear power appeared uncertain at best. Surveys suggested that nuclear power could be sold at prices competitive with those of coal-fired power plants, but many utility companies believed that the government was trying to spur a pace of nuclear power development that exceeded demand. Uncertainty over the reliability of an untested technology also made many companies reluctant to invest in nuclear power. Besides, coal and oil seemed abundant then, so the motivation for moving to another energy source was low. The federal government kept insisting, however, and made it so inviting that the utilities were drawn into "experiments" and "demonstrations" despite severe misgivings about liability in case of an accident. The AEC addressed the liability issue by placing a rather low cap of $560 million on any damages that could be collected from an accident.

Research and development moved at a quick pace, and in 1957 the first domestic nuclear power plant went on line as a demonstration at Shippingsport, Pennsylvania near Pittsburgh. Several other nuclear power plants were built as demonstration facilities between 1954 and 1962. The technical success of these facilities encouraged utility companies by showing that it was easier than expected to integrate nuclear power with electricity from other sources and by demonstrating the apparent reliability of nuclear power plants. The economic picture was not so flattering. Construction costs for the demonstration plants ran over budget, and the high operating costs made the nuclear plants uncompetitive with fossil fuel plants. Nevertheless, nuclear power began to flourish in a climate of optimism. A report commissioned in 1962 by President John Kennedy predicted a bright future. The nuclear power industry would grow to a capacity of 40 million kW by 1980 and produce half of the nation's electricity by the end of the century, according to the report. The estimates grew even rosier over the next few years and bumped the predicted 1980 capacity to 150 million kW. These numbers reveal an astonishing level of confidence in the feasibility and advisability of nuclear power given that it accounted for less than 0.5 percent of the electrical power produced in the United States during 1967.

Events following Kennedy's report seemed to bear out its predictions. Construction had begun on five major power plants by the end of 1963. The first of these to go on line was the 428,000-kW unit at San Onofre on the Southern California coast that reached full power near the end of 1967. Orders for nuclear power plants jumped sharply in 1966 as did their size. Plants with capacities of about 500,000 kW, like the one at San Onofre, gave way to designs for 1 million-kW stations. Nuclear power had suddenly become an attractive option to utility companies as the costs of transporting coal ceased to decline and public concern over air pollution and the ravages of coal mining steadily increased.

Doubts about the Nuclear Genie

The dream for abundant nuclear power in the United States appears to be over. No nuclear power plants have been ordered since 1978. The total power-generating capacity of nuclear power plants in this country stood at 116 million kW and accounted for only one-fifth of the total electricity actually generated as of January 1989. Although nuclear energy now ranks second only to coal in importance for electrical power production in the United States, there appears to be no way of meeting the predicted 50 percent level by the year 2,000. Nuclear power plants have become infamous for cost-overruns during construction, and some projects, such as the Shoreham nuclear power plant in New York, have been suspended indefinitely with huge bills yet to be paid. This rapid decline in the prospects for nuclear power in the United States parallels a major shift in public opinion.

Accidents at the Windscale nuclear reactor in England in 1957 and the Three Mile Island Reactor in Pennsylvania in 1979 alarmed a public already worried about risks stemming from nuclear power. People began to fear nuclear radiation and its association with cancer and death. Atomic bombs took nuclear radiation from the realm of medicine and made it terrible. Now that threat appeared to emanate from nuclear reactors developed on promises of safety and reduced pollution. Four years prior to the accident at Three Mile Island, more than 60 percent of American respondents to various surveys claimed to be in favor of nuclear power, while less than 20 percent

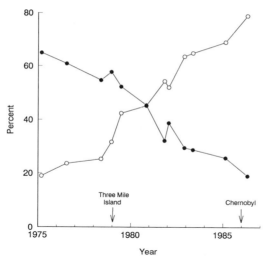

Figure 10.2 Proportion of people surveyed in the United States who approve (filled circles) or disapprove (open circles) of building more nuclear power plants. The years in which accidents occurred at the Three Mile Island and the Chernobyl nuclear power facilities are indicated. (Redrawn from Worldwatch paper #75.)

said they were opposed. Eleven years later those figures were more than reversed. Polls taken shortly after the Chernobyl accident in the Soviet Union in 1986 revealed that nearly 80 percent of respondents declared opposition to nuclear power. The Chernobyl disaster remains the worst accident in the history of nuclear power, but the shift in public opinion began years earlier (figure 10.2).

According to a recent poll, American college students and members of the League of Women Voters ranked nuclear power as the most threatening of thirty listed items, riskier than motor vehicles, hand guns, and smoking. (People commonly rate tobacco smoke as less risky than nuclear power despite the fact that 30 percent of the cancers in the United States can be traced to smoking.) People with expert knowledge of nuclear power, however, ranked it as the twentieth most risky item on the list. (They placed motor vehicles first.) What accounts for this striking discrepancy? The key lies in uncertainty. Activities like nuclear power are dreaded to the extent that the risks are uncertain. Our inability to feel low doses of nuclear radiation combined with a potentially fatal risk that seems beyond our control, the delayed effects of nuclear radiation, and the idea that the risk is imposed on us without our consent all help make nuclear power one of the most dreaded of modern technologies. This combination of factors may explain why X rays were perceived as much safer than nuclear power by people surveyed in the poll cited above. High doses of X rays can produce fatal cancers just as readily as radioactivity from uranium, yet college students ranked X rays seventeen positions below nuclear power—just riskier than hunting and not quite as risky as commercial aviation. The association of X rays with freely elected, medical diagnostic procedures may make the risk of X rays seem more acceptable than nuclear power, although they remain just as mysterious to most people. The "experts" who participated in this poll, however, rated X rays as a much more serious threat than nuclear power. Knowledge of radiation seems to exert a powerful influence on perceived risk. When the knowledge is extensive, X rays are perceived as more risky than nuclear power. When the knowledge is presumably less extensive, that perception is reversed.

The souring of public opinion on nuclear power in the United States has no doubt contributed to the decline of the industry from its heyday in the late 1960s. As a result, America is now proportionately less dependent on nuclear power for electricity than twelve of the twenty-seven countries that operate nuclear power plants. American utility companies produced 455 billion kWh of electricity from nuclear reactors in 1987, more than any other country, but we used so much electricity that nuclear power represented only 18 percent of our total production in that year. France, on the other hand, generated much less electricity from nuclear power in 1987, but what was produced amounted to 70 percent of its total electrical power production. Belgium followed close behind at 60 percent. Nuclear power enjoys a somewhat better reputation in these countries than in the United States. It is not clear whether favorable public opinion aided the development of nuclear power in France, for example, or followed a concerted governmental effort to promote nuclear power. Nevertheless, the policy regarding nuclear power in these countries reflects a notion that the benefits of nuclear power are worth the perceived risks.

Have France and Belgium made a better choice than the United States regarding nuclear power? If by reducing their dependence on fossil fuels, which in turn would lessen health risks from air pollution and threats of atmospheric warming from the greenhouse effect, then the decision to "go nuclear" seems correct. Nuclear power would not appear nearly so attractive if the benefits of reduced fossil fuel use are exceeded by risks of cancer and genetic damage from radiation. Just how large are these risks? That is the key question. In the remainder of this chapter, we try to sort through some of the key issues raised by that question. We start by briefly examining the nature of radioactivity and the scientific history that put us in a position to worry about it.

A Matter of Matter

The challenge facing nuclear scientists and engineers has been to find methods for controlling the atom's power and turning destructive energy to humanity's advantage. The discovery of atomic energy and the development of nuclear power provide one of the most significant and exciting chapters in scientific history. The roots of atomic power extend well back into the nineteenth century when, in 1820, the English scientist John Dalton proposed the modern version of the atomic theory. In doing so, he borrowed the Greeks' general notion of matter in the process: different materials result from different combinations of elements and each element consisted of specific types of atoms. Dalton's atomic theory did not immediately triumph, and it took nearly a century to become a pillar of science. The development of a periodic table of chemical elements by Mendeleev of Russia and Meyer of Germany in the 1860s helped clarify the situation by revealing an orderly progression in the characteristics of nature's elements. This achievement increased the urgency of questions about the nature and number of elements. Physicists, in particular, translated this query into: "What are atoms made of?" The pursuit of an answer to this apparently innocent question in the last decade of the nineteenth century and the first decade of the twentieth led to major discoveries that radically altered our perception of the natural world.

The Atom Yields Its Secrets

The super-sleuth scientist who cracked the atomic mystery had surprising origins because he arose not from the intellectual ferment

Figure 10.3 Marie Curie (seated) shown in her laboratory in 1912. (Photograph reproduced, with permission of the publisher, from: *Madame Curie* by Eve Curie, translation by Vincent Sheean, translation copyright 1937 by Doubleday, a division of Bantam, Dell Publishing Group, Inc.)

of continental Europe or even rapidly developing North America, but from New Zealand. Few places were more removed from the centers of western science in the 1880s than New Zealand. Ernest Rutherford, who grew up on a farm on the South Island of New Zealand, was undeterred and overcame both distance and the even more formidable atom with his genius for concepts and remarkable experimental skill.

Rutherford, after distinguishing himself as a physics and mathematics student at the University of New Zealand, was accepted in 1895 for doctoral work in physics at the prestigious Cavendish Laboratory at Cambridge University in England. He worked for several years on a project involving the newly discovered "radio waves" before shifting his interest in 1898 to another topic for his Ph.D. thesis. He turned from radio waves to questions regarding the nature of the mysterious "rays" that emanated from certain heavy elements like uranium and thorium. These rays had been discovered just two years earlier by Henri Becquerel, a French investigator who studied uranium, and by his former student, Marie Curie (figure 10.3), and her husband Pierre who had isolated the elements thorium, radium, and polonium (named after Marie Sklodowska Curie's homeland). These elements and their compounds surprised investigators by being able to expose photographic film sealed in a paper envelope. Becquerel and the Curies quickly realized that these particular elements must be spontaneously releasing something that penetrated the paper and exposed the film. Marie Curie termed this novel process "radioactivity."

The rays observed by Becquerel and the Curies bore an intriguing similarity to another type of ray reported in 1894 by the German physicist, Wilhelm Röentgen. Röentgen had observed that photographic plates held in front of a cathode ray tube became exposed although the tube emitted no visible light (see box 10.1). He suggested that the exposure must be produced by "X rays" emitted from the tube. Röentgen made the world's first X-ray photograph when he held his hand between the tube and a photographic plate.

Stimulated by these discoveries, the young and brilliant Rutherford entered the field eager to make his mark as a physicist. Were the rays emitted by radioactive elements another type of X ray or different altogether? Rutherford began to answer this question by watching the behavior of radioactivity rays in a magnetic field. These simple experiments led Rutherford to conclude that radioactivity is not a single entity but a composite of two types of rays made of particles differing in mass and charge: the more massive one was positively charged, whereas the very much smaller one carried a negative charge. Rutherford named the positive ones *alpha rays* and the others *beta rays* (see box 10.2). He eventually determined that the beta rays were actually electrons, and the alpha rays behaved like positively charged helium atoms. Each of these two particles contrasted sharply with X rays because they were charged and could be stopped by much less material than was needed to block X rays. Two years later, in 1900, the Frenchman Paul Villard discovered a third type of ray, one similar to X rays but even more energetic. Following the nomenclature sequence begun by Rutherford, he labeled these *gamma rays.*

Due in part to this work, Rutherford was offered a position in physics at McGill University in Montreal. There, he and his associates, including Frederick Soddy and Otto Hahn, proceeded to

Cathode Ray Tubes: Test Tubes for Atomic Particles

Late in the nineteenth century, physicists discovered that nearly evacuated sealed glass tubes gave "glows" or discharges when subjected to strong electric fields, with one end of the tube hooked to the positive terminal of a strong battery and the other end connected to the negative pole. When a metal object was placed in the tube, its shadow appeared on the positive end of the tube. These discharges were called *cathode rays* since the negative end of the tube was by convention called the cathode. The tubes were also known as *Crookes tubes* in honor of William Crookes, the English scientist, who had initially investigated the mysterious rays. The obvious question continued to vex the physicists who studied this phenomenon: what were the rays made of?

In 1897, J. J. Thomson of Cambridge University identified the source of the rays and in the process demonstrated that atoms were not indivisible—they were not the ultimate particles that the Greeks had speculated. Building upon the previous studies of cathode rays, he and his associates showed that these rays were actually extremely light particles with negative charges because they were attracted to a positively charged plate next to the tube. Thomson proposed that these particles or "electrons" were pulled off atoms of gas in the tube by the high-voltage electric fields use in these experiments.

They were also able to measure the "charge-to-mass" ratio of the electron and compare that to the same quantity measured for the positively-charged hydrogen ion (a hydrogen atom stripped of its one electron). Each of these should have the same charge but opposite in sign. The ratio for the electron was about two thousand times larger, indicating that it was a much smaller particle than the hydrogen ion (also called a "proton").

Some fifteen years later, Robert Millikan in America determined the actual value of the charge on the electron in his famous "oil drop experiment." Using Thomson's ratio, a simple calculation then gave him the mass of the electron—which turned out to be almost insignificant compared to the mass of the atom. However, be they ever so humbly small, the electrons determine the elements' chemical properties dear to the hearts of chemists; but physicists were after bigger game—the whole atom.

Today, cathode ray tubes (or CRTs) are used and enjoyed by billions of people in the form of television screens, computer monitors, oscilloscopes, and many other devices.

Box 10.1

The Type of Radiation Makes a Difference

Alpha (α) particles

An α-particle has the same configuration as a helium nucleus: two protons and two neutrons. Lacking electrons, it carries a double positive charge. This massive, highly charged particle moves relatively slowly and leaves a dense track of ionizations as it collides with atoms along its path. This makes α-particles damaging to cells, but it also means that they lose most of their energy in a short distance. Even high energy α-particles travel only a matter of millimeters in air and much less than that in living tissue such as skin. Only when α-emitting substances, such as radium, are ingested do they pose a serious health threat.

Beta (β) particles

Compared to α-particles, β-particles are small and fast. A β-particle is either a high energy electron kicked out of a disintegrating nucleus or the electron's opposite, a positron. Their low mass, combined with high energy, account for the tendency of β-particles to leave well-spaced ionizations along a zig-zag path as they carom off atoms. These particles travel distances of meters in air and penetrate the body's tissues easily, making β-emitters, such as radioactive sodium, a threat whether the source is outside or inside the body.

Gamma (γ) particles

Radioactive decay processes that produce α- or β-particles may also spew forth γ-rays. These photons behave like β-particles in traveling rapidly, penetrating living tissue easily, and leaving erratic tracks of well-spaced ionizations. At 1 MeV, γ-rays can penetrate more than 1 meter through soft body tissue compared to 7 mm for β-particles or 0.001 mm for α-particles at the same energy level.

Neutrons

The neutron's lack of charge allows it to penetrate the body's tissues with ease and to interact with nuclei along the way without interference from charged particles. High-energy neutrons, termed *fast neutrons*, impart enough energy through collisions with other nuclei to give those nuclei ionizing capabilities. Less energetic neutrons, called *slow neutrons*, are more likely to be captured by the nuclei they hit. This makes slow neutrons valuable in nuclear chain reactions because the absorbing uranium or plutonium nuclei become unstable and emit more neutrons as they decay. A similar process involving other types of atoms in living tissues produces γ-rays that can damage living cells.

Source: J. A. Pope. 1973. *Medical Physics*, 2nd ed. London: Heinemann Educational, pp. 165–171.)

characterize the radioactive processes of various radioactive elements. Rutherford and his coworkers began to suspect that radioactivity accompanies the transformation of atoms of one element into atoms of another element. This radical idea horrified those chemists who believed in the immutability of the elements. Despite this negative reaction, Rutherford rapidly gained international acclaim as an atomic physicist and was offered a prestigious position at Manchester in England in 1907. The following year, at the age of thirty-seven, he was awarded the Nobel Prize. To his chagrin, however, his prize was awarded in chemistry! Rutherford lamented that he still had not "really made it" in physics.

The Center of Attraction

In the year that Rutherford won the Nobel Prize in chemistry, the model of the atom was conceived as having its electrons and protons mixed together more or less like raisins and nuts in a dense plum pudding. But if various elements were bombarded with X rays, as was done by Philip Lenard in 1908, the X rays passed right through the material as if it consisted mostly of empty space. Rutherford became fascinated by this unorthodox idea and designed an experiment to test it (figure 10.4). Back in England, Rutherford and his two new associates, Hans Geiger and Ernest Marsden, found they could detect individual alpha particles by placing a small zinc sulfide-coated screen near the radioactive substance. The zinc sulfide advertised each alpha particle "hit" by emitting a tiny burst of greenish light. Using this device, they tested a number of radioactive substances to find those that were especially strong alpha emitters. These substances were fashioned into a crude alpha particle "gun" that they "fired" at targets made from gold and other metals.

As a first step, they used their new apparatus to repeat Lenard's strange observations. When the zinc sulfide detector was placed behind the target, glowing spots indeed showed that the alpha particles passed right through the gold foil just as if it didn't exist. Marsden then performed the crucial experiment following Rutherford's vague suggestion that he look for alpha rays scattered off the front of the foil (the side facing the alpha gun). Neither Rutherford nor Marsden expected to see anything, but it was the sort of blank

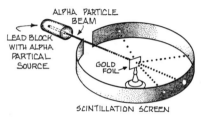

Figure 10.4 Rutherford's and Marsden's "great experiment." (Figure adapted from *Physical Science with Modern Applications*, Fourth Edition, by Melvin Merken, copyright © 1989 by Saunders College Publishing, reprinted by permission of the publisher.)

experiment required in meaningful research. The eighteen year-old Marsden, confined to a small dark room and squinting at the zinc sulfide screen, was astonished to observe faint pinpoints of light on the portions of the zinc sulfide screen *in front* of the target. That could only mean that a few alpha particles must be ricocheting off of the gold foil!

Rutherford shared Marsden's amazement but remained cautious about interpreting the new observation. Realizing what a tremendous force would be required to repel the massive, positively charged alpha particles, Rutherford likened the experimental result to firing "a 15-inch shell at a piece of tissue paper and [seeing] it came back and hit you" (Rhodes 1986). Rutherford also knew that Marsden's observation flew in the face of the prevailing "plum pudding" model of the atom. Almost incredibly, by today's standard of rush to publish, Rutherford left the world ignorant of these results for over a year. He had to be certain that his new model was supported by more experiments, and of course it was.

The first public interpretation of one of most important experiments in history was given in March 1911—not to a prestigious scientific society but to a local meeting of the Manchester Literary and Philosophical Society. At this gathering, Rutherford disclosed a radically new model for the atom, one in which the bulk of the atom's material was concentrated in a tiny but immensely dense, positively-charged nucleus surrounded by space sparsely dotted with negative electric charge. The nuclear atom was born. Rutherford's new model was not to be the last word in the story of modern physics by any means, but it did secure the role of the nucleus as containing most of an atom's mass.

Free of Charge

Nine years after presenting his discovery of the atomic nucleus, Rutherford made another startling announcement. In a lecture to the Royal Society of London in 1920, Rutherford, then director of the famous Cavendish Laboratory where he once studied, proposed another atomic particle in addition to the proton and the electron. It would be a particle, Rutherford claimed, that carried no charge, had a mass equal to that of the positively charged proton, and possessed "very novel properties" (Rhodes 1986). Because the neutral particle, termed the *neutron*, would not be affected by the intense electric fields produced by protons and electrons, it would not only be able to penetrate matter but it might also be captured by the nucleus of an atom or destroyed in the attempt. Rutherford had guessed at the neutron's existence in order to explain the observation that the atomic weights of the elements were, with the exception of hydrogen, larger than the atomic number. The atomic number indicates the number of protons. Why should oxygen have an atomic number of 8 but an atomic weight of 16? What else was there in the atom besides protons and electrons? The question of how all those positive protons could be packed together in the tiny nucleus also demanded explanation since like charges repel each other (see box 10.3). To Rutherford's way of thinking, the neutron would answer these questions. He knew it would also provide a tool for dissecting the fine structure of the atom. This novel idea would come to have vast consequences, but the neutron had to be detected first. The challenge of finding the phantom particle fell on the shoulders of Rutherford's assistant, James Chadwick. His search began in 1920, but the neutron proved to be an elusive prey. Finally, in 1931, one of Chadwick's students reported a curious event accompanying bombardment of a beryllium target with alpha particles. The beryllium atoms responded to the assault by releasing a particle. Soon afterward Irene Joliot-Curie, the daughter of Marie and Pierre Curie, and her husband Frédérick, tried to explain the odd phenomenon as a release of gamma rays, an explanation that failed to convince either Chadwick or Rutherford. Chadwick had used the radiation emitted from the beryllium to probe other elements and knew that the results could not be accounted for by a massless "par-

The Same but Different: Isotopes

One of the puzzles that bothered the pioneers in radioactivity research was the seemingly great variety of elements that underwent decay and were also produced in the radioactive decay processes. They wondered how they were going to fit into the known version of the periodic table of elements. Chadwick's discovery of the neutron supplied the missing piece of the puzzle. An amazingly simple but clever trick of nature allowed the same element, say carbon, to exist with different numbers of neutrons in the nucleus. Normally, carbon has 6 protons and 6 neutrons in the nucleus, plus 6 electrons outside to cancel the protons' charge. This is called *carbon-12* because its protons and neutrons add up to 12, its atomic mass number. This form of carbon is the basis of the system of atomic weights and has an atomic mass defined as exactly 12.0000. Its *atomic number* is 6, that is, the number of protons. The common forms of carbon are written as:

$$^{12}_{6}C \quad ^{13}_{6}C \quad ^{14}_{6}C$$

These forms are called *isotopes*, and almost every element exists as two or more isotopes. The decay of carbon-14, with 8 neutrons, is used in radiocarbon dating to authenticate the ages of ancient objects up to 40 thousand years old. Note that all carbon atoms have the same atomic number, which thus defines the element. Nitrogen-14, with an atomic number of 7, has 7 protons and 7 neutrons in its nucleus. Thus its atomic mass is the same as carbon-14, but it is a different element with different chemical properties.

The neutrons are a sort of atomic glue that holds the nucleus together. However, as more and more positively charged protons are crowded into the tiny nucleus, the nucleus may spontaneously disintegrate. For example, uranium-238 ($^{238}_{92}U$) slowly decays over many thousands of years, but others like iodine-131 ($^{131}_{53}I$) decay in a matter of days. Other isotopes are even shorter lived and may exist for only minutes or seconds. Radioactive tracers used in medicine for "imaging" are usually gone in an hour or so.

Many isotopes decay spontaneously, but these do not produce great amounts of power as is needed in an electric power station. Other applications, such as power plants for space probes like the Voyagers, can utilize spontaneous decay. The great scientific and technological challenge for the atomic physicists was to learn how to induce radioactive decay and then to create a situation in which the process was self-perpetuating—a chain reaction.

Box 10.3

ticle" like gamma rays but must indicate instead a large, uncharged particle, the neutron.

Chadwick won the Nobel Prize in 1935 for discovering the neutron, a discovery that eluded the Joliot-Curies by the slim margin of a misinterpreted experiment. Despite the stark objectivity sought by most scientists, the thrill of being first to glimpse one of nature's truths remains a powerful motivation, and the Joliot-Curies' loss in the search for the neutron must have been painful. Perhaps the 1935 Nobel Prize in chemistry that they shared for synthesizing new radioactive elements consoled them.

Working on the Chain

The path to atomic energy passed through the nucleus. In the 1930s, scientific understanding of the nucleus reached a level from which the technological breakthroughs of the next decade could be glimpsed. The eminent physicist, Hans Bethe, stated that "from 1932 on the history of nuclear physics [developed]" with the era before Chadwick's identification of the neutron being a sort of "prehistory" of the field (Rhodes 1986).

Many research groups and individuals worked intensively to ferret out the atom's further secrets in the 1930s, but one man had an idea for how those secrets could be put to use. Leo Szilard was a Hungarian émigré who combined a preference for theory over experiment with a penchant for the practical. At one point, he and Albert Einstein obtained a patent for a refrigeration system that had no moving parts. Szilard wandered about Europe for years without a permanent academic position before traveling to America and bringing with him a dream of a nuclear chain reaction. The key to unlocking this dream, Szilard realized, lay with the neutron. He thought, like Rutherford, that the uncharged neutron would be less likely to be repelled by the positive nucleus than a positively charged alpha particle. A nucleus that absorbed a neutron might become so unstable that it would fall into smaller pieces and release more neutrons. If this disintegration of the nucleus released at least two more neutrons, on average, and there was a "critical mass" of material available, then a self-sustaining chain reaction could be started. Szilard understood that the amount of energy released from a nuclear chain reaction could be enormous.

Szilard was so confident of his idea that, in the mid-1930s, he took out several patents in England on the "double neutron" process including the concepts of critical mass and nuclear power related to weapons. He relentlessly pushed these ideas among other physicists in large part because he feared the new German government under Hitler might embark on a program of atomic weapons research. Despite this apparent threat, Szilard had a hard time selling his idea.

One problem was that the disintegration of a nucleus had not been confirmed despite intensive studies of radioactivity. In addition, many prominent nuclear physicists did not consider it likely to occur. Rutherford called the idea "moonshine," Bohr thought that higher energy particles than neutrons were needed to bombard the nucleus in order to make it split apart, and the prominent American theoretician, J. Robert Oppenheimer, remained pessimistic about the whole idea.

The first great breakthrough that supported Szilard's hypothesis was achieved by the radiochemists Otto Hahn and Fritz Strassman in Berlin in mid-December, 1938, but the real significance of their experiment was explained a few days later by Lise Meitner and her nephew Otto Frisch, who had fled their native Austria to live in Scandinavia. Hahn and Strassman had bombarded uranium with neutrons and apparently found the much lighter element barium among the debris. How could this possibly happen? They had expected the neutrons to add to the uranium, not cause its demise, and so they were very cautious about their almost absurd result, which they asked their physicist friends, Meitner and Frisch, to review and interpret. Meitner soon realized the significance of the experimental result: the nucleus had broken apart. She also knew that a nuclear disintegration must be accompanied by the release of a very large amount of energy, more energy than released by any other physical or chemical process known. What was the source of all that energy? Meitner recalled a lecture given by Albert Einstein nearly thirty years before in which he described the conversion of matter into energy. A small part of the disintegrating atom's mass must be converted to the large amount of energy that forces the parts of the nucleus apart (see box 10.4). Frisch, in the process of writing up their interpretation, consulted an American biologist and applied the term *fission* to the breakup of the uranium nucleus. These re-

> ## Matter into Energy: $E = mc^2$
>
> Perhaps the most famous equation in science is the one above, which Einstein derived in his theory of relativity in 1905 to describe the conversion of matter into energy. This simple expression has enormous consequences because the mass (m) is multiplied by the square of the speed of light (3×10^8 meters per second times itself) to give a value of ten to the sixteenth power. Thus even a small amount of matter can potentially yield a great amount of energy.
>
> We can determine this matter lost to its energy equivalent in nuclear decay by carefully measuring the masses of the products and comparing that value to the reacting materials' masses. A tiny discrepancy will be found because energy is created. We are used to obtaining energy out of chemical reactions such as the combustion of gasoline, but the scale of energy production is increased a millionfold in the nuclear processes. This, of course, is what led to such excitement among those physicists who considered the possibility of atomic power, especially a bomb.
>
> A minor irony of history is that Einstein was not awarded the Nobel Prize for his development of relativity—it was too avant garde for the times. Instead, he received the prize for his other 1905 work, the analysis of the photoelectric effect, a phenomenon that may eventually replace nuclear power if efficient solar conversion technology is ever developed.

Box 10.4

sults astounded the physics community and gave Szilard half of his dream.

To finish the job, the genius of another great physicist was needed, one whose roots were not in northern Europe but in sunny Italy. Like Szilard and a number of other prominent scientists at the time, Enrico Fermi saw ominous clouds over Europe and chose to move from Italy to Columbia University in New York, where he linked up with, naturally, Leo Szilard. After the amazing report of nuclear fission, they quickly sought to answer the key question: did it proceed as a double neutron process (i.e., did the uranium emit more than one neutron when it split apart)? To do the crucial experiment, Szilard had to borrow $2000 to buy radium as a neutron source with which to bombard the uranium. Szilard and Walter Zinn conducted the experiment and found that approximately two neutrons were produced from each fission reaction confirming a

Fission and the Chain of Power

In order to continuously produce energy from a nuclear decay reaction, it is necessary to have a source of neutrons to carry out a constant bombardment of the target nucleus. An ingenious solution to this conundrum is to have each disintegrating atom supply those neutrons for the next round of collisions. However, if the atom provides just one neutron to replace the one that hit it, the reaction will proceed rather leisurely. Something with more zip is required to produce enough power to be useful. Remember that power is the amount of energy expended in a given time period: the shorter the time, the more powerful the process.

Now consider the following scenario embodied in this nuclear equation:

$$^1_0n + {}^{235}_{92}U \rightarrow {}^{137}_{52}Te + {}^{97}_{40}Zr + 2^1_0n + \text{Energy}$$

Here, U-235 is bombarded by a neutron that causes the uranium to undergo fission into "daughter" elements: tellurium and zirconium. Another decay path involves the creation of heavy isotopes of barium and krypton. Even more important is the production of two new neutrons that can collide with two more U-235 atoms and produce more neutrons that can continue the chain reaction.

If you are skeptical about this really working, do a little arithmetic and you will see that the number of neutrons increases as 2^n, where n is the number of successive steps in the decay process. Calculate the number of neutrons available after 10 successive decay generations. This process is called *exponential growth* and governs other processes such as population increase and financial investment.

The conversion of one element into one or more other elements is called *transmutation*. The alchemists of the Middle Ages sought to do this by changing lead into gold but, of course, never succeeded. It may be theoretically possible to do this now, but the cost would be prohibitively high.

Box 10.5

Figure 10.5 Leo Szilard (right) talking with Albert Einstein about a letter to President Franklin Roosevelt in 1939. (Photograph reproduced with permission of Gertrude Szilard.)

preliminary result obtained by Fermi and Herbert Anderson. Szilard had finally assembled the pieces he needed less than seven months before the outbreak of World War II (see box 10.5).

The all-important question, and experiment, now loomed: would a chain reaction occur? This issue, which Szilard had tried to hush up, became an open secret after French scientists, among them Frédérick Joliot, published a paper on uranium fission and the possibility of a chain reaction. The race was on to produce the first sustained chain reaction, but it turned out to be a stubborn problem.

Realizing the significance of their work in wartime, nuclear scientists on both sides of the Atlantic began informing government leaders of the importance of the recent remarkable discoveries. In the late summer of 1939, Szilard along with several other prominent immigrant physicists persuaded Einstein to write a letter to President Roosevelt describing the practical implications of nuclear research (figure 10.5). The president was delayed in considering it because his attention was focused on another drastic breakthrough, Germany's blitzkrieg in Poland. Once FDR understood the letter's implications, however, he authorized the establishment of a large-scale atomic research program, the Manhattan Project.

Final Link to the Bomb

Conclusive evidence for the chain reaction was three years in the making and came finally in December 1942, in a very unlikely setting: the bowels of the football stadium at the University of Chicago. The quarterback of the nuclear team was Fermi, who had been asked to move to Chicago in an effort to centralize the United States' war-related work on uranium fission.

One of the technical problems that had been conquered in the intervening years was the choice of a "moderator" for the nuclear "pile." Early on, the uranium was immersed in water, which served to slow down the neutrons and made them easier for uranium nuclei to capture. The problem with water as a moderator is that it also absorbs some neutrons, removing them from the reaction. Szilard proposed using very pure carbon as a moderator because it slowed the neutrons without absorbing them. This raised the possibility that the chain reaction within the pile's 40 tons of uranium oxide might proceed better than expected and increase the danger of a "meltdown." Fermi and Szilard reasoned that they could gain some control over the chain reaction by adding to the pile a series of "control rods" made from some neutron-absorbing material such as cadmium. Withdrawing or inserting the control rods would influence the number of neutrons available to be absorbed by the uranium nuclei and thus influence the chain reaction.

The historic experiment began on the morning of December 2, 1942. All the control rods but one were removed from the pile. The last control rod was slowly eased out of the pile inches at a time, precisely following Fermi's orders. At each new position, Fermi, Szilard, and a few other scientists monitored the production of neutrons by the pile and compared it to the number Fermi had predicted from his calculations. Their observations verified his results, so the experiment continued. When the control rod was removed further than ever before, the pile, as predicted, went "critical." The neutron production continuously increased, not leveling off as it had before, as Fermi let the reactor run. Four minutes later, a safety rod was then inserted and the experiment was over. The first self-sustaining chain reaction had been produced.

This demonstration of a working chain reaction validated the new scientific theories of the atomic nucleus and fission, but it fell far short of solving the myriad technical problems that would have to be confronted in converting scientific understanding into practical application. The onset of World War II elevated weapons applications over electrical power generation as the first priority for the new technology, although Japan is reported to have pointed its nuclear program in the direction of power sources for submarines and ships. The technical difficulties facing production of an atomic bomb appeared staggering. No one knew exactly how much uranium was required to make a bomb nor how a sufficient amount of uranium-235, the most reactive isotope of uranium, could be coaxed from uranium ore given that it constitutes only 0.7 percent of the total mass of uranium in the world. Once these problems were solved, the scientists and engineers had to determine how to bring the fissionable material together into a critical mass at exactly the right time. The device had to be small and light enough to be carried in an airplane, a constraint that Einstein intimated in his letter to Roosevelt might not be overcome. As if these problems weren't bad enough, scientists on the project feared that an atomic explosion might be intense enough to set the atmosphere on fire, nitrogen being burned by oxygen. To everyone's great relief, calculations showed this to be impossible. The leader of the German A-bomb effort, Nobel laureate Werner Heisenberg, had told Bohr in 1941: "I know that [a nuclear weapon] is in principle possible, but it would require a terrific technical effort, which, one can only hope, cannot be realized in this war." Germany, fighting a war on two fronts, could not sustain that effort, but the threat galvanized the Allies to be the first.

The Manhattan Project was a technological triumph, although its ultimate military and ethical significance will be debated forever. Never before or since have so many world-renowned scientists worked so closely together. Oppenheimer, who had shed his earlier skepticism, had become the scientific leader of the project. The task of keeping all the prima donna scientists focused on an engineering project fell to a gruff but practical Army man, Gen. Leslie Groves.

General Groves clearly recognized the profound difference between engineering and basic science and showed remarkable skill in coaxing the physicists to act less like scientists and more like engineers.

The gap between scientific understanding and engineering achievement exemplified by the Manhattan Project also appeared in the postwar development of nuclear power plants and continues to this day. The basic scientific principles underlying the operation of a commercial nuclear power plant, from nuclear fission to electrical power production, are well described in many textbooks, yet the nuclear power industry struggles to find a design for a safe, inexpensive, efficient nuclear power plant. Nuclear engineers face enormous and diverse practical problems encompassing all phases of plant design from public opinion about site selection, to safe reactor operation, and waste disposal. Large projects of such complexity are inevitably fraught with uncertainty. What procedures are needed to minimize error by plant operators? How will the reactor behave under various emergency situations? If radioactive materials are released from a reactor, how far will they spread in the environment, and by what routes will they affect people? What design features are needed to minimize the risk of accidents and release of radioactive materials? How can radioactive wastes be stored to ensure not only our own safety but the safety of subsequent generations? Engineers must cope with uncertainty in trying to answer these questions as well as assess the risks of nuclear power and make decisions about its future. Events have shown how difficult these decisions can be.

Chernobyl

In early 1986, the Soviet Union's nuclear power facility at Chernobyl contained four units of one million kW each, "enough to light Toledo, Ohio," and construction had begun on two more. When completed, it would take its place among the world's largest nuclear power generating stations. The location seemed ideal: rural yet close to major population centers in the Ukraine to the south and Byelorussia to the north. A tributary of the Dneiper River carried away the facility's waste heat. The water-cooled reactors used graphite as the moderator—not the material chosen for most commercial reactors in the United States but recognized as a good moderator by the pioneer nuclear physicists. The tall stack of graphite blocks in Unit 4 contained 1,659 metal rods, each packed with uranium oxide (UO_2) fuel pellets. A heavy steel plate covered the top of the "pile."

On April 26, 1986, a small group of technicians began testing Unit 4 to determine whether the plant's generators could supply emergency power to the Chernobyl facility during a blackout. In the course of their tests, they violated safety procedures by almost completely withdrawing the control rods from the reactor and shutting down some of the critical safety systems. The reactor responded to this unstable condition by developing a power surge. Within four seconds, the power generated by Unit 4 rose to a level a hundred times higher than normal—an energy release great enough to rupture some of the fuel rods. When water contacted the hot fuel

Figure 10.6 Spread of detectable radioactivity in air following the accident at the Chernobyl nuclear power facility on April 26, 1986. As winds shifted, radioactive material that originally headed northwest (plume A, April 26) began to spread over eastern and central Europe (plume B, April 27–28) and then over southeastern Europe (plume C, April 29–30). (Photograph reproduced, with permission, from: UNSCEAR, *Sources, effects, and risks of ionizing radiation, 1988 Report to the General Assembly*, with annexes. United Nations Publications, N.Y.)

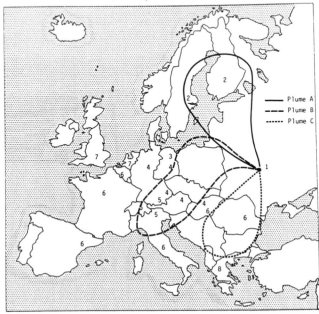

Risk

346

directly, it turned to steam with enough explosive force to shove the 1,000-ton steel lid onto the cooling channels. A second explosion moments later sent pieces of the reactor through the wall of the reactor's building, allowing air to enter. With its temperature driven up by the uncontrolled chain reaction and oxygen now freely available, the graphite core began to burn. Courageous firefighters bombed the reactor with loads of boron carbide, dolomite, clay, and lead to capture neutrons and choke off the chain reaction. Ten days later, the fire was finally extinguished, but not before the demolished reactor had spewed radioactive material high into the atmosphere where winds caught and distributed it over the western Soviet Union, Scandinavia, Europe, and eventually other parts of the Northern Hemisphere (figure 10.6).

Two of the more than 400 workers in and about Unit 4 died immediately as a result of the explosions. Thirty more workers died from burns or radiation sickness within ninety days after the accident. As tragic as these deaths were, they do not account for the millions of people exposed to the radioactive material released from the power plant. What about them? What risk do they face? People have been trying to answer similar questions ever since atomic bombs exploded over Japan.

Ionizing Radiation

The pioneers in nuclear physics learned early that nuclear radiation could damage a person's health or even kill. Henry Becquerel, the discoverer of nuclear radiation, carried tubes of radioactive material in his waistcoat pocket until he noticed a reddening of the skin near his navel. Being a good scientist, he did the control experiment and carried empty tubes for awhile to be sure the radioactivity was responsible. Apparently it was. The first death attributed to X rays came in 1900, only four years after the discovery of that high-energy radiation: a German "radiologist" enamored with the new device exposed his hands to X rays more than 1,000 times in public demonstrations before succumbing to cancer. The first international recommendations for protection against radiation and for methods of measuring exposure did not come until 1928, not soon enough to save Marie Curie who died in 1934 of radiation-induced anemia or possibly even Enrico Fermi who died of cancer in 1954. No one can be sure that radiation caused the deaths of these two scientists, but the doses they received over their professional careers certainly could have contributed.

Although the early nuclear scientists vaguely attributed risk to nuclear radiation, they did not know, indeed could not have known, how much is dangerous, what effects it produces, and the mechanism by which it acts on the human body. Aside from a uranium deposit in Gabon, Africa that apparently "went critical" 1.7 billion years ago, nuclear chain reactions on earth are an entirely modern, technological achievement. Just as the subtle risks of a new vaccine do not become detectable until large numbers of people use it, so the effects of nuclear radiation could not be known with any precision until large numbers of people were exposed to it. Detailed understanding of the biological effects of radiation began to take shape only after the bombings of Japan. Survivors of the attacks on Hiroshima and Nagasaki continue to be the major source of information about radiation's effects on people. In addition to this group, people exposed to radiation either accidentally or in the course of medical therapy provide most of the information now used to judge the risks of nuclear radiation. Long-term studies of Chernobyl's aftermath will no doubt add significantly to our understanding of these risks.

The danger in nuclear radiation springs from its high energy. Nuclear radiation is distinguished from visible light or heat by its very high energy and correspondingly short wavelength. Diagnostic X rays, for example, are about ten thousand times more energetic than blue light. Electromagnetic waves with such high energies dislodge electrons when they are absorbed by an atom. The battered atom now carries a positive charge, and the liberated electron quickly associates itself with another atom giving it a negative charge. These charged atoms, called *free radicals*, usually live short but highly reactive lives, regaining a more stable condition at the expense of atoms in nearby molecules. Radiation energetic enough to create ions is fittingly called *ionizing radiation*. Any electromagnetic radiation with a wavelength shorter than about 10^{-8} m has enough energy to be ionizing. This includes alpha radiation, beta radiation, gamma rays, and X rays.

Biological Effects of Ionizing Radiation

Radiation doesn't kill people—free radicals kill people. That simplistic statement disguises the fact that ionizing radiation creates free radicals, but it reveals that the link between radiation and biological damage is an indirect one. Free radicals are very reactive chemicals that are highly unstable and take any available opportunity to relieve themselves by passing their energy to a molecule or another atom. Given the prevalence of water in the human body, one of the most likely atoms to be converted to a free radical is oxygen, which has unlimited access to all parts of a cell. Serious biological damage begins when the negatively charged oxygen encounters an enzyme in the cell's cytoplasm or the genetic material, DNA, in the cell's nucleus. The interaction between the free radical and these types of molecules involves enough energy to damage the molecule and, in turn, impair its function.

Very high doses of ionizing radiation inflict so much damage through the actions of free radicals that the exposed cells die. A person can withstand only so much cell death before vital organs begin to fail and radiation sickness appears. A total of 145 of the firefighters and emergency workers at Chernobyl suffered radiation sickness. Typically in these victims, vomiting began within three hours after exposure as radiation-induced cell killing disturbed the gastrointestinal tract and central nervous system. Loss of appetite, diarrhea, and fatigue appeared about the same time. Fever set in about two to three days later. Bleeding and ulceration followed later still. Those who received the largest doses of radiation began losing hair and skin within several days of the accident as hair follicle and skin cells died. In the most serious cases, the death of bone marrow cells crippled the immune system, leaving the victim susceptible to disease. Heroic attempts to save lives with bone marrow transplants failed in eleven out of thirteen attempts.

The accident at Chernobyl confirmed what had been observed after the bombings of Hiroshima and Nagasaki: how far and how fast a person progresses through the symptoms of radiation sickness depends primarily on the radiation dose received. More than a hundred people at Chernobyl received at least one "gray" (Gy; see box 10.6) of radiation indicating that their bodies absorbed at least one Joule of energy for each kilogram of their body's weight. Chernobyl

A Confusion of Units

Rads, grays, rems, seiverts, curies, and bequerels are all units that crop up in discussions of radioactivity and ionizing radiation. This bewildering variety of terms is not a historical accident but reflects ongoing efforts to improve our description of ionizing radiation and its multiple effects.

The activity of a radioactive material, that is, the rate at which nuclear disintegrations occur, is measured in becquerels (Bq), a unit named in honor of Henri Becquerel who discovered the phenomenon of radioactivity in 1896. An activity of one disintegration event per second corresponds to 1 Bq. This unit is gradually replacing the curie as a measure of radioactivity. One curie, named for Marie and Pierre Curie, indicates 37 billion disintegrations in one second.

Radiation exposure is typically measured, when possible, by monitoring the ionizations produced by a radioactive source in a known mass of air. But the danger of radiation to a person depends more on how much energy is absorbed by the body's tissues than on the exposure level as measured in air. The absorbed dose, indicating the amount of energy imparted by ionizing radiation per kilogram of tissue, is frequently given in grays (Gy), where 1 Gy signifies 1 Joule absorbed by 1 kg of tissue. The gray replaces the rad (for radiation absorbed dose), where 1 Gy = 100 rad.

Although the gray is useful in measuring the absorption of all types of ionizing radiation by a variety of materials, it does not serve very well to describe damage to biological systems. Alpha (α) particles, for example, impart more damage to cells than beta (β) particles even when the absorbed dose is the same for each. This happens because α-particles leave a more dense track of ionizations in their wake than do β-particles. The seivert (Sv) serves as a measure of the "dose equivalent" by adjusting the absorbed dose according to the type of radiation and the duration of exposure. The seivert replaces the rem (for roentgen equivalent man) as the preferred unit for dose equivalent: 1Sv = 100 rem.

Source: J. A. Pope. 1973. *Medical Physics*, 2nd ed. London: Heinemann Educational, pp. 176–181.

Box 10.6

workers who received lower doses stood a greater chance of surviving; of the forty-three people at Chernobyl who received 2–4 Gy, one died, while seven of the twenty-one people who received 4.2–6.3 Gy died. The other deaths at Chernobyl resulted from either higher doses or from a combination of radiation sickness and burns.

The vast majority of people exposed to radiation from Chernobyl, including the millions of people living under the plumes of radioactive material that blew away from the accident site, received doses much smaller than those received by the firefighters and emergency workers. These low doses produced no immediate ill effects or fatalities, but that does not mean that the exposed people have escaped all harm.

Uncertainty: Radiation and Risk

Survivors of the atomic bomb blasts in Japan, in which radiation doses are assumed to have been received virtually instantaneously, continue to show a higher frequency of certain types of cancer than similar, nonirradiated groups of people (figure 10.7a). Increases in the frequency of cancers also occurred among women who received high doses of X rays over an extended period of time as treatment for breast cancer (figure 10.7b). Because each of these well-studied groups included tens of thousands of people (76,000 and 83,000 respectively), there remains little doubt that ionizing radiation, at least in high doses, can cause cancer. That is, cells lucky enough to survive high doses of ionizing radiation may eventually become transformed into cells that produce malignant tumors.

The key question for millions of people exposed to low doses of radiation from Chernobyl, as well as for the surviving emergency workers, concerns the amount of risk they face. How much radiation is dangerous? Put another way, what relationship exists between the amount of radiation exposure and the risk of cancer? Extensive investigations have not revealed an increased risk of hereditary, genetic damage among survivors of the atomic bombings of Hiroshima and Nagasaki, so this chapter ignores that feared effect of ionizing radiation.

The apparently simple question of the risk of cancer from radiation has proved extraordinarily difficult to answer. No cancer can

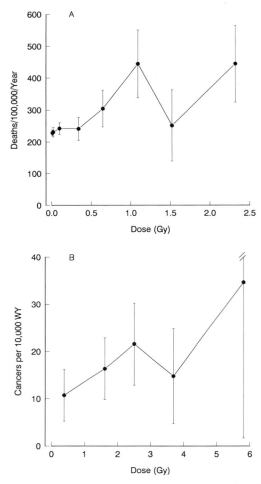

Figure 10.7 Cancer deaths or incidence related to ionizing radiation dose. A: Deaths from all types of cancer except leukemia among survivors of the atomic bomb blast at Hiroshima. B: Incidence of breast cancer among women exposed to X rays as treatment for lung cancer. WY indicates woman-year: ten women living for one year equals 10 WY. (Graph A redrawn from BEIR, 1980. Graph B redrawn from UNSCEAR, 1988.)

be linked unambiguously to ionizing radiation—radiation leaves no telltale sign. Ionizing radiation insidiously increases cancer-causing processes already at work in a population. This makes it extremely difficult, maybe impossible, to detect radiation-induced cancers except as statistically significant increases in cancer incidence within a large group of people. Several other factors further confuse the picture. The radiation dose that imparts a particular risk varies in a bewildering way with the type of radiation, the tissue that receives the most exposure, the duration of the exposure, and individual characteristics of the exposed person such as age, state of health, behavior, and genetic predisposition. Also, cancers do not appear immediately after exposure. Presumptive cancer cells lie dormant in the body for ten years on average before expressing their malignant properties. Virtually everything seems to obscure whatever link might exist between radiation exposure and cancer.

Despite these difficulties, information about the relationship between radiation dose and cancer incidence has emerged over the last thirty years. The single most important source for this information remains the group of people who survived the atomic bombs at Hiroshima and Nagasaki. As of 1988, 76,000 people in this group had been examined, their likely radiation doses determined, and their medical records compared to a similar group of nonirradiated people. The atomic bomb survivors continue to suffer an increased incidence of leukemia as well as cancers of the breast, stomach, ovary, esophagus, colon, lung, and bladder, but mysteriously not cancers of the gallbladder, pancreas, uterus, or prostate. Cases of leukemia have drawn the most attention because that type of cancer appears especially sensitive to ionizing radiation. The incidence of leukemia clearly rises with increasing doses of radiation among the atomic bomb survivors (figure 10.8). Other cancers, taken together, show a similar but less striking relationship.

The arrangement of the data points in figure 10.8 no doubt reflects some underlying "dose/response" relationship based on the mechanism by which ionizing radiation induces cancer, but exactly what that relationship is remains a mystery because the underlying biological events linking ionizing radiation with cancer are not yet completely understood. For the time being, several mathematical models must take the place of biological knowledge (table 10.1). Each of the expressions in table 10.1 seems to describe the data

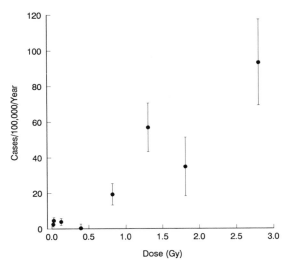

Figure 10.8 Incidence of leukemia among survivors of the atomic bomb blast at Nagasaki. Vertical bars represent 50 percent confidence limits. (Redrawn from Land, 1980.)

Table 10.1
Dose/Response Models for Ionizing Radiation and Cancer

Name	Equation
linear	$R = a + b_1 D$
quadratic	$R = a + b_2 D^2$
linear-quadratic	$R = a + b_1 D + b_2 D^2$

R, response (cancer incidence); D, radiation dose; a and b indicate coefficients that vary with the type of radiation and cancer. These simplified expressions do not include factors that account for radiation-induced cell-killing.

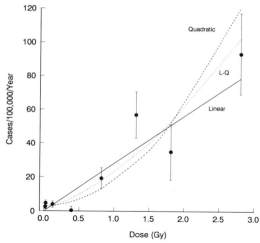

Figure 10.9 As in figure 10.8 but with various curves fitted to the data in an attempt to describe the underlying relationship between ionizing radiation dose and cancer incidence. The three models used to generate these dose/response curves are linear, linear-quadratic (L-Q), and quadratic. The cell-killing effect of high doses is not included in the models represented here. (Redrawn from Land, 1980.)

reasonably well (figure 10.9), but different models are better suited to some cancers. The linear model, for example, fits the incidence of breast cancer in atomic bomb survivors better than the other two models. This means that the incidence of breast cancer increases at a steady rate with dose until the dose becomes so high that cells begin to die before they can turn malignant. At that point, of course, cancer incidence begins to tail off (not shown in figure 10.8). Thyroid cancer and leukemia, however, seem to be better described by the linear-quadratic model although the other two models can not be ruled out statistically.

Once a dose/response relationship has been selected, it can be used to estimate the number of cancers that will appear in a population throughout the lifetime of its members. This information plays a major role in estimating the magnitude of the health problem created by an accident at a nuclear power plant, for example, and in helping governments and health care agencies plan for future health problems. Selecting the wrong dose/response relationship may lead to mistaken estimates of radiation's health effects. A linear model predicts a larger number of cancers at doses between about 0.07 to 2.0 Gy than the other two models (figure 10.10). If the linear model

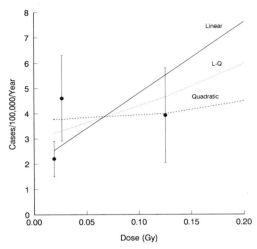

Figure 10.10 Detail of figure 10.9 showing how the different models predict different cancer incidences at very low doses. (Redrawn from Land, 1980.)

proves incorrect, then individual risk assessments and public policies based upon it will overestimate the dangers of ionizing radiation. The other two models would underestimate the danger if they were wrong. In the face of this uncertainty, most investigators assume that the linear dose/response model applies because it projects a worst case scenario over a large portion of the dose range.

The best way to get information regarding the correctness of the various dose/response models for humans would be to track radiation-induced cancers in a population for an entire lifetime. No one has been able to complete a study like that yet: 60 percent of those who survived the atomic bomb blasts in Japan are still alive. We must rely instead on educated guesses or models of how much risk an irradiated group of people will face in the future.

Models of Risk: More Uncertainty

According to one view, an irradiated population faces an additional risk of cancer, but the magnitude of that added risk remains constant during a lifetime (figure 10.11*a*). That is, radiation-induced "excess" cancers are added to the spontaneous cancers, and all age groups experience the same incremental increase in risk. This is called the *absolute risk* or *additive model*. It might be, however, that radiation does not simply add a fixed increment to the spontaneous cancer risk

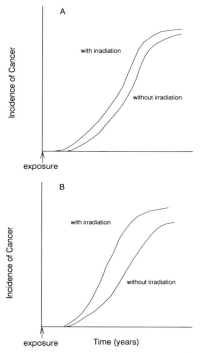

Figure 10.11 Risk models used to project how the incidence of cancer in a population will change after exposure to ionizing radiation. A: Absolute risk model. B: Relative risk model. In each model, there is a latency period between the time of exposure and when excess cancer cases begin to appear in the population.

---

but acts as a multiplier of that risk. This model, called the *relative risk* or *multiplicative model*, predicts that the radiation-induced risk of cancer will increase along with the risk of spontaneous cancer as people grow older (figure 10.11*b*).

When the various dose/response relationships are combined with the two risk models, an unsettling range of risk estimates emerges (figure 10.12). The number of excess cancers, cancers over and above the number expected from spontaneous cancer formation, predicted to follow a 0.1 Gy irradiation of one million people varies from 120 when the quadratic relationship is coupled with the absolute risk model to 5,300 when the linear model is combined with the relative risk model. Note that the higher value is only about one-thirtieth of the normally expected cancers in a group of one million people as discussed below.

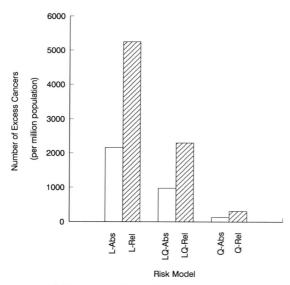

Figure 10.12 Projections of cancer incidence that arise from various combinations of dose/response model and risk model. L-Abs = linear dose response model combined with the absolute risk model. L-Rel = linear combined with relative. LQ-Abs = linear quadratic combined with absolute. LQ-Rel = linear quadratic combined with relative. Q-Abs = quadratic combined with absolute. Q-Rel = quadratic combined with relative.

Table 10.2
Lifetime Excess Cancer Deaths Following Exposure to 1 Gy (expressed as % risk per person per Gy)*

Cancer	Relative Risk Model	Absolute Risk Model
red bone marrow	0.97	0.93
bladder	0.39	0.23
breast	0.60	0.43
colon	0.79	0.29
lung	1.5	0.59
ovary	0.31	0.26
esophagus	0.34	0.16
stomach	1.3	0.86
other	1.3	1.1
total	7.5	4.9

*Data from survivors of atomic bomb blasts in Japan based on a linear dose/response relationship (UNSCEAR 1988, p. 39).

In general, the relative risk model yields higher risk estimates than its alternative (table 10.2). For the workers at Chernobyl, exposure to 1 Gy added 7.5 percent to their risk of fatal cancer according to the relative risk model but only 4.9 percent according to the absolute risk model. These increased risks imply four to eight excess cancers for every hundred people at Chernobyl exposed to 1 Gy. The difference between these numbers may seem small, but it translates into a difference of 40,000 fatal cancers when that level of exposure is spread over a million people. This disparity between the two projections is important to health authorities, especially those in the most heavily affected areas, who must plan according to the expected number of cancer cases. Which model provides the better estimate? There is, as yet, no way to tell, although the unfolding story of the atomic bomb survivors seems to conform more closely to the relative risk model, unfortunately.

How about the Risk of Low Doses?

The discussion to this point about radiation and cancer has dealt with people who have been exposed to fairly large amounts of ionizing radiation. But the number of heavily irradiated people at Chernobyl pales in comparison to the number of people who received low doses. One accepted convention defines low doses as less than or equal to 0.2 Gy, intermediate doses as lying between 0.2 and 2.0 Gy, high doses as 2.0 to 10 Gy, and very high doses as more than 10 Gy. Approximately one hundred workers at the Chernobyl nuclear power plant received doses of 1 Gy or larger, but millions of people from the Middle East to Scandinavia and North America received much smaller doses as they came in contact with radioactive materials ejected by the burning reactor.

The explosions and fire at the Chernobyl facility sent a variety of radioactive materials into the atmosphere including radioactive isotopes of iodine (e.g., I-131) and cesium (e.g., Cs-137). These materials, called *radionuclides*, rose thousands of feet into the atmosphere where winds carried them initially northwest over Leningrad and Scandinavia then over eastern Europe and, later still, over the Balkan Peninsula and western Europe. The first increase in radiation outside the Soviet Union was detected in Sweden on April 27, one

> ## Half-Lives
>
> A radioactive atom is inherently unstable, and at any moment it may lose a particle to achieve a more stable condition. The high-energy particle given off in this "decay" carries the energy that may lead to ionizations when it interacts with other molecules. The amount of radiation decreases over time at a rate characteristic of the radioactive element. The rate at which this probability decreases is described by the element's half-life: the amount of time needed to see a 50 percent reduction in the probability of decay. I-131, for example, has a half-life of about eight days: half the existing radioactive iodine is converted to another element in eight days. In another eight days, half is again lost so that only one-fourth of the original I-131 is left. This is called *exponential decay*. In seven half-lives, fifty-six days for I-131, more than 99 percent of the radioactivity will be lost:
>
> Fraction remaining $= (1/2)^7 = 1/128 = 0.008$ or 0.8%
>
> I-131 disappears fairly quickly, but Cs-137 has a half-life of 30 years. At that rate it will take 210 years (seven half-lives) for the amount of radioactive cesium to be reduced by 99 percent. This is one of the factors that makes Cs-137 a more worrisome product of nuclear fission than I-131.

Box 10.7

day after the accident. Atmospheric turbulence eventually spread very small amounts of radionuclides over most of the Northern Hemisphere.

Despite the small amount of material involved, detectable increases in radioactivity appeared in early May in Israel, Kuwait, Japan, India, Canada, and the United States. Radioactive gases and deposited materials exposed people externally, but the major threat came when people ingested radionuclides that had been deposited by rain or fallout on crops. The body rapidly concentrates ingested iodine, including I-131, in the thyroid where irradiation can lead to thyroid cancer. Radioactive cesium finds its way into muscle and other soft tissues where it can lead to a variety of cancers. Both of these radionuclides provided cause for concern, but Cs-137 is of greater concern over the long run because it retains its radioactivity for a much longer time (see box 10.7). Short-lived isotopes provide local cause for concern, but long-lived ones have the potential to become global problems.

Table 10.3
Effective Dose Equivalents for the First Year After the Chernobyl Accident*

Country	Effective Dose Equivalent (mSv)
Byelorussia (Belarus)	2.00
Switzerland	1.30
Greece	0.96
Bulgaria	0.72
Finland	0.49
Sweden	0.44
Israel	0.09
Denmark	0.03
Japan	0.008
Portugal	0.002
United States	0.002

*Data represent whole body exposure averaged over the most exposed region of a country (UNSCEAR 1988).

Local and federal governments took several steps in the weeks following the Chernobyl accident to reduce the radiation exposure to people in the areas around the power plant and in neighboring countries. The most drastic measures included evacuation of 115,000 people from several towns around the accident site. Many people took government-supplied iodine to reduce the chances of I-131 uptake by swamping the thyroid gland with a nonradioactive form. Contaminated fields were plowed deeply to bury the fallout and prevent people from directly contacting radioactive materials. Similarly, roads were covered with asphalt. Many houses and buildings had to be completely buried.

These defensive measures undoubtedly helped, but they did not reduce the exposure to zero. Surveys show that the most heavily exposed area was in Byelorussia, north of Chernobyl, where the effective dose equivalent for the first year after the accident rose to 2 mSv. People in other regions received considerably less radiation exposure. People in the United States, among the least exposed, received about 0.002 mSv during the first year after the accident (table 10.3).

What cancer risk do the people exposed in these areas face? Problems in finding an accurate answer to this question appear immediately in trying to establish the doses of radiation to which people have been exposed. Discouragingly few circumstances allow a person's radiation dose to be known precisely. The best cases occur in medical diagnostic and therapeutic procedures, but even there the dose is somewhat uncertain outside the center of the irradiated zone. People who have studied the aftermath of the atomic bombings or accidents involving radioactive materials had to estimate doses after the fact according to what they knew about the source of the radiation, the types of radiation emitted, the radionuclides to which people had been exposed, the pathways of exposure, the distance separating people from the radiation source, and the shielding effects of any materials that happened to lie between a person and the radiation source at the time of exposure. Changes in any one of these factors force alterations in estimations of dose. Recent reviews of the Hiroshima bombings, for example, suggest that there was less neutron radiation emitted and that wood provided more protection than previously thought. The encouragement offered by these changes was more than offset by an increased estimated yield of the bomb (15,000 tons instead of 12,500 tons) and the realization that gamma rays penetrate human tissue more effectively than previously believed. The net effect of these changes was to increase the estimates of risk associated with ionizing radiation by a factor between two and four over previous estimates.

Once the exposure is estimated, the likely effects of such exposures must be determined. In the absence of a better way, the carcinogenicity of low doses of ionizing radiation has been estimated by extrapolating from high doses. This immediately raises the problem, however, of which dose/response equation to assume. Again, the number of cancers expected per unit dose of radiation will be higher at low doses if the linear model is chosen than if either of the other two are used.

Next, the effects must be projected by using either the additive or the relative risk model. This decision injects more uncertainty into the process because the "real" risk model has not yet been identified. Given the various combinations of dose/response relationship and risk model, in addition to the assumptions and simplifications that go into estimating radiation exposure, it is no wonder that predictions for the number of cancers caused worldwide by the Chernobyl accident varied from 5,000 to 100,000! The most widely quoted estimate, however, is 28,000 excess cancers worldwide, but we still do not have a definitive number.

Is It Safe?

Twenty-eight thousand fatal cancers: that is a lot of grief. It would seem that the Chernobyl accident will leave a horrible legacy of disease. But consider that cancer kills approximately 16 percent of the population even in the absence of nuclear power accidents or atomic bombs. This means that approximately 160,000 out of every million people in the United States die of cancer. It is the second leading cause of death after heart disease among Americans. Assuming that this also applies to the world's population of approximately five billion, we can expect approximately 800 million cancer deaths worldwide. Radiation-induced cancers remain hidden within this thicket of expected cancers unless they are so common that they increase the cancer risk substantially.

The prevalence of spontaneous cancers will probably disguise the major health impact of the Chernobyl accident. This is hard to conceive—the worst nuclear power plant accident to date may leave no detectable trace in the cancer statistics of the general population. If the Chernobyl accident induces 100,000 cancers in the Northern Hemisphere, an upper estimate that hardly anyone now accepts, that will represent an increase of less than 0.02 percent above the spontaneous rate, an imperceptibly small elevation. If a yardstick represents the number of spontaneous cancers expected in the Northern Hemisphere's population, then a 0.02 percent increase would mean lengthening the yardstick by seven thousandths of an inch, approximately the diameter of a good-sized *Amoeba*. The numbers become no more revealing when applied to a smaller region. Of the 28,000 fatal cancers more commonly predicted to follow Chernobyl, half could well occur in the former Soviet Union. Even if these "excess" cancers appeared only within the 60 million people in the western reaches of the former Soviet Union, that would mean 14,000 excess cancer fatalities compared to about 10 *million* expected cancers! This translates into a change in the cancer risk from 16 percent (due to "background") to 16.03 percent. Such a small increase is not likely to be detected in a group of this size. Most of the radiation-related cancers will be expected to appear among the much smaller group of people who lived near the plant at the time of the accident. As the size of the "target" group decreases, however, the difficulty of detecting an excess cancer risk increases.

Several factors conspire to blur whatever damaging health effects might be produced by low levels of ionizing radiation. The major one concerns statistical limitations to epidemiological studies. As the radiation dose decreases, the probability of contracting cancer decreases, but the difficulty of detecting that cancer increases. The only way to "see" these radiation-induced cancers is to increase the number of people surveyed in the irradiated and nonirradiated groups. This is easier said than done. As the dose decreases by a factor of ten, the sample size must increase by a factor of 100 in order to maintain reliability. If 1,000 people are required to detect the effects of 1 Gy, then 100,000 people would be needed to see the effects of 0.1 Gy, and the sample size would have to balloon to ten million to reveal the effects of 0.01 Gy!

Furthermore, the number of people required for a study depends on the spontaneous frequency of the cancer being studied. Studies of breast cancer, for example, would require a larger number of people than studies of leukemia because breast cancer occurs in approximately 250 out of every million women in the United States per year while leukemia is less common at 44 cases per million population per year. Detecting the effects of low doses of ionizing radiation on breast cancer in a retrospective study would require participation by 600,000 women with breast cancer, but only 25,000 cases of breast cancer appear each year in the United States. If the study does not include a "matched control group," one of equal size and makeup as the experimental group, then the required sample size could run upwards of 100 million women. Even if the required number of participants could be found, such a study would take decades to complete, and the costs would be enormous.

This discussion of the difficulty in accurately assessing the health consequences of the Chernobyl accident should not dilute concern for the people who lived in contaminated areas. There is no denying the traumatic effect produced by the accident on these people apart from whatever cancer risk they may face. An increasingly large number of citizens have had to abandon their farms and homes, perhaps permanently in some cases. As late as four years after the accident, Byelorussia, the most heavily contaminated Soviet republic, proposed relocating two million of its citizens, one fifth of its population. Estimated doses of radiation exposure have been increasing as the area contaminated by the Chernobyl accident continues to be studied in detail. In addition to lives lost and the dislocation of survivors, there remain billions of dollars in costs for cleaning up the countryside.

Radiation Is Nothing New

Up to this point, only one technological source of ionizing radiation has been mentioned. But even if nuclear technology ceased to exist today, our exposure to ionizing radiation would continue. Not only do other industries, such as coal mining, expose people to ionizing radiation, but it has been with us for our entire evolutionary history. We can not escape it. This so-called background radiation comes from a variety of sources most of which trace their ancestry to the origin of the planet (table 10.4).

Table 10.4
Background Radiation

Source	Annual Effective Dose Equivalent (mSv)*
cosmic rays	0.35
potassium-40	0.33
radon-222	1.1
radon-220	0.16
uranium series (without radon)	0.14
other	0.32
TOTAL NATURAL SOURCES	2.4
MEDICAL EXPOSURES	0.6
OCCUPATIONAL EXPOSURE	0.002
NUCLEAR POWER PRODUCTION	0.0002
TOTAL	≈ 3

*Data from UNSCEAR 1988.

The average background radiation exposure from natural sources amounts to 2.4 mSv per person per year, but exposures vary on either side of that average depending on a person's behavior and location. People who live in Denver, Colorado, for example, receive about twice the dose of cosmic rays as people living at sea level because the atmospheric shield against cosmic rays is thinner at high altitudes. The average exposure of 2.4 mSv from background radiation sources provides a crude but interesting bench mark against which other exposures may be compared (table 10.4). According to the best available estimates, the people most heavily exposed to radionuclides from the Chernobyl facility received 2.0 mSv in the first year after the accident, nearly doubling their annual exposure from background radiation. The danger these people face is not yet clear, but a study in China turned up no differences in cancer mortality between groups of people whose background radiation exposure differed by a factor of three. In the United States, fallout from the Chernobyl accident amounted to about 0.06 percent of background. The loss-of-coolant accident at the Three Mile Island nuclear power facility in 1979 released only 2 percent the amount of radioactive gases and 0.00002 percent as much I-131 as the Chernobyl accident, leading to an average exposure less than 1 percent of background for people within a 50 mile radius of the plant and less than 10 percent of background for workers in the plant. (No radioactive cesium escaped from the Three Mile Island facility.) Rosalind Yalow, Nobel laureate, has commented that reporters who flew more than three hours above 33,000 feet to cover the Three Mile Island story received as much extra radiation as a person living near the plant during the accident.

Uncertainty Breeds Controversy

Our blindness to the effects of low doses of ionizing radiation creates bitter controversy over health problems blamed on nuclear industries. In one of the more bizarre episodes in the early history of atomic bombs, more than 200,000 American soldiers took part in atmospheric tests of atomic bombs conducted in Nevada between 1946 and 1962. The Army wanted to know how atomic blasts would affect the performance of combat soldiers positioned near a detonation site. At that time, the dangers of ionizing radiation were not as well understood as they are today, and government officials claim that they did not anticipate a great risk to the soldiers. Eight out of 3,224 soldiers who had participated in the 1957 test codenamed "Smokey" eventually died of leukemia. Spontaneous cancers should have accounted for about three or four deaths from that cause in a group of that size. No other type of cancer appeared more often than expected in the Smokey group. The highest dose received by a soldier who later died of leukemia was 36 mSv, according to the film badge he wore, but most soldiers in these tests received less than 5 mSv, about twice the annual exposure from natural background radiation. Understandably, relatives of these soldiers and survivors of the other tests demanded that the federal government compensate them for their trauma and losses. Late in 1990, the federal government signed into law the National Atmospheric Testing Compensation Act to give cash awards to some uranium miners and residents downwind of atomic bomb test sites. This act was passed despite evidence that an apparent increase in the incidence of cancer among Utah "downwinders" following the atomic bomb tests simply reflected an unusually low incidence of cancer in the years immediately preceding the testing period.

As the preceding comment suggests, not everyone agrees that the atomic bomb tests posed a significant threat to the soldiers involved. A subsequent, more extensive survey of 46,186 participants in the atomic bomb tests confirmed the increased risk of leukemia within the Smokey group but failed to detect an increased risk overall for leukemia or any other cancer. Forty-six soldiers from this larger group have died of leukemia whereas fifty-two such deaths would be expected from the spontaneous development of leukemia. The total number of cancer deaths in the larger group is also slightly less than the number expected in the absence of excess radiation (1,046 vs 1,243). As the authors of the original Smokey study cautioned, the unusually high incidence of leukemia observed in the initial study may signify nothing more than a statistical quirk.

The conclusion that no detectable excess risk exists among veterans of atomic bomb tests raised hackles, as might be expected. Critics of the larger study complained that comparing the soldiers' health to that of civilians, who are presumably less healthy than soldiers on average, may have misled the investigators into overestimating the number of expected cancers and overlooking the excess cancers within the irradiated group. The National Research Council that supervised the larger study responded that a sufficient number of veterans could not be found to serve as a control group, and even if they could have been found the extra time and cost would have been prohibitive.

The scenario of a small study indicating risk followed by a larger study showing no excess risk has been played several times. When six out of 146 "nuclear workers" at the Portsmouth Naval Shipyard died of leukemia, people became alarmed that work around nuclear propulsion systems might carry a risk of radiation-induced cancer. But a subsequent, larger study found no increased risk of fatal cancer among 7,615 shipyard workers exposed to an average of 30 mSv of radiation compared to 15,585 nonexposed shipyard employees. A group of Navajo Indians who mined uranium for the nuclear weapons industry between 1949 and 1961 have demanded that the United States government compensate them for the lung cancers they blame on the radiation to which they were exposed. Workers from these mines have developed lung cancer at five times the expected rate, but the Justice Department has cited scientific reports in arguing that there is no evidence for an increased cancer risk among people who participated in atomic bomb tests or worked in American uranium mines. On the other hand, workers in Czechoslovakian uranium mines who received intermediate doses of radiation, mainly from inhaled radon gas, show a substantial increase in the risk of fatal cancers, especially lung cancers.

The conflicting reports tempt us to ask in exasperation, "Well, are low doses of ionizing radiation dangerous, or aren't they?" Without doubt, high doses of ionizing radiation increase the frequency of malignancies, but the probability decreases so much at low doses that the dangerous effects can no longer be detected against the backdrop of spontaneous cancers. If the presumed carcinogenic effects of low doses can not be detected, does that mean that ionizing radiation is safe at those levels? Uncertainty hides the scientific answer we seek from the sorts of epidemiological studies we have conducted in the past. It may be necessary to adopt a completely different approach to learn what risks of cancer attend low doses of ionizing radiation.

Cells and Understanding Radiation's Risks

Scientists at the Eleanor Roosevelt Institute for Cancer Research in Colorado have begun to view the effects of ionizing radiation—not from the long-term perspective of effects on human populations but from the short-term perspective of how it affects a peculiar type of cell. These scientists combined a human connective tissue cell with an ovary cell from the Chinese hamster to form a hybrid "cell" containing all of the hamster's genetic information but only a small portion of the human genetic material, the portion contained on chromosome 11. (Humans have 23 visually distinct pairs of chromosomes that are numbered 1 to 23.) This ingeniously designed cell provides a living package within which the effects of radiation on a human chromosome can be studied. Damage to the human chromosome does not interfere with the cell's ability to divide because cell division in these hybrids is governed entirely by the hamster's genes. Damage to the human genes *does* interfere with the production of certain proteins that become attached to the cell's outer membrane and jut out into the surrounding fluid. The presence or absence of these human proteins in the hybrid cell's membrane can be taken as a measure of the damage inflicted by radiation on the human chromosome. The trick lies in detecting the proteins' presence. If the proteins are manufactured and inserted in the cell's membrane, the hybrid cell can be killed by certain chemicals from the human immune system that bind selectively to the human proteins. Cells that are killed by this treatment are judged to carry a normal, undamaged human chromosome 11. Cells that survive the treatment must have suffered damage to chromosome 11 that prevents them from making the human proteins or incorporating them into their membranes.

The scientists evaluated the mutagenic effects of ionizing radiation by exposing thousands of their hybrid cells to X rays then simply counting the number that survived exposure to the immune system chemicals. What they found ran counter to all expectations. Low doses of ionizing radiation proved much more damaging to human chromosomes than previously thought (figure 10.13). A dose of 0.25 Gy produced a hundred times more mutant cells than predicted from previous experiments on the mutational effects of ionizing radiation. Other scientists had seriously underestimated the

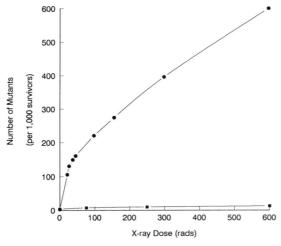

Figure 10.13 Number of mutants produced by irradiating hybrid cells with various doses of X rays. When the experiment takes into account the cell-killing effects of X rays (solid circles), low doses of radiation are revealed to be much more mutagenic than when the cell-killing effects are not taken into account (solid squares). (Redrawn from Waldren et al., 1986).

effects of radiation because their methods did not account for the cell-killing effects of radiation.

Do these results mean that low doses of radiation are more likely to cause cancer than currently believed? A quick comparison of figure 10.13 with any of the curves in figure 10.9 suggests the answer is a resounding "yes." But maybe not. The answer depends on the mechanism by which radiation induces cancer. Because genetic damage appears in so many cancerous cells, it seems that the carcinogenic effects of radiation must arise from the damage it inflicts on the genetic material, DNA. The sequence of events leading from irradiation to cancerous cells, however, remains frustratingly unclear.

Cancer

Cancerous tumors seem to develop in two stages: initiation and promotion. Initiation may be triggered either by turning on dormant "oncogenes" or by turning off other genes that normally suppress initiation. Oncogenes appear to be genes that are normally active during embryonic life but quiescent during adult life. Turning them on restores embryonic-like qualities to the cell, allowing it to multiply uncontrollably. A cell that has undergone initiation typically lies dormant for a period of time, often years. Initiation appears to be frequent, but most initiation events do not lead to malignant tumors. During the early part of this lag time, the cell's own chemical machinery may repair the damaged DNA and eliminate the initiated state, or the body's immune system may recognize and destroy the transformed cell.

"Promotion" begins when the transformed "initiation cells" start dividing. Various hormones, including some of the steroids like estrogen, assist the transition to the promotion stage. When the tumor grows to a diameter of about 0.2 mm, it needs an abundant blood supply or its cells will die from starvation and lack of oxygen. Tumors, in their insidious way, ensure their own blood supply by secreting chemicals that cause blood vessels to develop fingerlike projections toward the group of cancerous cells. It is at about this time that tumors in soft tissues become detectable with current techniques. If the tumor grows unchecked, it robs nutrients from the blood stream, and the rest of the body wastes away. Tumor cells may also begin to migrate about the body, damaging tissues or taking up residence along the way. We are fortunate that 90 percent of all tumors never gain their own blood supply and remain small and harmless.

Aside from the initial genetic damage, it is not clear at which points ionizing radiation influences this two-stage process. Radiation probably affects the activity of oncogenes and "suppressor" genes, and it may spark the promotion stage, but the ways in which it affects these events or other points in the process upon which it acts are not yet understood. Without this knowledge, the cellular view of radiation-induced mutations can not be meshed with the epidemiological view of cancer in human populations, and the cancer risk from low doses of ionizing radiation remains unknown.

Nuclear Power: Decisions Despite Uncertainty

The question was posed earlier: "Is it safe?" No single scientific answer can be given when that question is directed at ionizing radiation. The term "safe" has no clear, universally accepted definition, only operational definitions. Designations of safe levels of ionizing radiation have steadily declined since the discovery of X rays about a hundred years ago, and there is, as yet, no detectably low level at which ionizing radiation ceases to be dangerous. The danger of radiation, however, does not show itself clearly until the dose rises to a high level. We are challenged in the atomic age to act reasonably in this highly uncertain zone between what we know and what we fear. We do not always succeed.

A poll published in *Newsweek* magazine shortly after the Chernobyl accident revealed that, while 49 percent of the 762 respondents oppose "nuclear generation of electricity," 70 percent would oppose construction of a nuclear power plant within five miles of their home. Apparently, a sizable number of people who are undecided or in favor of nuclear power draw the line at living close to a power plant, preferring that it be located in unpopulated areas or that other people assume whatever risk may exist.

Small amounts of radionuclides, including isotopes of hydrogen and iodine and carbon, escape from nuclear power plants during their normal operation. These materials either form in the coolant or seep into the coolant from corroded fuel rods. All reactors scavenge radionuclides from the coolant that leaves the plant, but some release is inevitable. Nuclear reactors normally release such tiny amounts of these materials that exposures to people can not be measured but must be inferred from a model of radionuclide transfer in the environment. According to this model, the most exposed people in and around operating pressurized water-cooled nuclear power plants, such as the ones in the United States, receive about 4.5 mSv (0.0045 Sv) in one year, about 0.2 percent of background radiation and 200 times less than the average annual exposure from medical diagnostic procedures. This tiny excess exposure has not measurably increased the cancer risk for people living near nuclear power plants. Surprisingly, even those people who worked with plutonium on the Manhattan project have not shown an increased incidence of cancer, nor have employees at the Hanford nuclear facility that produces plutonium.

Normal operation of a nuclear power plant is one thing, but many people cite fear of accidents as a reason for shying away from these facilities. The staggering complexity of nuclear power plants, a general lack of knowledge about how radionuclides behave in the environment, the relative youth of the nuclear power industry, and the unpredictability of human behavior all contribute to tremendous uncertainty in any risk estimates for accidents. One risk estimate, published in 1975, evaluated the likelihood of a serious accident—one that released radionuclides into the environment—as lying between one every hundred thousand reactor-years and one every billion reactor-years. (One reactor operating for one year equals one reactor-year while a hundred reactors operating for one year equals a hundred reactor-years.) The maximum risk of a worst-case accident has recently been reduced from one in a hundred thousand reactor years to one in a million reactor years. This implies that we might expect a radiation-emitting accident about once every 2,500 years given the world's current number of reactors! A risk estimate such as this does not preclude three radionuclide-releasing accidents from occurring within thirty years, as happened between 1957 and 1986, but such a sequence of events makes many people uneasy. Scientific studies on small-scale models of nuclear power plants are not likely to improve risk estimates very much. Small models are simply too far removed from complex, full-scale, human-operated power plants to be helpful in a detailed way. The Chernobyl accident exposed at least one flaw in small-scale models. The destructive effects and even the possibility of steam explosions in nuclear power plants had been discounted prior to the Chernobyl accident because they did not pose a problem in small-scale models.

Nuclear power thus embodies a great irony. The power of scientific research and technological innovation is nowhere more evident than in our ability to control the release of energy from an atomic nucleus. Yet that energy release carries with it risks that seem beyond the complete understanding of science and technology. Despite incomplete knowledge, we must act. Policies must be formulated, decisions must be made according to what we know even when the information is painfully incomplete. What do we do in the meantime? Abandoning nuclear power may increase air pollution as we rely more heavily on fossil fuels until alternative energy sources and energy conservation policies can be developed further. Many people, including members of the Bush administration, responded to the war with Iraq in 1991 by calling for expansion of the nuclear power industry to reduce our dependence on imported oil. Increasing our reliance on nuclear power, however, will likely increase the risk of exposing large parts of our population to low doses of ionizing radiation. Our decision will depend heavily on our making the best possible assessments of those risks and weighing them against the alternatives.

Are we willing to live with an uncertain risk while we work to learn more about nuclear power? A Polish woman interviewed after the Chernobyl accident confessed that "At first I was frightened, but now that I have heard all the explanations, I am still frightened." Our perception of safety may be guided by quantitative assessments of risk, but our final judgment grows out of a much larger and more nebulous set of beliefs and attitudes regarding our sense of control over a risk and the degree to which we think nuclear power is needed.

Exercises

1. (a) Explain how a self-sustaining nuclear chain reaction is started and then propagated.
   (b) Show how U-235 plus a neutron can yield barium-145 (atomic number 56) and krypton-89 (atomic number 36).
   (c) If fission occurred with three neutrons emitted by the material undergoing fission, how many neutrons would be produced at the end of the sixth set of collisions?

2. Suppose you borrowed $25 from a friend who said that it wasn't necessary to pay it back quickly. In fact the friend, knowing that you might not have much money for a while, is very generous and patient and gives you at least six months to repay. However, in order to avoid having the loan come due all at once, the only stipulation is that you pay back 1 cent the first week, 2 cents the second week, 4 cents the third week, and so on. How much will you have paid back after four months?

3. Calculate the amount of energy (in Joules) that arises from one gram (0.0010 kg) of matter being transformed into energy. Assume that this happens over a period of 10 seconds in a nuclear reactor—how much power is produced?

4. In the Chernobyl accident a large quantity of radioactive cesium-137 was released over Europe and the rest of the Northern Hemisphere. Estimate what fraction of the cesium is now still in the biosphere. (This isotope of cesium has a half-life of ten thousand days.)

5. Why does the type of mathematical model determine the level of risk of getting cancer from radiation? If you were a public health official, which one of the models would you choose to apply in determining public policy regarding radiation? Defend your choice.

6. Why can't a nuclear reactor explode like an atomic bomb?

Conclusion

IV

An Intricate Web

11

The Pace of Innovation

A witness to the Wright brothers' historic flight in 1903 might well have lived to see television images of the first footprint on the moon. A life across that time span in the United States changed drastically as telegraphs gave way to telephone networks, vaccines brought many infectious disease under control, small isolated electrical power plants grew into immense power grids, automobiles replaced horses as gasoline became abundant, prestressed concrete altered the landscape, and physicists tapped the vast reservoir of energy in the atomic nucleus. That is a great deal of innovation for one lifetime, and this brief list is by no means exhaustive.

Such rapid change can be an invigorating source of hope that conditions will improve continuously within our lifetimes. From this optimistic point of view, scientists may appear as brilliant stars guiding our progress. But some people detect a sinister, or at least dehumanized, element in modern technology, an element that casts a different, less flattering light on scientists, who are blamed for modern ills. Just after World War II, Aldous Huxley bitterly lamented that "Thanks to the genius and cooperative industry of highly trained physicists, chemists, metallurgists and mechanical inventors [i.e., engineers], tyrants are able to dragoon larger numbers of people more effectively, and strategists can kill and destroy more indiscriminately and at greater distances than ever before."

Both views, in the extreme, are too simplistic to be correct. Each makes the mistake of considering science to be the fount from which all technology springs. This highly ordered view of technological progress is so appealing that it has become the predominant, modern view of how science relates to technology. It seems logical, for one thing, that we must understand nature before we can control it. Francis Bacon argued this point more than three hundred years ago in trying to justify scientific study: scientists discover how nature works and engineers apply that knowledge for human benefit. Accordingly, scientists and engineers deserve credit for projects that turn out well but must accept blame for those that increase human suffering.

Technology as Applied Science?

Much of our technological progress over the past 150 years supports the view of technology as applied science. The rapid growth of telegraphy out of the discoveries of Oersted and Henry makes a case in point, especially given that Henry turned his scientific work into a prototype telegraph. Henry, in particular, actively sought to buttress the view that basic scientific research pays off in terms of practical applications. Similarly, Faraday's scientific investigations of electromagnetism formed the direct antecedents of the electrical generator. The telegraph and the generator almost certainly would not have been developed without a prior understanding of electricity and magnetism because the phenomena involved could not have been predicted from everyday experience. The same holds true for the relationship between nuclear physics and the control of nuclear power. The success of the Manhattan Project and the feasibility of commercial nuclear power plants depended entirely on a basic physical understanding of the atomic nucleus. As that understanding increased, the ability to control that aspect of nature improved. Vaccines make an especially interesting example in this regard because, although Jenner developed the original vaccine without understanding the underlying biology, further progress in vaccination stalled until a great deal of basic biological research had been completed. The shot-in-the-dark approach, simply did not pay off in vaccine development.

This sort of parent-offspring link between basic science and technology became institutionalized during this century as many corporations, especially oil companies, established R&D divisions and staffed them with research scientists. These companies gambled that investments in basic research would pay big dividends in innovation and market share. One of the great early successes that showed the wisdom of the gamble was development of techniques for recovering large amounts of gasoline from a barrel of oil. As another act of institutionalizing the relationship between science and technology, the Massachusetts Institute of Technology was founded in 1861 to encourage the merger of science with engineering.

Faith that science begets technology underlies federal policy in the United States regarding the funding of scientific research. Representatives of the National Science Foundation (NSF), for example, have argued forcefully and convincingly over the years that feeding the basic research community is necessary to maintain our technological edge and economic vitality. Policymakers no doubt took encouragement from the successes following the government's investment in scientific research during World War II, research that led not only to the atomic bomb but also to penicillin and radar, all of which contributed significantly to our war effort.

Cracks in the Facade

This optimistic view of basic science was bolstered by an NSF-sponsored investigation into the direct ancestry of several modern technologies. The final report, entitled *Tracers*, claimed that 70 percent of the ideas considered critical for the development of these technologies could be traced by the Foundation to basic scientific research. The conclusion that basic scientific research fosters technological advancement seemed obvious. The Department of Defense (DOD), however, came to a very different conclusion. In the DOD study, appropriately named *Hindsight*, the recent histories of twenty weapons technologies were combed for evidence that basic scientific research had led directly to crucial technological achievements. In contrast to the NSF report, the DOD ascribed only 0.3 percent of the crucial achievements to basic scientific research. One study suggests that science is indispensable for technological achievement, while another study decides basic science is nearly irrelevant. What gives?

One important difference between *Tracers* and *Hindsight* may be the time period surveyed by each: *Tracers* extended its search back fifty years before the technical innovation under study, while the *Hindsight* study covered fewer years. It takes time to convert a scientific discovery into a useful application. Even in the cases of the telegraph and the generator where the applications seemed born almost fully formed from the science, considerable thought and effort had to be invested by Morse, Siemans, Edison, and others to solve the myriad *technical* problems of building a useful, marketable device. A study that does not reach back beyond the development phase of a technology may miss its important scientific antecedents.

The difference between the conclusions reached in the *Tracers* and *Hindsight* studies, however, may reveal flaws in the common perception of science as the mother of invention. As the amount of time and the difficulty in developing a technology increase, the link to scientific antecedents, if it existed at all, becomes fuzzier and fuzzier. Soon, the scientific discoveries begin to look less like initiators of technology and more like enablers. In this condition, they become part of the broad foundation on which the technology rests and cease to be direct-line ancestors to innovation.

Such might be the case with flight. The general scientific theories established by Newton, Euler, and Bernoulli had existed for more than a century before the Wrights began to look seriously at the flying machine problem. These theories, while beautiful generalizations of nature's behavior, proved completely inadequate when it came to designing the details of a flying machine. The necessary details emerged only through the painstaking laboratory research and field trials undertaken by Orville and Wilbur. The Wright brothers' link to science appeared more clearly in their methodology than in the prior scientific information they used. As a result, their work reflects a different relationship between science and technology than the one so commonly cited today. The discoveries that led directly to their success came not from unfettered, basic scientific research, but from research conducted according to the scientific style in the service of a specific technical goal, powered flight. The fusion of scientific methodology with practical goals has been termed engineering science and characterizes the emergence of modern engineering practice.

Prestressed concrete offers another case where the scientific underpinnings are so remote from the innovation that they can hardly share the credit. Freyssinet thought to combine stretched steel cables with concrete after pondering the deficiencies of reinforced concrete and studying the behavior of bridges he built. Even the development of high-strength steel, so crucial to prestressed concrete, depended as much, if not more, on trial-and-error as on a scientific understanding of materials. The technology of prestressing sprang from other, preexisting technology, not from science. The same can be said of the battery or the steam turbine that runs most modern electrical power plants. The direct-line ancestor to the turbine was a cream separator! A technological advance may even precede and initiate a scientific advance as was the case with the steam engine and thermodynamics.

Yet another type of relationship between science and technology is one in which we use scientific methodology to investigate technology-related problems such as the greenhouse effect or excess ionizing radiation. The substitution of careful measurements, laboratory and field experiments, simulations, and modeling for the trial-and-error approach not only helps define the problem precisely but also promises to suggest the best solutions.

These examples do not deny that basic scientific research can form a powerful stimulus to technical innovation. The amazing variety of applications suggested by recent advances in molecular biology testify that science can indeed serve as the mother of invention. But all of modern technology can not comfortably fit within that relationship. The varied sources of technical advancement suggest a much more complex relationship between science and technology, one more of equal partnership than of parent-offspring.

The Enterprises of Science and Engineering

As science emerged from natural philosophy and engineering developed from craftsmanship and other roots, the two disciplines took on distinct characteristics. Most scientists claim that their ultimate goal is to see nature's patterns, understand how nature works. Engineers, on the other hand, strive to control nature for human benefit; fundamental understanding takes a back seat to utility.

These sharply differing goals affect the constraints under which scientists and engineers work. In its purest form, basic science is an unfettered exploration of nature. Any question can be asked, as long as it will submit to scientific methodology, and any answer is acceptable. In actual practice, of course, the availability of money, equipment, and appropriate techniques influence the selection of research topics as do less tangible factors such as the desire for professional recognition and advancement and a sense of public acceptability. These constraints are perturbations on the basic scientific enterprise, however, not necessary features of it.

Engineers, however, are not interested in *any* answer to *any* question. They must solve pressing human problems or correctly anticipate human needs, and their solutions must not only be useful but also affordable and acceptable. Political, psychological, and economic constraints are just as much a part of engineering problems as technical constraints. Designing for human use means taking into account all of our quirks as well as our institutions and ways of living. Devices, structures, or systems that fail to address one or more of these constraints are not very likely to succeed, or at least we will not be very happy with them.

The difficulty in designing under these myriad and sometimes unpredictable constraints accounts in large part for the lag time between a scientific discovery and a technical application. Faraday was satisfied to measure the small currents produced by his crude generator; he could use it to continue his investigations of electromagnetic phenomena in nature. A commercially useful generator, however, must not only work but work well. It must generate large amounts of electrical power at low cost if the company that builds it hopes to survive. But that is not all. The type of electrical current it produces must match the needs of the devices that will depend on it and yet allow low-cost transmission within the constraints of the power plant's environment. Alternating current was not as desirable as direct current for running elevators in the early days of that uplifting technology because it caused jerky changes in speed rather than smooth transitions. Direct current, however, proved too costly to transmit over long distances. No wonder sixty years passed between Faraday's discovery of electromagnetic induction and installation of the first commercial three-phase AC generator.

Social pressures, whatever form they might take, shape technology even when scientific discoveries seem to be driving new technical developments. The atomic bomb, for example, grew out of nuclear physics only because of World War II. Hitler's rise to power coincided with the discovery of nuclear fission and set the imperative for the transformation of that knowledge into a weapon. Had that war (or some other war) not provided the impetus for the bomb, it may never have been built because many prominent physicists, including Robert Oppenheimer at first, thought it could not be done and did not want to waste their time on a futile project. There was no *scientific* imperative to build the bomb. The nature of technology means that, inevitably, the technical expression of scientific knowledge depends heavily on the social context within which the technology is being developed.

The tight constraints confining technology create problems for engineers that are entirely foreign to most basic scientists. The difference in their perspectives can produce important consequences when these more nebulous problems are confronted. Again, the atomic bomb supplies a fitting example. The astonishingly rapid progress made in the Manhattan Project, for example, must be credited to the ability of Brigadier General Leslie Groves, the military commander of the project, to recognize when the work should turn from scientific research to engineering design and production. In an action that particularly galled the scientists on the project, Groves enforced secrecy to the extent that different groups of scientists were not allowed to know what other groups were doing unless absolutely necessary. Groves was aware of the scientist's penchant for wandering from one question to another according to personal interest and felt that this attribute of science had to be squelched in favor of a task-oriented engineering approach if the bomb was to be finished in time to decide the war. Richard Hewlett, after reviewing the early history of the nuclear power industry, suggested that the failure to include at least one engineer on the original Atomic Energy Commission resulted in unnecessary delays in the search for a useful nuclear power plant design.

The view of technology that treats engineering as applied science fails to recognize the profound distinctions between science and engineering and as a result overlooks the complex interactions between science, engineering, and society that produce modern technology. In the simplified view, science is likely to receive too much credit for technological achievements as well as too much blame for technology-related problems, and the creative role of engineering is likely to be underestimated with possible consequences for policies that try to influence innovation. Science and engineering are "mirror image twins" (Layton 1971) that function independently yet interact extensively in complex ways. The work of science and engineers continues inescapably within our social institutions and so is subjected to all the forces created by those institutions. By working in this intricate, imperceptible web, we have produced our modern world.

Abbreviations, Symbols, and Units

Chapter 2

amp	Ampere, unit of electrical current
N	number of coil turns
ohm	electrical resistance
V	volt, unit of electrical potential
ρ	electrical resistivity (per unit length)

Chapter 3

AC	alternating current
B	magnetic field strength
d	density of water
DC	direct current
ft-lb	foot-pound
HP	horsepower
I	current
KE	kinetic energy
kW	kilowatt
kWh	kilowatt-hour
MW	megawatt
PE	potential energy
P	power

q	rate of water flow
R	resistance
V	voltage
W	watt
ω	angular velocity

Chapter 4

C_d	coefficient of drag
C_l	coefficient of lift
D	drag
k	Smeaton coefficient
L	lift
R	force perpendicular to the wing's surface
S	surface area
V	velocity of fluid
ρ	fluid density
θ	angle of attack

Chapter 5

A	area of piston
AC	alternating current
DC	direct current
ft-lb	foot-pound
hp	horsepower
I	current
KE	kinetic energy
kW	kilowatt
kWh	kilowatt-hour
L	length of piston
MW	megawatt
N	strokes per minute of piston
P	pressure
PE	potential energy

psi	pounds per square inch (pressure)
R	resistance
V	voltage
W	watt

Chapter 7

C	compression (stress)
d	depth of beam
F	force
I	inertia
k	kip
kft	kip-ft (kilopounds × length)
M	bending moment
psi	pounds per square inch
S	stress
T	metric tons
Tm	ton-meter
α	angular acceleration
Γ	torque

Chapter 10

Bq	becquerel, a unit of radioactive decay rate
c	speed of light
Ci	curie, a unit of radioactive decay rate
E	energy
Gy	gray, unit of radiation energy absorbed
m	mass
rad	(radiation absorbed dose) unit of radiation energy absorbed
rem	(roentgen man equivalent) unit of radiation dose
Sv	seivert, unit of radiation dose
λ	wavelength of light

Glossary

absorption band
a part of a spectrum that shows that the transmittance of radiation is significantly less than 100 percent; a valley or downward peak that is opaque or partly transparent.

absorption window
a part of the spectrum in which the transmittance is near 100 percent; a "clear" part.

aerosol
very fine droplets suspended in air like those emitted from a spray can of paint.

ailerons
movable surfaces hinged to the trailing edges of an airplane's wings that allow the pilot to control the airflow over the wings and hence the lift; offer the pilot control over roll.

albedo
the reflectance of the earth; percentage of sun's radiation that is reflected by clouds and the surface of the planet.

aliphatic hydrocarbons
compounds of carbon and hydrogen (and their derivatives), which may be straight-chain, branched-chain, or cyclic. The saturated aliphatic compounds have the general formula, C_nH_{2n+2}, (butanes, pentanes, hexanes, etc.), but the unsaturated compounds have fewer hydrogens per carbon than given by the formula for saturated aliphatics (butylenes, pentylene, hexylene, etc.)

alkane
see aliphatic hydrocarbons.

alpha particle
a helium nucleus.

alternating current
a form of electricity in which the current rapidly reverses direction; current oscillates sixty times per second in American commercial electricity.

angle of attack
the angle between the front-to-back axis of a wing and the direction of airflow; determines the amount of lift and drag on a wing.

antibody
protein molecule secreted by a select group of immune cells in the blood stream in response to exposure to an antigen; binds to specific antigens and prepares them to be expelled from the blood.

antigen
any substance treated as foreign by the body such that the body develops an immune response against it.

aromatic hydrocarbons
compounds of carbon and hydrogen containing benzene and/or its derivatives.

atomic number
the number of protons in the nucleus of an atom.

bacterium
a microorganism that is a few thousandths of a millimeter in size; characterized by the presence of a cell wall, in addition to a membrane, and lack of a nucleus.

barrel (bbl)
the unit of production in petroleum processing; one barrel is equivalent to 42 U.S. gallons or 5.61 ft^3 or 158.8 liters.

battery (electrical)
one or more cells connected either in series (positive terminal of one connected to negative terminal of another, except for those at the ends), or in parallel (all positive terminals connected to each other, all negative terminals also connected to each other).

bending moment
a moment or torque internal to a structure that is created in response to loads on the structure; its value is equal to the weight of the load times the distance between the point at which the load is assumed to act and the axis of rotation.

Bernoulli's principle
the scientific principle stating that the pressure of a fluid decreases as its flow rate increases such that the product of pressure and flow rate is constant according to the equation: $p + \rho V^2/2 =$ constant (where p is pressure, ρ is fluid density, and V is velocity).

beta particle
an electron or a positron emitted by some radioactive decay processes.

biosphere
the part of the earth's surface that is inhabited by living organisms.

British thermal unit

a unit of heat (thermal energy) in the "English system," as opposed to the metric system; it is the amount of heat needed to raise the temperature of one pound of water one degree Fahrenheit.

camber

in aeronautics, the upward curvature of a wing; measured as a fraction of the wing's width.

cantilever

a structure with one end fixed to a support and the other end free to move.

carcinogen

any factor that causes cancer.

catalyst

a substance that alters the rate of a chemical reaction but remains unchanged itself so that it can recycle through the process many times.

cell (voltaic, electrolytic)

a device for converting chemical energy directly to electrical energy. It consists of an electrolyte in which a pair of electrodes are immersed.

Celsius

temperature scale used in science and by most of the world's people; each degree Celsius (°C) equals 9/5°F.

center of pressure

the single point in a wing's cross-section representing the location of lift's net effect after the actions of lift all over the wing have been averaged. This concept is analogous to the concept of center of gravity.

CFC

chlorofluorocarbon; a class of synthetic organic compounds that have been shown to lead to depletion of the earth's ozone layer as well as be greenhouse gases.

chain reaction

in atomic fission, a sequence of reactions in which the first fission event produces neutrons that then trigger subsequent fission events, each of which produces more neutrons with the same effect.

chord

the imaginary straight line running from a wing's leading edge through the wing's trailing edge.

clinical trial

experiment involving human subjects in which the value of a new medical treatment or drug is tested.

coefficient of lift or drag

numbers that summarize the combined effect of angle of attack and camber on lift or drag; used in equations to determine amount of lift or drag produced by a wing under various conditions.

compression
a force or stress that tends to push the particles of a material together.

condenser
a cooling device used to cause gases to revert to their liquid state.

conductor (electrical)
a substance that allows the flow of electrons; copper, graphite, and silver are good conductors.

continuous beam
a beam in which the ends extend beyond the supports.

control group
a group of subjects in an experiment that is treated exactly the same as another group (the experimental group) except with respect to the variable of interest in the experiment; control groups are essential in experimental science to provide the background against which observed effects of a treatment are measured.

control rods
rods of a material that absorbs neutrons, inserted into the core of a nuclear reactor to control the chain reaction.

cracking
process of breaking down larger hydrocarbon molecules into smaller ones.

creep
the tendency of concrete to shrink as it ages and cures.

critical mass
the amount of fissionable material needed to make a chain reaction self-sustaining.

culture
in addition to being the collective works and sensibilities of a society, a culture is a controlled growth of a microorganism such as a bacterium or a virus.

current (electrical)
the rate of flow of electrical charges through a conductor.

deforestation
systematic removal of trees for their wood, which may be burned or used for products, or for clearing the land for agriculture, cattle and sheep grazing, building homes, and so forth.

direct current
current with a constant voltage and constant direction of flow.

distillate
any fraction of a mixture removed during the distillation process; usually characterized by its boiling point range.

distillation
a physical separation process in which a mixture of chemicals in the liquid state is vaporized and condensed according to various boiling point ranges; the resulting distillate (or condensate or fraction) is a new mixture that contains

fewer numbers of chemicals in higher purity because the vapor phase has a different ratio of concentrations of the components than does the original liquid.

drag
the force that opposes an object's forward movement through a fluid. Induced drag is the resistance that inevitably accompanies lift, it changes with velocity and angle of attack. Parasitic drag is the resistance created when an object pushes fluid aside as it moves forward.

DTP
acronym for the combination vaccine, diphtheria-tetanus-pertussis.

effective dose equivalent
a standard measure of radiation absorbed by an organism.

efficiency
the ratio of the energy output to the initial input of energy; for example, electrical energy compared to potential energy in water used to generate it.

electrolyte
usually a solution containing dissolved charged particles, called ions, that allow the passage of current; a salt solution is an electrolyte and a good conductor. In some batteries, pastes are made from electrolytes.

electromagnet
a metal magnet whose magnetic field strength depends upon the current in the coil of wire wrapped around it; often has an iron core or, more recently, is made of alloys, mixtures of metals.

electromagnetic induction
process by which an electrical current is made to flow (induced) in a wire by varying the strength of a magnetic field around the wire, for example, by moving the magnet relative to the wire; one of the fundamental processes of electromagnetism.

electromagnetism
in general, the idea that electricity and magnetism are physically related—one can lead to the other; the science of electricity and magnetism and their interactions (see also electromagnetic induction).

elevator
in aeronautics, a horizontal surface, usually placed at the tail of an aircraft, that the pilot angles upward or downward in order to change the wings' angle of attack, that is, it offers the pilot pitch control. Upward deflection of the elevator, for example, pushes the tail down, thus increasing the wings' angle of attack and causing the aircraft to rise.

energy
the capacity to do work; units of calorie, Btu, kWh, and others.

epidemiology
the study of the cause and control of epidemic diseases.

fission
the process by which an atomic nucleus splits into smaller fragments.

fluid catalytic cracking
a refining process in which a very finely divided catalyst is sprayed into the hydrocarbons being cracked.

foot-pound
a unit of work; work done in raising a one-pound object one foot; also a unit of energy.

foot-pound per second
rate at which work is done; a unit of power.

force
defined by Isaac Newton as mass times acceleration. Forces cause bodies to accelerate; weight is a force because it is the product of an object's mass and the acceleration due to gravity.

force couple
a pair of forces, equal and opposite, but with different lines of action such that they generate a torque.

fossil fuel
carbon-based substances like coal, peat, or petroleum that have been produced over millions of years from the decomposition of vegetation under high pressure and temperature.

fractionation
separation of a mixture into components (fractions) based upon their relative boiling points; see distillation.

Francis turbine
a turbine used in hydroelectric plants with high flow rates; it is totally immersed in the flowing water.

frequency
number of times an event occurs in a particular time period—cycles per second, heartbeats per minute, and so forth.

fuel oil
one of the higher boiling components of crude oil, diesel fuel.

galvanometer
an instrument for measuring a small electric current by the movement of a magnetic needle or of a coil in a magnetic field.

gamma ray
one type of very high energy electromagnetic radiation; released by many radioactive elements during fission.

gas oil
a distillate of crude oil that boils higher than kerosene (450–800°F) and is the main feedstock to the catalytic cracking units. Its name comes from its use as an enriching agent in the production of city, or manufactured, gas.

gasoline
a volatile, flammable, liquid hydrocarbon mixture that is produced in the refining of crude oil and used as the fuel in the internal combustion engine.

generator
a device using the relative motion of wire coils and magnets to "induce" the flow of electric current in the wire; one that produces direct current is often called a dynamo.

geosphere
the inorganic or geologic part of the earth's surface.

germ theory
the now widely accepted theory, initially proposed by Louis Pasteur and Robert Koch, that infectious diseases are caused by microorganisms such as bacteria.

greenhouse effect
the warming of a planet's atmosphere due to the presence of gases, such as carbon dioxide, that absorb heat.

greenhouse gas
a gaseous substance that stays in the atmosphere and absorbs infrared radiation (heat).

half-life
the period of time required for a quantity of material to be reduced by half through the process of radioactive decay.

head
the vertical height through which water falls at a hydroelectric plant; the vertical distance between the surface of the water behind the dam and the surface of the tailrace.

horsepower (hp)
a unit of power originally defined by James Watt as 550 ft-lbs/sec, somewhat greater than the experimentally measured power of one horse; 1 hp = 33,000 ft-lbs per minute = 746 W.

host-specific
characteristic of being restricted to living on a small number of available animals or plants that act as hosts; applied to a variety of animals and plants in addition to viruses.

hydrocarbon
any chemical compound consisting of the elements of hydrogen and carbon; a compound in which these two are the main elements.

hydroelectric
refers to electrical power generated by flowing water; also hydropower or hydro.

immune system
the collection of substances, cells, tissues, and organs in the body that generate protective responses, involving cells or antibodies, against foreign materials or cells.

immunity
the state of being immune, or protected.

immunology
the study of the immune system and its responses to foreign cells and substances.

immune response
protective reaction of the immune system to invasion of the body by foreign microorganisms, cells, or materials.

immunogenic
having the capacity to stimulate an immune response.

impulse
the product of a force and the period of time over which it is applied to an object. It represents a rapid transfer of momentum to an object; for example, as a baseball bat hits a ball, the change in the ball's momentum equals the impulse.

infrared
the portion of the radiation spectrum just below visible light in energy and frequency but longer in wavelength; heat such as that used in cooking.

inoculation
the intentional exposure of a person or animal to a disease-causing microorganism; this word has largely been replaced by "vaccination" when referring to prophylactic application of the microorganism.

insulator
a material that does not allow current (or heat) to flow through it readily; for example, ceramics, rubber, and wood are poor conductors but good insulators that have a high resistance.

iso
a prefix attached to a name of a hydrocarbon, often to indicate branching.

isomers
molecules having the same chemical formulas but different structures.

isotopes
atoms that have the same number of protons but different numbers of neutrons; different forms of the same element.

kerosene
a fraction of crude oil that boils above gasoline; used for heating and lighting.

kilowatt
a unit of electrical power equal to 1000 watts; also equivalent to 1.34 horsepower.

kilowatt-hour
a measure of total electrical energy; power multiplied by time elapsed.

kinetic energy
energy an object possesses by virtue of its motion.

kip (k)
a unit of weight indicating 1,000 pounds; a kilopound.

law of large numbers
the concept, initially developed by Jacques Bernoulli, that an estimate of a quantity approaches the true value for that quantity as the number of samples on which the estimate is based increases.

lift
a force that opposes gravity; created by wings as one component of an object's resistance to movement through a fluid.

Manhattan Project
one name given to the project conducted in the United States during WWII to build an atomic bomb.

megawatt
a unit of power equal to one million watts.

meltdown
term applied to the consequences of having a nuclear reaction proceed so rapidly that the heat generated can not be dissipated before the metal in the reactor's core melts.

moderator
a material or element, such as boron, that absorbs neutrons and is used to reduce the progress of nuclear chain reactions.

molecule
an assembly of two or more atoms.

moment
torque; the product of a force and the distance separating the point at which the force is applied from the axis of rotation; analogous to force in translational motion.

momentum
a property of an object in motion; product of the mass of an object and its velocity; a change in direction is a change in velocity, hence a change in momentum.

Montreal Protocol
1987 international agreement to limit the use and production of chlorofluorocarbons and halons, which lead to the destruction of the ozone in the stratosphere.

mutagen
a substance that induces mutations or changes in genetic material.

naphtha
a fraction (or distillate) from the distillation of crude oil with a boiling point range lower than 400°F–500°F (occasionally up to 650°F); it includes the fractions used to formulate gasoline and the lighter grades of fuel oils such as kerosene and diesel fuel oil. But, as a finished product, naphtha denotes a product with a much narrower boiling point range, for example, varnish makers' and painters' (VM&P) naphtha with a 200°F–300°F boiling point range.

naphthenes
cyclic aliphatic hydrocarbons.

negative feedback
a control process that involves an initiator, an action, and a component that is sensitive to the action and can shut down the initiator at an appropriate stage; a room thermostat is an example.

neutron
one of the major subatomic particles that forms the nucleus of atoms; it is electrically neutral (i.e., has no charge).

Newton's third law
when two bodies (A and B) interact, the force exerted by A on B is exactly equal and opposite to the force exerted by B on A; for example, the floor counteracts the force of gravity acting on your body; also known as the "action, reaction" law.

Ohm's law
in an electrical circuit, the voltage equals the product of current and resistance.

oncogene
genes that, when activated, induce cells to become cancerous.

ozone layer
the region in the stratosphere that contains relatively high concentrations of ozone, a triatomic molecule of oxygen (O_3), which absorbs ultraviolet light.

pathogen
any organism that causes disease or pathological effects.

Pelton wheel
a water wheel with scoop-shaped buckets used to generate electricity under low or moderate flow rates.

permeability (magnetic)
a measure of the magnetization produced in a material by an external magnetizing agent, which could be a current-carrying wire or a permanent magnet.

phagocytosis
the act by cells of taking in particles from the surrounding fluid.

phase
for two sinusoidally varying quantities, such as AC voltage and current, the relative displacement of their peak values in a cycle.

photochemical process
chemical process initiated by light such as photosynthesis in plants.

pier
the pillar or column supporting a bridge.

pile
a name given by Enrico Fermi to the accumulation of radioactive material and moderator in a nuclear reactor.

pitch
in aeronautics, movement in the vertical plane that results in an aircraft's nose either pointing up or down.

polyphase AC
term applied to electric current generated so that more than one alternating current is produced at a time, each out of phase with the other; three-phase AC is one example.

positive feedback
cyclical process in which a disturbance to a system produces responses that exacerbate the original disturbance.

potential difference (electrical)
difference in the potential energy of a unit electrical charge between two points. It is numerically equal to the work needed to move a unit charge from one point to the other.

power
the rate at which work is done, or energy is used.

ppm
part per million, a unit of concentration in chemistry; equivalent to one penny in 10 thousand dollars.

prestressed concrete
a building material composed of concrete and steel with the steel stretched to generate an amount of tension such that, when the steel is secured to the concrete, it creates compression stresses in the concrete sufficiently large to overcome most or all of the tension in the concrete induced by loads.

prospective study
a study in which the effects of an experimental treatment are observed as they appear rather than after the fact.

radiation
energy emitted by various sources such as the sun, flames, hot objects, decaying atoms, and so forth; types include radio waves, microwaves, infrared, visible light, ultraviolet light, X rays, gamma-rays. (Particle radiation is discussed in chapter 10.)

radioactivity
the spontaneous fission of an atom accompanied by the release of high-energy particles or electromagnetic radiation.

radionuclides
radioactive elements.

reciprocating steam engine
a device run by steam pressure causing a piston to move back and forth; the piston is then connected to a lever or wheel that performs some task or transmits power.

resistance
the relative ability of certain substances to conduct electricity; the lower the resistance, the better the conductor (e.g., copper has a smaller resistance than graphite).

roll
in aeronautics, rotation of an aircraft around its longitudinal axis.

rotor
the part of a turbine that contains the blades that spin; the moving part of a generator that, in older designs, contained the coils, but now is more likely to consist of the magnet.

rudder
in aeronautics, the flat vertical surface, usually attached to the tail of the plane, that the pilot can angle to the left or right to "steer" the aircraft, that is, it offers the pilot yaw control; leftward deflection of the rudder, for example, causes the nose of the plane to swing to the left.

run-back
a device used on a still to allow uncracked petroleum to return to the still to be reheated so that it undergoes cracking.

scroll casing
the spiral-shaped pipe that surrounds a Francis turbine and feeds water to it.

Second Industrial Revolution
the period roughly between 1870 and 1950 in which the industrialized nations shifted to a more scientific approach to technological development; many of the common objects such as the auto and telephone were introduced.

semaphore
visual signaling coded by means of the arrangement of rods, arms, or flags.

series
a pattern of electrical connections, among lights, for example, that establishes a single pathway for current to flow.

short-circuit
connecting the terminals of a battery or other source of electrical potential difference by a path of negligible resistance.

simple beam
a beam supported at each end such that one end of the beam is free to rotate while the other end is fixed.

sink
in the context of the greenhouse effect, this term refers to those processes that can absorb carbon dioxide and other gases; for example, trees remove carbon dioxide from the air.

span
in this context, the portion of a bridge between adjacent supports.

spectroscopy
the study of spectra or patterns of wavelengths that make up light.

spectrum
an infrared spectrum refers to the pattern of absorbance (or transmittance) by a particular molecule over the range of wavelengths for infrared radiation.

steam plant
a facility for generating electricity where steam is used to power an engine or turbine that then drives a generator; heat is derived from burning fossil fuels or from decay of radioactive material.

steam turbine
a cylindrical machine in which rotation is caused by steam flowing along the rotor axis and across blades mounted on the rotor.

stratosphere
the upper part of the atmosphere.

streamline
in aeronautics, a concept developed by Leonhard Euler to describe the lines traveled by air "particles" over the surfaces of a wing.

stress
the amount of force applied per unit area.

tailrace
the stream of water after it has passed the water wheel or turbine in a hydropower plant.

tension
the force or stress that tends to pull apart the particles of a material.

thermal efficiency
result obtained from dividing the final energy output by the initial energy in the fuel.

thermodynamics
the study of the interconversion of different forms of energy; the first law states that when energy is expended, the sum of the heat used and work done is equal to the initial energy available—this is called "conservation of energy"; the second law states that not all the available energy can be transformed into work but that some is "lost" as waste heat—therefore, 100 percent efficiency is not a practical goal.

three-phase AC
one type of polyphase AC; involves three distinct alternating currents.

thrust
in aeronautics, the force generated by the propulsion system that counteracts the effects of drag and allows the aircraft to move forward through the air.

toxoid
inactivated toxin; suitable for use in a vaccine to induce antibody formation.

transformer
a device used to convert electricity from a low voltage to a high voltage or vice versa.

Glossary

troposphere
the lower part of the atmosphere where the weather, and thus climate, occurs.

tube still
a pipelike device in which petroleum is heated in order to undergo cracking.

turbine
a device in which a fluid such as water, air, or steam supplies the power to turn blades mounted on a shaft called a rotor.

turbogenerator
a machine for generating electricity in which the generator is powered by a turbine; usually applied to cases where the turbine is driven by steam.

ultraviolet
light or radiation which is higher in energy than visible light; it can damage skin cells and possibly lead to cancer and is partly absorbed by the ozone layer in the stratosphere.

unit operation
a single chemical or physical change, such as distillation, in the sequence of processes involved in refining or other chemical production.

vaccination
the term originally used by Edward Jenner in referring to the intentional exposure of people to the cowpox microorganism in order to induce a protective response against smallpox, later applied by Pasteur more broadly to include prophylactic application of any microorganism.

vaccinia virus
virus related to cowpox virus and smallpox virus; used to make smallpox vaccine.

virus
a microorganism typically consisting of a protein coat surrounding genetic material (either DNA or RNA depending on the type of virus).

velocity
distance traveled per unit time in a specified direction; similar to speed except that the concept of speed does not involve the direction of movement.

volt
a unit of electromotive force that exists between two conductors; it designates the amount of work needed to move a quantity of electric charge from one point to another.

voltage
measure of electrical difference in between two points, such as a transmission cable and the ground or between two cables at different potential; voltage is higher at the point which has more electric charge than the other point, analogous to water at a high elevation and its tendency to flow downward.

watt
a unit of electric power named after James Watt; equal to 0.74 ft-lbs/sec.

wavelength
property of radiation (light) that is analogous to the property of water waves, which have specific distances between the peaks of the waves.

wing-warping
a technique for controlling roll of an aircraft in which flexible wings are twisted across their longitudinal axis, like grabbing a shoe box at the ends and rotating your hands in opposite directions; first developed by the Wright brothers to adjust the angle of attack of each wing in opposite directions and induce a roll; replaced by ailerons in modern aircraft.

work
(v.) the process of applying a force on an object over a distance; for example, picking up a book.

work
(n.) the application of a force on an object that produces movement of the object; the amount of work is the product of the force and the distance over which the object moves.

X ray
one form of electromagnetic radiation with a very high energy.

yaw
movement in the horizontal plane, that is, leftward or rightward movements of a plane.

Bibliography

Chapter 2

Abbot, C. G. 1932. *Smithsonian Scientific Series*, vol. 12. New York: Smithsonian Institution Press.

Andrews, F. T. 1989. The heritage of telegraphy. *IEEE Communications Magazine*, August 1989, 23–30.

Coulson, T. 1959. *Joseph Henry, His Life and Work*. Princeton: Princeton University Press.

Derry, T. K. and T. I. Williams. 1961. *A Short History of Technology from the Earliest Times to A.D. 1900*. New York: Oxford University Press.

Dibner, B. 1967. Communications. In *Technology in Western Civilization*, vol. 1, edited by M. Kranzberg and C. W. Pursell. New York: Oxford University Press.

Dizard, W. P. Jr. 1985. *The Coming Information Age*. New York: Longman.

Harlow, A. F. 1971. *Old Wires and New Waves*. New York: Arnow Press and New York Times.

Hindle, B. 1983. *Emulation and Invention*. New York: Norton.

Mabee, C. 1943. *The American Leonardo: A Life of Samuel F. B. Morse*. New York: Alfred A. Knopf.

Magie, W. F. 1935. *A Source Book in Physics*. New York: McGraw-Hill.

May, K. O. 1972. Wilhelm Friedrich Gauss. In *Dictionary of Scientific Biography*, vol. 5, edited by American Council of Learned Societies Devoted to Humanistic Studies. New York: Scribner and Sons.

Oliver, J. W. 1956. *History of American Telegraphy*. New York: The Ronald Press.

Pope, F. L. 1884. *Modern Practice of the Electric Telegraph: A Handbook for Electricians and Operators*. New York: D. van Nostrand.

Rheingold, N. 1972. Joseph Henry. In *Dictionary of Scientific Biography*, vol. 6, edited by American Council of Learned Societies Devoted to Humanistic Studies. New York: Scribners and Sons.

Romer, A. 1988. *Materials and Technology in History*. New York: Clarkson College.

Smith, A. 1980. *The Geopolitics of Information: How Western Culture Dominates the World*. New York: Oxford University Press.

Stephens, C. E. 1989. The impact of the telegraph on public time in the United States, 1844–1893. *IEEE Technology and Society Magazine*, March 1989, 4–10.

Trowbridge, J. 1901. *Samuel Finley Breese Morse*. Boston: Small, Maynard and Co.

Chapter 3

Boyle, W. 1936. *The City That Grew*. Los Angeles: Southland Publishing Co.

Coleman, C. E. 1952. *PG and E of California: The Centennial Story of the Pacific Gas and Electric Company, 1852–1952*. New York: McGraw-Hill.

Dunsheath, P. 1962. *A History of Electrical Power Engineering*. Cambridge, MA: The MIT Press.

Faraday, M. 1851. *Faraday's Diary: Being the Philosophical Notes of Experimental Investigation Made by Michael Faraday*, vols. 1 and 5. London: G. Bell and Sons.

Hughes, T. P. 1983. *Networks of Power: Electrification in Western Society 1880–1930*. Baltimore: The Johns Hopkins University Press.

Hughes, T. P. 1989. *American Genesis: A Century of Technological Enthusiasm*. New York: Viking.

Layne, J. G. 1952. *Water and Power for a Great City: A History of the Department of Water and Power of the City of Los Angeles to December 1950*, vols. 1 and 2. Los Angeles: Los Angeles Department of Water and Power Publications.

Layton, E. T. Jr. 1979. Scientific technology, 1845–1900: The hydraulic turbine and the origins of American industrial research. *Technology and Culture*, 20:64–89.

Logan, E. Jr. 1981. *Turbomachinery: Basic Theory and Application*. New York: Marcel Dekker.

Monition, L., M. Le Nir, and J. Roux. 1984. *Microhydroelectric Power Stations*. New York: J. Wiley and Sons.

Myers, W. A. 1986. *Iron Men and Copper Wires: A Centennial History of the Southern California Edison Company*. Glendale: Trans-Anglo Books.

Rolle, A. 1981. *Los Angeles: From Pueblo to City of the Future*. San Francisco: Fraser.

Snyder, G. M. and A. M. Whitsett. 1982. *Development of Small Hydroelectric Generating Plants on Water Distribution Systems: Report to the Metropolitan Water District of Southern California.* Los Angeles: Metropolitan Water District.

Southern California Edison Company. 1936–1950. *Annual Reports.* Los Angeles: Southern California Edison Company.

Stevens, J. E. 1988. *Hoover Dam: An American Adventure.* Norman: University of Oklahoma Press.

United States Department of Interior and Bureau of Reclamation. 1948. *Boulder Canyon Project. Final Report. Part 1. Introduction. Bulletin 1. General History and Description of the Project.* Washington, D.C.: Government Printing Office.

United States Department of Interior and Bureau of Reclamation. 1950. *Boulder Canyon Project. Final Report. Part 1. Introduction, Bulletin 2. Hoover Dam. Power and Water Contracts.* Washington, D.C.: Government Printing Office.

United States Department of Interior and Bureau of Reclamation. 1985. *Hoover Dam: Fifty Years.* Washington, D.C.: Government Printing Office.

Wise, G. 1988. William Stanley's search for immortality. *American Heritage of Invention & Technology*, Spring/Summer 1988, 4:42–49.

Chapter 4

Anderson, J. D. Jr. 1984. *Fundamentals of Aerodynamics.* New York: McGraw-Hill.

Anderson, J. D. Jr. 1987. *The Wright Brothers: The First True Aeronautical Engineers and The Wright Flyer: An Engineering Perspective.* Washington, D.C.: Smithsonian Institution Press.

Anderson, J. D. Jr. 1988. The historical development of aerodynamics prior to 1810 and its application to flying machines. Preprint. Department of Aerospace Engineering, University of Maryland, College Park, Maryland.

Cayley, G. 1809. On aerial navigation. *A Journal of Natural Philosophy, Chemistry, and the Arts* 24:164–174.

Cayley, G. 1810. On aerial navigation. *A Journal of Natural Philosophy, Chemistry, and the Arts* 25:81–87.

Cayley, G. 1810. On aerial navigation. *A Journal of Natural Philosophy, Chemistry, and the Arts* 25:161–169.

Combs, H. 1979. *Kill Devil Hill: Discovering the Secrets of the Wright Brothers.* Englewood: Ternstyle.

Crouch, T. D. 1978. *Engineers and the Airplane, The Wright Brothers, Heirs of Prometheus.* Washington, D.C.: Smithsonian Institution Press.

Crouch, T. D. 1989. *The Bishop's Boys: A Life of Wilbur and Orville Wright.* New York: W. W. Norton and Co.

Dalton, S. 1977. *The Miracle of Flight.* New York: McGraw-Hill.

Giacomelli, R. and E. Pistolesi. 1976. Historical sketch. In *Aerodynamic Theory: A General Review of Progress*, vol. I, edited by W. F. Durand. Maryland: Peter Smith.

Laurence, J. 1961. *Sir George Cayley: The Inventor of the Aeroplane*. London: Parrish.

McFarland, M. W. 1953. *The Papers of Wilbur and Orville Wright, vol. 2: 1906–1948*. New York: McGraw-Hill.

Miller, R. and D. Sawers. 1970. *The Technical Development of Modern Aviation*. New York: Praeger.

Moolman, V. 1980. *The Road to Kitty Hawk, The Epic of Flight*. Alexandria: Time-Life Books.

von Kármán, T. and J. M. Burger. 1976. General aerodynamic theory: Perfect fluids. In *Aerodynamic Theory: A General Review of Progress*, vol. 2, edited by W. F. Durand. Gloucester: Peter Smith.

Wolfson, R. and J. M. Pasachoff. 1987. *Physics*. Boston: Little, Brown and Company.

Wolko, H. S. (ed). 1987. *The Wright Flyer: An Engineering Perspective*. Washington, D.C.: Smithsonian Institution Press.

Wright, O. and W. Wright. 1908. The Wright brothers' aeroplane. *The Century Magazine* 76(5):641–650.

Chapter 5

Billington, D. P. 1987. *Structures and Machines in Urban Society, Lecture Notes*. Princeton: Department of Civil Engineering and Operations Research.

Coleman, C. E. 1952. *PG and E of California: The Centennial Story of the Pacific Gas and Electric Company, 1852–1952*. New York: McGraw-Hill.

Daumas, M. and P. Gille. 1979. The steam engine. In *A History of Technology and Inventions*, vol. III, edited by M. Daumas. New York: Crown.

Dickinson, H. W. 1963. *A Short History of the Steam Engine*. London: F. Cass & Co.

Dunsheath, P. 1962. *A History of Electrical Power Engineering*. Cambridge, MA: The MIT Press.

Hodgins, E. and F. A. Magoun. 1938. Behemoth: *The Story of Power*. New York: Garden City.

Hughes, T. P. 1983. *Networks of Power: Electrification in Western Society 1880–1930*. Baltimore: The Johns Hopkins University Press.

McWilliams, C. 1973. *Southern California: An Island on the Land*. Santa Barbara: Peregrine Smith.

Myers, W. A. 1986. *Iron Men and Copper Wires: A Centennial History of the Southern California Edison Company*. Glendale, CA: Trans-Anglo Books.

Newmark, M. H. and M. R. Newmark (eds). 1970. *Sixty Years in Southern California: 1853–1913*. Los Angeles: Zeitlin and Ver Brugge.

Rolle, A. 1981. *Los Angeles: From Pueblo to City of the Future*. San Francisco: Boyd and Fraser.

Sandfort, J. 1962. *Heat Engines: Thermodynamics in Theory and Practice*. Garden City: Doubleday Anchor.

The Brush system of electric lighting. 1882. *Scientific American Supplement* 11:4359–4363.

van Riemsdijk, J. T. and K. Brown. 1980. *The Pictorial History of Steam Power*. London: Octopus Books.

Weaver, J. D. 1973. *El Pueblo Grande: A Nonfiction Book About Los Angeles*. Los Angeles: The Ward Ritchie Press.

Workman, B. 1936. *The City That Grew*. Los Angeles: Southland.

Chapter 6

Aramco Handbook. 1968. *Oil and the Middle East*. Dhahran: Arabian American Oil Company.

Bryant, L. 1967. The origin of the automobile engine. *Scientific American* 216:102–112.

Bryant, L. 1967. The origin of the four-stroke cycle. *Technology and Culture* 8:178–198.

Enos, J. L. 1962. *Petroleum, Progress and Profits: A History of Process Innovation*. Cambridge, MA: The MIT Press.

Flink, J. L. 1981. Henry Ford and the triumph of the automobile. In *Technology in America*, edited by C. W. Pursell, Jr. Cambridge, MA: The MIT Press.

Gerding, M. 1986. *Fundamentals of Petroleum*, 3rd edition. Austin: Petroleum Extension Service, University of Texas.

Grayson, M. (ed.) 1985. Gasoline and other motor fuels. In *Kirk-Othmer Concise Encyclopedia of Chemical Technology*, 3rd edition, vol. 11, edited by M. Grayson. New York: Wiley and Sons.

Grayson, M. (ed.). 1985. Petroleum: nomenclature in the petroleum industry, origin of petroleum, composition, resources, drilling fluids, enhanced oil recovery, refinery processes, and products. In *Kirk-Othmer Concise Encyclopedia of Chemical Technology*, 3rd edition, vol. 17, edited by M. Grayson. New York: Wiley and Sons.

Hancock, E. G. (ed.). 1985. *Technology of Gasoline*. Oxford: Blackwell Scientific Publications.

Nevins, A. 1954. *Ford: The Times, the Man, the Company*. New York: Charles Scribner and Sons.

Rae, J. B. 1971. *The Road and the Car in American Life.* Cambridge, MA: The MIT Press.

Schackne, S. and N. D. Drake. 1960. *Oil for the World.* New York: Harper and Brothers.

Spitz, P. H. 1988. *Petrochemicals: The Rise of an Industry.* New York: John Wiley and Sons.

Thompson, E. V. and W. H. Ceckler. 1977. *Introduction to chemical engineering.* New York: McGraw-Hill.

Williamson, H. F., R. L. Andreano, A. R. Daum, and G. C. Klose. 1981. The age of illumination. In *The American Petroleum Industry,* vol. 1. Westport: Greenwood Press.

Williamson, H. F., R. L. Andreano, A. R. Daum, and G. C. Klose. 1981. The age of energy. In *The American Petroleum Industry,* vol. 2. Westport: Greenwood Press.

Chapter 7

Albano, A., W. P. Case, and N. H. Copp. 1990. *Forces and Forms in Large Structures.* Stony Brook: Research Foundation of the State University of New York.

Beaufait, F. W. 1977. *Basic Concepts of Structural Analysis.* New Jersey: Prentice Hall.

Billington, D. P. 1976. Historical perspective on prestressed concrete. *Journal of the Prestressed Concrete Institute* 21 : 2–25.

Billington, D. P. 1979. *Robert Maillart's Bridges: The Art of Engineering.* Princeton: Princeton University Press.

Billington, D. P. 1983. *The Tower and the Bridge: The New Art of Structural Engineering.* Princeton: Princeton University Press.

Brodsly, D. 1981. *L. A. Freeway: An Appreciative Essay. Berkeley:* University of California Press.

Cerny, L. 1981. *Elementary Statics and Strength of Materials.* New York: McGraw-Hill.

Fisher, D. A. 1963. *The Epic of Concrete.* New York: Harper and Row.

Ghali, A. and A. M. Neville. 1977. Structural Analysis: *A Unified Classical and Matrix Approach,* 2nd edition. London: Chapman and Hall.

Gordon, J. E. 1978. *Structures: or Why Things Don't Fall Down.* New York: Plenum Press.

Leonhart, F. 1964. *Prestressed Concrete Design and Construction.* Berlin: Wilhelm Ernst and Sohn.

Magnel, G. 1954. *Prestressed Concrete,* 3rd edition. New York: McGraw-Hill.

Menn, C. 1985. Aesthetics in bridge design. *Bulletin of the International Association for Shell and Spatial Structures* August 1985, 53–62.

Menn, C. 1976. Projectgrundlagen und Entwurf. In *Felsenaubrücke-Projectund Ausführung*, edited by H. Rigendinger. Zurich: Gasser and Eggerling.

Portland Cement Association. 1980. *Principles of Quality Concrete*. New York: J. Wiley and Sons.

Rigendinger, H., and W. Maag. 1976. Die projektierung des überbaus. In *Felsenaubrücke-Projectund Ausführung*, edited by H. Rigendinger. Zurich: Gasser and Eggerling.

Chapter 8

Bennenson, A. A. S. 1989. Smallpox. In *Viral Infections in Humans: Epidemiology and Control*, 3rd edition, edited by A. S. Evans. New York: Plenum Medical Book Co.

Boylston, Z. 1726. *An Historical Account of the Small-Pox Inoculated in New England Upon All Sorts of Persons, Whites, Blacks, and of all Ages and Constitutions. With Some Account of the Nature of the Infection in the Natural and Inoculated Way, and their different Effects on Human Bodies. With some short Directions to the Unexperienced in this Method of Practice*. London: Chandler.

Bulloch, W. 1938. *The History of Bacteriology*. New York: Oxford University Press.

Chase, A. 1982. *Magic Shots: A Human and Scientific Account of the Long and Continuing Struggle to Eradicate Infectious Diseases by Vaccination*. New York: W. Morrow and Company.

Cherry, J. D. 1989. Pertussis and the vaccine controversy. In *Immunization*, vol 8. in *Contemporary Issues in Infectious Diseases*, edited by R. K. Root, J. M. Griffiss, K. S. Warren, and M. A. Sande. New York: Churchill-Livingston.

Cherry, J. D. 1990. 'Perttussis vaccine encephalopathy': It is time to recognize it as the myth that it is. *Journal of the American Medical Association* 263:1679–1680.

Cody, C. L. J. Baraff, J. D. Cherry, S. M. Marcy, and C. R. Manclark. 1981. Nature and rates of adverse reactions associated with DTP and DT immunizations in infants and children. *Pediatrics* 68: 650–660.

Creighton, C. 1965. *A History of Epidemics in Britain, Volume 2: From the Extinction of the Plague to the Present Time*. New York: Barnes and Noble.

Frauenthal, J. C. 1981. *Smallpox: When Should Routine Vaccination be Discontinued?* Boston: Birkhäuser.

Golden, G. S. 1990. Pertussis vaccine and injury to the brain. *Journal of Pediatrics* 116:854–861.

Gonzalez, E. R. 1982. TV report on DTP galvanizes US pediatricians. *Journal of the American Medical Association* 248:12–22.

Hinman, A. R. and J. P. Koplan. 1984. Pertussis and pertussis vaccine: Further analysis of benefits, risks and costs. *Developmental and Biological Standards* 61:429–437.

Hinman, A. R. and J. P. Koplan. 1984. Pertussis and pertussis vaccine: Reanalysis of benefits, risks, and costs. *Journal of the American Medical Association* 251:3109–3113.

Jenner, E. 1798. An inquiry into the causes and effects of the variolae vaccinae, a disease discovered in some of the western counties of England, particularly Gloucestershire, and known by the name of the cow pox. In *Scientific Papers—Physiology, Medicine, Surgery, Geology*, vol. 38, The Harvard Classics, edited by C. W. Eliot. New York: P. F. Collier and Sons.

Kempe, C. H. 1960. *Studies on smallpox and complications of smallpox vaccination.* Pediatrics 26:176–188.

Kendrick, P. and G. Eldering. 1939. A study of active immunization against pertussis. *American Journal of Hygiene* 29(B):133–153.

Lane, J. M., F. L. Rubin, J. M. Neff, and J. D. Miller. 1969. Complications of smallpox vaccination, 1968: National surveillance in the United States. *New England Journal of Medicine* 281:1201–1208.

Langer, W. L. 1976. Immunization against smallpox before Jenner. *Scientific American* 234:112–117.

Lapin, J. H. 1943. *Whooping Cough.* Baltimore: Charles C. Thomas.

Miller, D. L., R. Alderslade, and E. M. Ross. 1982. Whooping cough and whooping cough vaccine: The risks and benefits debate. *Epidemiological Reviews* 4:1–24.

Morbidity and Mortality Weekly Report. 1989. Death rate: Leading causes, 1986–1988. 38(8):117–123.

Mortimer, E. A., and P. K. Jones. 1978. Pertussis vaccine in the United States: The benefit-risk ratio. In *International Symposium on Pertussis, Nov. 1–3, 1978*, edited by C. R. Manclark and J. C. Hill. Washington, D. C.: DHEW publication no. NIH 79-1830.

Pizza, M., A. Covacci, A. Bartoloni, M. Perugini, L. Nencioni, M. Teresa De Magistris, L. Villa, D. Nucci, R. Manetti, M. Bugnoli, F. Giovannoni, R. Olivieri, J. T. Barbieri, H. Sato, R. Rappuoli. 1989. Mutants of pertussis toxin suitable for vaccine development. *Science* 246:497–500.

Rhodes, A. J. and C. E. Van Royen. 1968. *Textbook of Virology*, 5th edition. Baltimore: The William and Wilkin Co.

Robbins, A. and P. Freeman. 1988. Obstacles to developing vaccines for the third world. *Scientific American*, November 1988, 126–133.

Sauer, I. 1937. Whooping cough: A study in immunization. *Journal of the American Medical Association* 100:239–241.

Sekura, R. D., J. Moss, and M. Vaughan (eds.). 1985. *Pertussis Toxin.* New York: Academic Press.

Silverstein, A. M. 1988. *A History of Immunology.* New York: Academic Press.

Stewart, G. T. 1978. Pertussis vaccine: The United Kingdom's experience. In *International Symposium on Pertussis, Nov. 1–3, 1978,* edited by C. R. Manclark and J. C. Hill. Washington, D.C.: DHEW Publication no. NIH 79-1830.

Stewart, G. T. 1979. Toxicity of pertussis vaccine: Convulsions are not evidence of encephalopathy. *Journal of Epidemiology and Community Health* 33: 150.

Stuart-Harris, C. H. 1978. Experiences of pertussis in the United Kingdom. In *International Symposium on Pertussis, Nov. 1–3, 1978,* edited by C. R. Manclark and J. C. Hill. Washington, D.C.: DHEW Publication no. NIH 79-1830.

Taylor, F. K. 1979. *The Concepts of Illness, Disease, and Morbus.* Cambridge: Cambridge University Press.

Tortora, G. J., B. R. Funckl, and C. L. Case. 1985. *Microbiology: An Introduction,* 3rd edition. Redwood city: Benjamin Cummings.

Winslow, O. E. 1974. *A Destroying Angel: The Conquest of Smallpox in Colonial Boston.* Boston: Houghton Mifflin.

World Health Organization. 1980. *The Global Eradication of Smallpox: Final Report of the Global Commission for the Certification of Smallpox Eradication.* Geneva: World Health Organization.

Chapter 9

Abrahamson, D. E. (ed.). 1989. *The Challenge of Global Warming.* Covelo, CA: Island Press.

Beardsley, T. 1989. Not so hot. *Scientific American,* November 1989, 17.

Dye, L. and M. Dolan. 1989. Greenhouse threat to sea level scaled back. *Los Angeles Times,* December 8, A3.

Fulkerson, W., D. B. Reister, A. M. Perry, A. T. Crane, D. E. Kash, and S. I. Auerbach. 1989. Global warming: An energy technology R&D challenge. *Science* 246: 868–869.

Herzog, A. 1977. *Heat.* New York: Simon and Schuster.

Hileman, B. 1989 Global warming. *Chemical & Engineering News,* March 13, 25–44.

Kellogg, W. W. and R. Schware. 1981. *Climate Change and Society: Consequences of Increasing Carbon Dioxide.* Boulder: Westview Press.

Maranto, G. 1986. Are we close to the end? *Discover,* January, 28–50.

Mathews, J. T. 1987. Global climate change: Toward a greenhouse policy. *Issues in Science and Technology,* Spring, 58–68.

McKean, K. 1983. Hothouse Earth. *Discover,* December, 99–102.

National Research Council. 1989. *Ozone Depletion, Greenhouse Gases, and Climate Change*. Washington, D. C.: National Academy Press.

Patursky, B. 1988. Dirtying the Infrared Window. *MOSAIC* 19:24–37.

Raloff, J. 1988. CO_2: How will we spell relief? *Science News* 134:411–414.

Schneider, C. 1989a. Preventing climate change. *Issues in Science and Technology*, Summer, 55–62.

Schneider, S. H. 1989b. The changing climate. *Scientific American*, Special Issue. September, 70–79.

Schneider, S. H. 1989c. The greenhouse effect: Science and policy. *Science*, February 10, 771–781.

Schneider, S. H. 1989. *Global Warming: Are We Entering the Greenhouse Century?* San Francisco: Sierra Club Books.

Seidel, S. and D. Keyes. 1983. *Can We Delay a Greenhouse Warming?* Washington, D.C.: Environmental Protection Agency.

Shands, W. E. and J. S. Hoffman, (eds.). 1987. *The Greenhouse Effect, Climate Change, and U.S. Forests*. Washington, D.C.: The Conservation Foundation.

Chapter 10

Adelstein, S. J. 1987. Uncertainty and relative risks of radiation exposure. *Journal of the American Medical Association* 258:655–657.

Alberts, B., D. Bray, J. Lewis, M. Raff, K. Roberts, and J. D. Watson. 1989. *Molecular Biology of the Cell*. New York: Garland.

Committee on the Biological Effects of Ionizing Radiation. 1980. *The Effects on Populations of Exposure to Low Levels of Ionizing Radiation*. Washington, D.C.: National Academy Press.

Edwards, M. 1987. Chernobyl—One year after. *National Geographic* 171:633–653.

Fuentes, G. 1986. 'Atomic veterans' push claims campaign. *Los Angeles Times*, August 28.

Goldman, M. 1987. Chernobyl: A radiobiological perspective. *Science* 238:622–623.

Hogerton, J. F. 1968. The arrival of nuclear power. In *Scientific Technology and Social Change: Readings from Scientific American*, edited by G. I. Rochlin. San Francisco: W. H. Freeman and Company.

Land, C. F. 1980. Estimating cancer risks from low doses of ionizing radiation. *Science* 209:1197–1203.

Medvedev, Z. 1990. *The Legacy of Chernobyl*. New York: W. W. Norton.

Pochkin, E. 1982. *Nuclear Radiation: Risks and Benefits*. London: Oxford University Press.

Raloff, J. 1985. Source terms: The new reactor safety debate. *Science News* 127:250–253.

Rhodes, R. 1986. *The Making of the Atomic Bomb.* New York: Simon and Schuster.

Rovin, A. 1990. 600 Navajos at hearing on aid for ill uranium miners. *Los Angeles Times*, March 14.

United Nations Scientific Committee on the Effects of Atomic Radiation (UNSCEAR). 1988. *Sources, Effects, and Risks of Ionizing Radiation: 1988 Report to the General Assembly, with Annexes.* New York: United Nations Publications.

United States Nuclear Regulatory Commission. 1985. *Reassessment of Technical Bases for Estimating Source Terms.* Washington, D.C.: U.S. Government Printing Office.

von Winterfeldt, D. and W. Edwards. 1984. Patterns of conflict about risky technologies. *Risk Analysis* 4:55–68.

Waldren, C., L. Correll, M. A. Sognier, and T. Puck. 1986. Measurement of low levels of x-ray mutagenesis in relation to human disease. *Proceedings of the National Academy of Science* 83:4839–4843.

Chapter 11

Jevons, F. R. 1976. The interaction of science and technology today, or, is science the mother of invention? *Technology and Culture* 17:729–742.

Gibbons, M. and C. Johnson. 1970. Relationship between science and technology. *Nature* 227:125–127.

Hewlett, R. G. 1976. Beginnings of development of nuclear technology. *Technology and Culture* 17:465–478.

Huxley, A. 1946. *Science, Liberty and Peace.* New York: Harper and Brothers.

Layton, E. T. Jr. 1971. Mirror-image twins: The communities of science and technology in 19th century America. *Technology and Culture* 12:562–580.

Layton, E. T. Jr. 1976. American ideologies of science and engineering. *Technology and Culture* 17:688–701.

McKelvey, J. P. 1985. Science and technology: The driven and the driver. *Technology Review* 88:38–47.

Index

Addition law, 250, 260
Aerodynamics, theory, 122
Airplane, 7, 266. *See also* Flight
 comparison of Wright's, 113
 military uses in early history of, 117
 progress in cruising speed of, 121
 rapid improvement of, 120
 and Wright's 1900 glider, 93, 95, 103
 and Wright's 1901 glider, 100
 and Wright's 1902 glider, 105
 and Wright's 1903 Flyer, 107–115
 and Wright's 1950 Flyer, 117
Albedo, 295, 300
Alcohols, 165
Alkanes, 162–164
Alkenes, 164
Alternating current (AC), 52
 generators, 56
 versus direct current (DC), 53–54
 and long distance transmission, 54, 55
 and three-phase AC generators, 4, 40, 41, 55–58
 and transformers, 58–59
Ampere, Andre, 22
Angle of attack, wing, 85, 94, 96, 109
Angular acceleration, 202
Antarctica and global warming, 306, 312
Antibody, 264, 269
Antigen, 264
Archimedes, 201

Arc lamps, 38, 50, 52, 141
Armour, J. Ogden, 178
Arrhenius, Svante, 291, 293
Atlantic cable, 16
Atmosphere
 carbon dioxide content of, 292, 295
 greenhouse gases and, 294, 309
 infrared absorption profile of, 308
Atmospheric engine, 131
Atom. *See also* Nuclear Fission
 Dalton model of, 329
 development of, 329–339
 and isotopes, 333
 nuclear model of, 334, 335
 plum pudding model of, 334, 335
 radiation and, 333
Atomic bomb, 321–324, 366–367
Atomic energy commission, 324
Atomic fission. *See* Nuclear fission
Automobile, 7
Automobile industry, 151, 152, 174, 175
Aviation fuel, 165, 185

Balance used by Wright brothers, 102, 104
Banning-Alameda steam plant, 135, 139, 141
Battery, 19–20, 24, 27
Battle of the currents, 53
Beam as a bridge form, 193, 194, 235

Becquerel, Henri, 330–331, 347
Behring, Adolph von, 269
Belknap, Frank, 178
Bending moment
  in cantilevered beams, 220
  created by prestressing cable, 212
  definition of, 206–208
  in Felsenau Bridge, 228
  and force couple, 208–210
  in the Southeast Connector, 235–236
  in Walnut Lane Bridge, 207, 208, 215
Benzene, 165
Bernoulli, Daniel, 67, 87, 380
Bernoulli, Jacques, 249, 250
Bernoulli's principle, 69, 87, 89, 122–123
Bissell, George, 153
Black, Joseph, 292
Boulder Canyon Project Act, 61. See also Hoover Dam
Boulton, Matthew, 134
Boundary layer on aircraft wings, 124
Boylston, Zabdiel, 248–251
Brewer, Francis, 153
Brush, Charles, 166
Brush generator, 53
Bureau of Reclamation, 61. See also Hoover Dam
Burton process. See also Thermal cracking
  compared to Dubbs process, 178, 179
  development of, 168–173
  first installation of, 173
  improvements of, 174
  profits from, 175
  yield of gasoline from, 175
Burton, William, 166–175, 181

California Department of Transportation (Caltrans), 231
Callendar, G. S., 293
Camber of aircraft wings, 89, 93, 94, 101, 106
Cancer, 367–371. See also Ionizing radiation
Cantilever construction, 218–228
Carbon cycle, 301
Carbon dioxide
  atmospheric concentration of, 301, 302
  correlation with temperature, 293, 303, 305
  from decay of vegetation, 300
  infrared spectrum of, 298
  measurement of, 292
  oceans as sinks for, 305
  properties of, 292
  sources of, 298–300, 305
Carnot, Sadi, 142
Catalysis (catalysts)
  and cracking of petroleum, 183, 184
  description of, 182
  in fluid bed process, 185
  in Houdry's fixed bed process, 185
  and metals as catalysts, 181
  and Suspensoid process, 186
Catalytic cracking. See Fluid bed catalytic cracking; Houdry process; Suspensoid process
Cathode rays, 332
Cayley, George, 89, 108
Cell (voltaic), 20. See also Battery
Cellular immunity. See Immunity
Center of pressure, 95, 96, 101
CFCs (Chlorofluorocarbons)
  atmospheric levels of, 307
  as greenhouse gases, 306
  and Montreal Protocol, 310
  and ozone layer, 310
  sinks for, 311
  sources of, 308
Chadwick, James, 336–338
Chain reaction, 338–349. See also Nuclear fission
Chanute, Octave, 90, 108
Chemical energy, 108
Chernobyl accident, 345–347, 348, 349, 359–361, 363–365
Chord line in aircraft wing, 93
Circulation in aerodynamics, 123
Clausius, Rudolph, 142
Clean circulation process. See Dubbs process
Cleveland refinery, 158, 159
Clouds, 305, 311
Coal, 38
Coke, 168, 169, 173, 180
Colorado river, 60, 61–62
Communications, 14–16, 19

Index 418

Compression. *See* Stresses in bridges
Concrete, as a building material, 194
Concrete beams and prestressing, 196. *See also* Prestressed concrete
Conductor, electrical, 21
Congress and telegraph, 28
Conservation of energy, law, 87
Conservation of matter (mass), 187, 299
Continuous beam, 232, 235
Control of aircraft, 97, 98, 106, 217. *See also* Flight
Cooke, William, 16, 29
Cornell, Ezra, 28
Cowpox, 253–255. *See also* Smallpox vaccine
Cracking (of petroleum). *See* Burton process; Dubbs process; Fluid bed catalytic cracking; Houdry process; Suspensoid process; Thermal cracking
Creep and behavior of concrete, 198
Crookes tubes. *See* Cathode ray tube
Crowe, Frank, 66
Crude oil (petroleum). *See* Oil; Petroleum
Curie, Irene Joliet-, 336–338
Current (electric), 21. *See also* Alternating current; Electricity
Curtiss, Glenn Hammond, 119

Da Vinci, Leonardo, 34
Davis, A. P., 61
Dayton-Wright Co. and early commercial flight, 120
Decker, Almarian W., 41, 53
Deforestation, 298, 300
Direct current electricity, 52
Disease, concepts of, 266–267, 274
Distillation, 147, 160
Doheny, Edward, 138
Double neutron process, 339. *See also* Nuclear fission
Drag, 86, 93–94, 107
 coefficient of, 110
 induced, 108, 110
 parasitic, 108, 110
 ratio to lift, 103
 total, 111

Drake, Edwin, 154, 155, 156
Drilling (for oil), 154–156
Dubbs, Carbon Petroleum, 177–181
Dubbs, Jesse, 177–179
Dubbs process, 179–181

Edison, Thomas A., 33, 52, 83
Efficiency
 in bridge design, 221
 definition of, 47
 of Hoover Dam, 69
 hydro versus steam power and, 59–60
 at Mill Creek Station number 1, 59
 of Pelton wheel, 45, 47, 57
Ehrlich, Paul, 270
Einstein, Albert, 17, 323, 338, 340, 342
Electrical generator, 48, 131, 266. *See also* Alternating current, and three-phase AC generators
Electrical power, 3, 38, 54
 transmission, 4, 40, 54, 55, 58, 70
Electricity, 7, 38, 39. *See also* Alternating current; Direct current; Hydroelectric power; Los Angeles
Electric trolleys, 39, 71–72
Electromagnet, 23–27
Electromagnetic induction, 48–49
Electromagnetism, 17, 19, 23
Elevator as a control surface on aircraft, 96, 112
Energy, 37, 108
 conversion in a hydropower plant, 42
 conversion of fossil fuel to electrical, 133
 sources for work, 148
 versus power, 132
Engine, 111, 116. *See also* Steam engine
Engineering, 5, 50, 381–382. *See also* Science and engineering
Equilibrium, 92, 202
Euler, Leonhard, 87, 380
External loads on bridges, 203

Fall, Albert, 61
Faraday, Michael, 7, 17, 25, 48–51, 58, 378
Felsenau bridge. *See also* Bending moment; Cantilever construction; Prestressed concrete

Felsenau bridge (cont.)
 aesthetic value of, 222, 230
 bending moment diagram for, 228
 cantilever construction and, 223–228
 Christian Menn and, 222
 display of cross section of, 223
 efficiency of, 230
 stresses in, 226, 227, 229
Fermi, Enrico, 324, 340
Finsterwalder, Ulrich, 218–220
First Industrial Revolution, 6, 131, 134, 290
Fission. *See* Nuclear fission
Flagler, Henry, 158
Flight, 380. *See also* Airplane; Control of aircraft
 airman's approach to, 91
 chauffeur's approach to, 91
 control of aircraft in, 117
 criteria for success in, 115
 first human-powered, 4, 80, 115
 and Otto Lillienthal, 83, 90, 91–92, 96–97
 problem of, 116
 progress of, 81
 scientific theories of, 88
 stability and control in, 94–98, 106–107
Fluid bed catalytic cracking, 186–189
Fluid dynamics, 87, 96
Flyer, the Wright, 80, 107–117. *See also* Airplane
Ford, Henry, 147, 148, 150, 151, 152, 174
Fossil fuels and steam plants, 72, 129, 131, 139, 140
 carbon dioxide production by, 298, 299
 efficiency of, 140–143
 and electricity in Los Angeles, 130
 environmental problems of, 144
 and scale of recent plants, 143
Fourier, Jean Baptiste, 291
Francis, James, 67
Francis turbine, 67, 68, 136
Freeways, 230–233, 237–241
Freons. *See* CFCs
Freyssinet, Eugene, 197, 199, 381

Gale, Leonard, 27
Gasoline, 4, 146, 147. *See also* Refining
 as carbon dioxide source, 299
 and combustion equation, 299
 components of, 162
 consumption of, 149
 demand for, 175, 184
 energy content of, 161
 and octane number, 176–177
 production of, 191
 seasonal variation in composition of, 190
 from thermal cracking of petroleum, 171
 yields of, 148, 173, 189, 191
Gauss, Karl Friedrich, 16
Generator, electric, 47, 51, 52, 378. *See also* Alternating current
 coil and, 51
 Edison and DC, 52
 magnetic field and, 51
 steam turbine and, 135, 137
 three-phase AC, 54
Germ theory of disease, 266–267, 274
Gilliland, E. R., 189
Global warming. *See also* Greenhouse effect
 consequences of, 311–313
 description of, 293–295
 role of clouds in, 305
Goldilocks problem, 295
Great Aerodrome, the, 79–80
Greenhouse effect
 adaptation strategy for, 315–316
 carbon dioxide's role in, 298–301
 computer models of, 304
 consequences of, 311
 and energy conservation, 316
 evidence for, 301–305
 and excess warming, 295
 gases' roles in, 294, 306–309
 historical background and theory of, 291, 293–295
 and infrared radiation, 297
 mechanism and radiation's role in, 296
 and ocean levels, 304, 312
 political concerns about, 313–315

Index   420

prevention strategy for, 316
and science and technology, 318
and solar fluctuation, 305
and water vapor, 311
and weather patterns, 312–313
Greenhouse gases, 294, 306–309
Greenland ice shelf, 312

Head, defined for hydroelectric
 power, 43, 67
Hennebique, François, 195
Henry, Joseph, 7, 17, 22–26, 49, 378
Hiroshima, 323, 348, 351–352
Hoover Dam
 consideration of environmental
  effects of, 74
 construction of, 64
 cost of electricity from, 63
 head of, 67
 operation of, 66
 power produced by, 66
 and production of electricity for
  Los Angeles, 63, 70, 71
 total electrical generating capacity
  of, 68
Hoover, Herbert, 61
Horsepower, 43, 134
Humoral immunity. *See* Immunity
Humphreys, Robert, 168, 169, 171, 172
Huntington, Henry, 71
Houdry, Eugene, 181, 183
Houdry process, 183–185
Hydrocarbons, 161, 162, 168
Hydroelectric power, 72, 73, 125,
 130. *See also* Alternating current;
 Electricity

Ice ages and carbon dioxide, 293, 303
Ice cores, 302
Illumination, 131, 147, 153, 156
Immunity, 269–270
Immunology, origin of field, 269–270
Impulse turbines, 47
Incandescent light bulbs, 39
Induction. *See* Electromagnetic
 induction
Industrial revolution. *See* First
 Industrial Revolution; Second
 Industrial Revolution

Information age, 14
Infrared radiation, 297, 298
Infrared spectrum, 297, 298, 308, 309
Inoculation, 247–252
Internal combustion engine,
 108–109, 147, 150, 151, 175
Internal forces in bridges, defined, 203
Ionizing radiation. *See also* Atom;
 Nuclear Fission
 background level of, 365–366
 biological effects of, 347–365
 and cell death, 349
 definition of, 348
 discovery of, 331
 dose-response models for effects of,
  353–358
 and free radicals, 348
 and risk of cancer, 351–371
 risk of low doses of, 359–366
 risk models for, 356–359
 types of, 333, 348
 units of, 349–350
Isomers, 162, 163
Isotope, defined, 337

Jenner, Edward, 252–257, 258, 261,
 263–265
Jenner, Henry, 255
Joukowski, Nikolai, 123
Joule, James, 142

Keeling, Charles, 293
Kelvin, Lord, 142
Kerosene, 147, 149, 153, 156
Kill Devil Hills, 92, 113
Kilowatt, defined, 44
Kinetic energy, 43, 46–47, 108
Kitty Hawk and the Wright
 Brothers, 80, 92. *See also* Flight;
 Wright Brothers
Knock (engine), 175, 176–177
Koch, Robert, 266–267
Kutta, M. Wilhelm, 123

Lanchester, Friedrich, 122
Langley, Samuel Pierpont, 78, 79,
 95, 105
Laval, Gustav, 135
Law of large numbers. *See* Probability

Le Sage, George, 15
Lewis, Warren, 188
Lift. *See also* Airplane; Flight
  and aircraft wings, 90, 93, 107, 109, 122
  and Bernoulli effect, 86–87
  calculating, 123–124
  and coefficient of lift, 93, 109
  definition of, 86–87
  and Francis turbines, 67
Lillienthal, Otto, 83, 90, 91–92, 96–97
Loads on bridges, 203
Los Angeles
  arc lamps for streets of, 131
  early history of, 36
  early population growth in, 37, 39
  electrification of, 36–75, 129–144
  fossil fuels and electricity for, 129–144
  and freeways, 232
  and hydroelectricity, 36–75
  and interurban rapid transit system, 71, 230
  and petroleum industry, 138, 139
  and prestressed concrete bridges, 233

Magnetic field equation, 24
Maillart, Robert, 195
Manhattan project, 343–345, 383
Maqnel, Gustave, 200, 206, 211, 215
Mars, 295
Mass production, 150, 152
Mather, Cotton, 247–248
Materials and bridge design, 192–193
Mauna Loa, 293, 302
Maxwell, James Clerk, 17, 51
Menn, Christian, 220–223, 228
Metchnikoff, Elia, 269–270
Methane
  Arctic sources of, 313
  as cleaner fuel, 317
  effects of, 307
  infrared spectrum of, 298
  sinks for, 309
  sources of, 306
Mill Creek Station #1 and hydroelectric power, 41–42, 44, 45, 48, 56, 68
Millikan, Robert, 332
Model T Ford, 152, 174

Moment, 202–203. *See also* Bending moment
Momentum, 45–46, 47, 86
Montreal Protocol, 310, 314, 317
Moon, 294
Morse, Samuel
  background of, 18–19
  and electromagnetism, 19
  and lawsuits over telegraph, 31
  personal views of, 32
  and telegraph, 16, 25, 26, 28
  and telegraph code, 26, 27, 33
Multiplication law, 250

Nagasaki and the atomic bomb, 323. *See also* Cancer; Nuclear fission
Neutron, 333–338
Newcomen, Thomas, 131
Newton, Isaac, 87, 201, 380
  aerodynamic lift and, 88
  and concept of resistance, 85–86
  and the Second Law of Motion, 67, 84, 86
  and the sine-squared law, 86
  and the Third Law of Motion, 84, 85, 202
  and the three laws of motion, 48, 84
Nitrous oxide, 307, 311
Nuclear fission, 338, 339, 340–344
Nuclear physics, 329–345
Nuclear power
  and Chernobyl accident, 345–347, 349, 359, 361 (*see also* Chernobyl)
  decisions regarding, 371–374
  dependence on, 328
  perceived risk of, 327–328
  public opinion of, 326–329
  and submarines, 324
  and utility companies, 325–326
Nuclear reactor, 324, 343–344. *See also* Chernobyl; Nuclear power

Oceans, 301, 304, 312
Octane number, 175, 176–177, 186
Oersted, Hans, 7, 22, 48, 378
Ohm, Georg, 21
Ohm's Law, 21, 24, 52, 54
Oil, 36, 149, 155–157, 159. *See also* Petroleum

Oil industry. *See* Petroleum industry
Oppenheimer, J. Robert, 322
Otto engine, 147, 150, 151
Otto, Nicolaus, 147, 150
Ozone, 307, 308, 311
Ozone layer, 294, 310

Parsons, Charles, 135, 136
Pasteur, Louis, 266–267, 268–269
Pelton, Lester, 45
Pelton wheel, 45–47, 135
Penstock and hydroelectric plants, 43
Pertussis, 271–272
Pertussis vaccine. *See also* Vaccine
  benefits of, 276–278
  decisions to use, 281–283
  development of, 273–275
  improvements of, 283–285
  and injury compensation, 283
  and large-scale vaccination program, 275
  and mouse test, 274
  risks of, 275–276, 279–283
Petrochemicals, 190
Petroleum. *See also* Refining
  deposits of, 147
  distribution of, 156
  fractions of, 160
  gas oil from, 169
  heavy oil from, 169
  products from refining of, 169
Petroleum industry, 156, 157, 168, 189–191
  in Southern California, 130, 138, 139
Pony Express, 13
Positive feedback, 312, 313
Potential energy and hydroelectricity, 42–43
Power, 42, 43, 108, 111, 132
Prandtl, Ludwig, 124
Prestressed concrete
  advantages in bridge design of, 198
  and continuous beams, 232
  definition of, 197
  and design, 200
  development of, 197–200
  and eccentricity, 212
  and the Felsenau Bridge, 224–227
  in freeway bridges, 236–237
  in the Walnut Lane Bridge, 211–213
Probability, 249, 250
Propellers and flight, 112, 114
Propulsion system for flight, 107, 114
Pumped storage, 42

Radiation, 296, 297, 309. *See also* Ionizing radiation
Radioactivity, 330–331, 360. *See also* Ionizing radiation
Radionuclides, 359
Railroad, 13, 30, 34
Reaction force and lift, 85, 107
Redondo Beach Plant, 139–142
Refining, 147, 156, 160
  and batch process, 175, 178
  and catalytic cracking (Houdry process), 184, 185
  and continuous process (Dubbs process), 175, 178
  and fluid bed catalytic cracking, 186
  and material balance, 187
  products from, 167, 186, 190
  and Suspensoid process, 186, 187
  and thermal cracking (Burton process), 167–169
  yields of gasoline from, 148, 167, 175, 179, 180, 189
Reinforced concrete, 195
Research and development laboratory, 166, 167
Resistance. *See* Reaction force
Resistance (electrical), 21
Revelle, Roger, 293
Risk, 9. *See also* Ionizing radiation; Pertussis vaccine; Smallpox Vaccine
Rockefeller, John D., 147, 157–159
Röentgen, Wilhelm, 331
Rudder on aircraft, 106
Rumford, Count. *See* Thompson, Benjamin
Rutherford, Ernest, 330–331, 334–335

Salginatobel Bridge, 197
Schilling, Paul, 16
Science and engineering, 5, 7, 377–384
  and electromagnetic induction, 50
  and flight, 90, 124
  and greenhouse effect, 318

Science and engineering (cont.)
  and hydroelectric power, 74–75
  and prestressed concrete, 240–241
  and vaccines, 286
Second Industrial Revolution, 7, 37, 290
Seuss, Hans, 293
Shear force, 205
Siemens, Werner, 52
Signaling methods before telegraph, 15
Silliman, Benjamin, 154, 161
Simple beam, 203, 231
Sine-squared law, 87
Sinks for greenhouse gases, 300, 301, 309, 311. *See also* Greenhouse effect; Greenhouse gases
Smallpox
  description of, 236, 247
  epidemic in Boston, 247–251
  eradication program, 261–263
  risks of, 249–250
  virus, 264
Smallpox vaccine. *See also* Vaccine
  benefits of, 260
  discovery of, 253–257
  early clinical trial of, 257
  opposition to, 256–257
  related injury, 259
  risk estimates for, 258–260
  and vaccinia virus, 264–265
Smeaton, John, 89
Smeaton's Coefficient, 89, 93, 103, 105, 110
Smith, "Fog," 28
Smith, William, 154, 155
Southeast connector, 234–241
Southern California Edison Co., 139
Standard Oil Company, 158, 159, 173
Stanley, William, 58
Steam engine, 4, 8, 108, 266
  and electricity, 129, 133
  and First Industrial Revolution, 131
  and fossil fuels, 290
  and global warming, 290, 317
  Watt's improvements of, 134
Steam plants. *See* Fossil fuels and steam plants
Steam power, 6, 131, 134. *See also* Fossil fuels and steam plants
Steam turbine, 130, 135, 136, 137, 138
Steel, properties of as a building material, 195, 231
Stresses in bridges
  compression and tension, 192–196
  definition of, 209
  in the Felsenau Bridge, 224–229
  in a freeway bridge, 235, 237, 239
  in the Walnut Lane Bridge, 209–210, 213, 214, 215–216
Structural art, criteria, 221, 228
Sun Oil Company, 184
Suspensoid process, 186, 187. *See also* Fluid bed catalytic cracking
Szilard, Leo, 338–340, 342, 343

Taylor, Charles, 111
Technology in relation to science, 377–388. *See also* Science and engineering
Telecommunications, 3, 7, 14
Telegraph, 3, 19, 25, 48, 125, 378
  controversy over inventor of, 31
  first message over, 29
  key, 27
  and Morse code, 26, 27, 33
  network, 33, 34
  poles, 29, 31
Telephone, 34, 48
Temperature
  correlation with atmospheric carbon dioxide, 303, 305
  and energy of a gas, 294, 295
Tension. *See* Stresses in bridges
Tesla, Nikola, 55. *See also* Alternating current
Thermal cracking, 168
  and Burton's still, 172
  chemical reactions of, 170
  fractions of crude oil used in, 169
  pressurized, 171
  yield of gasoline from, 175
Thermodynamics, 8, 142, 299
Thompson, Benjamin (Count Rumford), 142
Thomson, William (Lord Kelvin), 142
Three Mile Island, radiation exposure from accident, 366. *See also* Cancer; Ionizing radiation

Three-phase AC generator. *See* Alternating current, and three-phase AC generators
Thrust and flight, 108, 110, 111
Time zones and railroad, 30
Titusville, 153–155
Torque. *See* Moment
Transformer, 55, 58–59
Transmission of electricity. *See* Electrical power, transmission
Trolleys. *See* Electric trolleys
Turbines and generation of electricity, 44–47, 67
Turbogenerator, 135, 137
Tyndall, John, 291

Unit operations, 189
Universal Oil Products, 178, 180

Vaccination, 255, 265, 269
Vaccine. *See also* Pertussis vaccine; Smallpox vaccine
 and Adolph von Behring, 269
 for pertussis, 272–285
 for rabies, 269
 and science versus engineering, 286
 for smallpox, 252–265
 types of, 265
Vail, Alfred, 28
Venus, 295, 305
Vertical forces and bridge design, 204
Virchow, Rudolph, 266
Volta, Alessandro, 20
Voltage, 21, 24

Walnut Lane Bridge, 201. *See also* Bending moment; Prestressed concrete
 cross section of beam in, 210
 external forces on, 203
 and the Felsenau Bridge, 224
 internal forces in, 204
 key lessons from, 217
 stresses in, 209–210, 214, 216
 use of materials in, 213, 215
 vertical forces in, 204
Water as an energy source, 40
Water vapor, 311

Water wheels, 45. *See also* Turbines and generation of electricity
Watt, James, 6, 131, 133, 134, 290
Wayss, G. A., 195
Weber, Wilhelm, 16
Wenham, Francis H., 89
Westinghouse, George, 52–53
Wheatstone, Charles, 16, 29
Whooping cough. *See* Pertussis
Wind tunnel, 90, 102–103
Wings and flight, 102, 124. *See also* Airplane; Flight
Wing warping, 98, 99, 102
Wire services, 14
Wright Brothers, 82, 92. *See also* Airplane; Flight
 business and manufacture of airplanes and, 128
 as engineers, 111
 keys to success of, 81–82, 122
 and the 1903 Flyer, 81, 107–115, 116
 patents and, 107